VOLUME FIFTY FIVE

 DEVELOPMENTS IN PETROLEUM SCIENCE

HYDROCARBON EXPLORATION AND PRODUCTION

2ND EDITION

DEVELOPMENTS IN PETROLEUM SCIENCE 55

Volumes 1-7, 9-18, 19b, 20-29, 31, 34, 35, 37-39 are out of print.

VOLUME FIFTY FIVE

DEVELOPMENTS IN PETROLEUM SCIENCE

HYDROCARBON EXPLORATION AND PRODUCTION

2ND EDITION

By

Frank Jahn, Mark Cook and Mark Graham
TRACS International Consultancy Ltd.
Aberdeen, UK

ELSEVIER

Amsterdam • Boston • Heidelberg • London • New York • Oxford
Paris • San Diego • San Francisco • Singapore • Sydney • Tokyo

Elsevier
Radarweg 29, PO Box 211, 1000 AE Amsterdam, The Netherlands
The Boulevard, Langford Lane, Kidlington, Oxford OX5 1GB, UK

First edition 1998
Second edition 2008
Reprinted 2009, 2010 (twice), 2011

British Library Cataloguing in Publication Data
A catalogue record for this book is available from the British Library

Library of Congress Cataloging-in-Publication Data
A catalog record for this book is available from the Library of Congress

ISBN–13: 978-0-444-53236-7

For information on all Elsevier publications
visit our website at www.elsevierdirect.com

Printed and bound by CPI Group (UK) Ltd, Croydon, CR0 4YY

Transferred to Digital Print 2011

Working together to grow
libraries in developing countries

www.elsevier.com | www.bookaid.org | www.sabre.org

ELSEVIER   BOOK AID
             International   Sabre Foundation

CONTENTS

Principal Authors

Frank Jahn has worked as a Petroleum Geologist in Brunei, Thailand, the Netherlands, the UK and Australia. After 11 years with a multinational oil company he co-founded TRACS International in 1992. He has designed and teaches multidisciplinary training courses related to oil and gas field exploration and development worldwide, particularly graduate development programmes. He is now based in Perth, Australia where he works as a petroleum consultant.

Mark Cook joined the oil industry in 1981 as a reservoir engineer, and worked for 11 years in a multinational oil company in the Netherlands, Oman, Tanzania and the UK. In 1992 he co-founded TRACS International, where he is currently CEO, based in Scotland. His specific technical interests lie in petroleum engineering, risk analysis and economics, and he remains involved in training course development and delivery.

Mark Graham has 29 years of oil industry experience, working initially as a wireline logging engineer with Schlumberger in the Middle East, followed by more than 10 years with a multinational oil company in both operations and petroleum engineering positions in the Far East and North Sea. He co-founded TRACS International, where he is a Director, currently responsible for general business development in addition to working as a petroleum engineer, economist and project manager.

We thank Fiona Swapp for support in the graphics and publishing of this book, and for technical updates provided by TRACS International team members, particularly Liz Chellingsworth, Bjorn Smidt-Olsen, Jonathan Bellarby and Jenny Garnham.

TRACS International Training Ltd., and TRACS International Consultancy Ltd., can be contacted through:-

Tel:	+44 (0)1224 321213
Fax:	+44 (0)1224 321214
email:	tracs@tracsint.com
website:	http://www.tracs.com
Address:	Falcon House,
	Union Grove Lane,
	Aberdeen, AB10 6XU,
	United Kingdom

Introduction: About This Book

'Hydrocarbon Exploration and Production' takes the reader through all the major stages in the life of an oil or gas field, from gaining access to opportunity, through exploration, appraisal, development planning, production and finally to decommissioning. It straightforwardly explains the fiscal and commercial environment in which oil and gas field development takes place.

This comprehensive and current introduction to the upstream industry, is useful to industry professionals who wish to be better informed about the basic technical and commercial methods, concepts and techniques used. It is also intended for readers who provide support services to the upstream industry.

It draws together the many inter-disciplinary links within the industry in a clear and concise manner, while pointing out the commercial reason for the activities involved in the business – each chapter is introduced by pointing out the commercial application of the subject. The many illustrations are clear and plentiful, and are designed to maximise the learning while containing the detail necessary to preserve technical authenticity.

The authors are all practising consultants in the business, and have included the major advances in the industry in this latest edition, including technical methods for field evaluation and development and techniques used for managing risk within the business.

TRACS International has provided training and consultancy in Exploration and Production related issues for many clients worldwide since 1992. This book has gradually developed from course materials, discussions with clients and material available in the public domain.

F. Jahn
M. Cook
M. Graham

CHAPTER 1

THE FIELD LIFE CYCLE

Introduction and Commercial Application: This section provides an overview of the activities carried out at the various stages of field development. Each activity is driven by a business need related to that particular phase. The later sections of this book will focus in more detail on individual elements of the field life cycle (Figure 1.1).

1.1. GAINING ACCESS PHASE

The first step an oil company will undertake in hydrocarbon exploration and production is to decide what regions of the world are of interest. This will involve evaluating the technical, political, economic, social and environmental aspects of regions under consideration. Technical aspects will include the potential size of hydrocarbons to be found and produced in the region, which will involve *scouting studies* using publicly available information or commissioning regional reviews, and a consideration of the technical challenges facing exploration and production, for example in very deep offshore waters.

Political and economic considerations include political regime and Government stability, the potential for nationalisation of the oil and gas industry, current embargoes, fiscal stability and levels of taxation, constraints on repatriation of profits, personnel security, local costs, inflation and exchange rate forecasts. Social

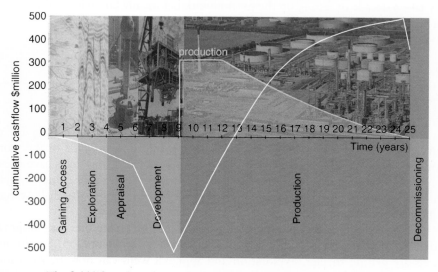

Figure 1.1 The field life cycle and typical cumulative cash flow.

considerations will include any threat of civil disorder, the availability of local skilled workforce and local training required, the degree of effort which will be required to set up a local presence and positively engage the indigenous people. The company will also consider the precautions needed to protect the environment from harm during operations, and any specific local legislation. There may also be a reputational issue to consider when doing business in a country whose political or social regime does not meet with the approval of the company's home Government or share-holders. Finally, an analysis of the competition will indicate whether the company has any advantage. It may be that if the company has an existing presence in-country from another business interest, such as downstream refining or distribution, the experience from these areas could be leveraged.

Some 90% of the world's oil and gas reserves are owned and operated by *National Oil Companies* (NOCs), such as Saudi Aramco (Saudi Arabia), Petronas (Malaysia), Pemex (Mexico). For an independent oil company to take a direct share of explo-ration, development and production activities in a country, it first needs to develop a suitable agreement with the Government, often represented by the NOC.

The invitation to participate may be publicly announced, in the form of a *licensing round*, as discussed in Chapter 2. Alternatively an arrangement for participation may be privately agreed with the NOC. In order to gain an advantageous position on this process, an oil company will expend effort to understand the local conditions, often by setting up a small presence in-country through which relationships are formed with key Government representatives such as the Oil and Gas Ministry, Department of Environmental Affairs and local authorities.

The understanding of local conditions and the requirements of the country, along with the relationships built, may result in a direct agreement for participa-tion in the country or at least an advantageous position when a public bidding round occurs. The investment made during the *Gaining Access* phase may be considerable, especially in terms of time and the commitment of representatives – it may take a decade of setting up the groundwork before any tangible results are seen, but this is part of the investment process of hydrocarbon exploration and production.

1.2. EXPLORATION PHASE

For more than a century petroleum geologists have been looking for oil. During this period major discoveries have been made in many parts of the world. However, it is becoming increasingly likely that most of the 'giant' fields have already been discovered and that future finds are likely to be smaller, more complex, fields. This is particularly true for mature areas like the North Sea and the shallow water Gulf of Mexico (GoM).

Fortunately, the development of new exploration techniques has improved geologists' understanding and increased the efficiency of exploration. So although targets are getting smaller, exploration and appraisal wells can now be sited more accurately and with greater chance of success.

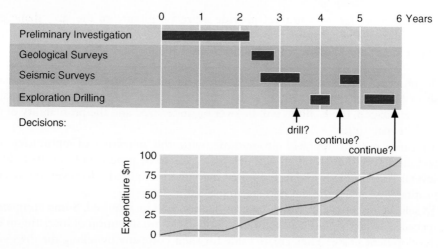

Figure 1.2 Phasing and expenditure of a typical exploration programme.

Despite such improvements, exploration remains a high-risk activity. Many international oil and gas companies have large portfolios of exploration interests, each with their own geological and fiscal characteristics and with differing probabilities of finding oil or gas. Managing such exploration assets and associated operations in many countries represents a major task.

Even if geological conditions for the presence of hydrocarbons are promising, host country political and fiscal conditions must also be favourable for the commercial success of exploration ventures. Distance to potential markets, existence of an infrastructure and availability of a skilled workforce are further parameters which need to be evaluated before a long-term commitment can be made.

Traditionally, investments in exploration are made many years before there is any opportunity of producing the oil (Figure 1.2). In such situations companies must have at least one scenario in which the potential rewards from eventual production justify investment in exploration.

It is common for a company to work for several years on a prospective area before an exploration well is 'spudded' – an industry term for starting to drill. During this period the geological history of the area will be studied and the likelihood of hydrocarbons being present quantified. Prior to spudding the first well a work programme will have to be carried out. Field work, magnetic surveys, gravity surveys and seismic surveys are the traditional tools employed. 'Exploration' in Chapter 3 will familiarise you in some more detail with the exploration tools and techniques most frequently employed.

1.3. APPRAISAL PHASE

Once an exploration well has encountered hydrocarbons, considerable effort will still be required to accurately assess the potential of the find. The amount of data acquired so far does not yet provide a precise picture of the size, shape and producibility of the accumulation.

Four possible options have to be considered at this point

- To proceed with development and thereby generate income within a relatively short period of time. The risk is that the field turns out to be larger or smaller than envisaged, the facilities will be over or undersized and the profitability of the project may suffer.
- To carry out an appraisal programme with the objective of optimising the technical development. This will delay 'first oil' to be produced from the field by several years and may add to the initial investment required. However, the overall profitability of the project may be improved.
- To sell the discovery, in which case a valuation will be required. Some companies specialise in applying their exploration skills, with no intention of investing in the development phase. They create value for their company by selling the discovery on, and then move on with exploration of a new opportunity.
- To do nothing. This is always an option, although a weak one, and may lead to frustration on behalf of the host nation's Government, who may force a relinquishment if the oil company continues to delay action.

In the second case, the purpose of *appraisal* is therefore to reduce the uncertainties, in particular those related to the producible volumes contained within the structure. Consequently, the purpose of appraisal in the context of field development is not to find additional volumes of oil or gas! A more detailed description of field appraisal is provided in Chapter 8.

Having defined and gathered data adequate for an initial reserves estimation, the next step is to look at the various options to develop the field. The objective of the *feasibility study* is to document various technical options, of which at least one should be economically viable. The study will contain the subsurface development options, the process design, equipment sizes, the proposed locations (e.g. offshore platforms) and the crude evacuation and export system. The cases considered will be accompanied by a cost estimate and planning schedule. Such a document gives a complete overview of all the requirements, opportunities, risks and constraints.

 ## 1.4. DEVELOPMENT PLANNING

Based on the results of the *feasibility study*, and assuming that at least one option is economically viable, a field development plan (FDP) can now be formulated and subsequently executed. The plan is a key document used for achieving proper communication, discussion and agreement on the activities required for the development of a new field, or extension to an existing development.

The FDP's prime purpose is to serve as a conceptual project specification for subsurface and surface facilities, and the operational and maintenance philosophy required to support a proposal for the required investments. It should give management and shareholders confidence that all aspects of the project have been identified, considered and discussed between the relevant parties. In particular, it should include

- objectives of the development
- petroleum engineering data
- operating and maintenance principles
- description of engineering facilities
- cost and manpower estimates
- project planning
- summary of project economics
- budget proposal.

Once the FDP is approved, there follows a sequence of activities prior to the first production from the field

- *FDP*
- *Detailed design* of the facilities
- *Procurement* of the materials of construction
- *Fabrication* of the facilities
- *Installation* of the facilities
- *Commissioning* of all plant and equipment.

1.5. PRODUCTION PHASE

The production phase commences with the first commercial quantities of hydrocarbons (first oil) flowing through the wellhead. This marks the turning point from a *cash flow* point of view, since from now on cash is generated and can be used to pay back the prior investments, or may be made available for new projects. Minimising the time between the start of an exploration campaign and 'first oil' is one of the most important goals in any new venture.

Development planning and production are usually based on the expected *production profile* which depends strongly on the mechanism providing the driving force in the reservoir. The production profile will determine the facilities required and the number and phasing of wells to be drilled. The production profile shown in Figure 1.1 is characterised by three phases

1. *Build-up period*	During this period newly drilled producers are progressively brought on stream.
2. *Plateau period*	Initially new wells may still be brought on stream but the older wells start to decline. Production facilities are running at full capacity, and a constant production rate is maintained. This period is typically 2–5 years for an oil field, but longer for a gas field.
3. *Decline period*	During this final (and usually longest) period, all producers will exhibit declining production.

1.6. DECOMMISSIONING

The *economic lifetime* of a project normally terminates once its net cash flow turns permanently negative, at which moment the field is decommissioned. Since towards the end of field life the capital spending and asset depreciation are generally negligible, economic decommissioning can be defined as the point at which gross income no longer covers operating costs (and royalties). It is of course still technically possible to continue producing the field, but at a financial loss.

Most companies have at least two ways in which to defer the decommissioning of a field or installation

(a) reduce the operating costs, or
(b) increase hydrocarbon throughput

In some cases, where production is subject to high taxation, tax concessions may be negotiated, but generally host Governments will expect all other means to have been investigated first.

Maintenance and operating costs represent the major expenditure late in field life. These costs will be closely related to the number of staff required to run a facility and the amount of hardware they operate to keep production going. The specifications for product quality and plant up-time can also have a significant impact on running costs.

As decommissioning approaches, *enhanced recovery,* for example chemical flooding processes are often considered as a means of recovering a proportion of the hydrocarbons that remain after primary production. The economic viability of such techniques is very sensitive to the oil price, and whilst some are used in onshore developments they can less often be justified offshore.

When production from the reservoir can no longer sustain running costs but the technical operating life of the facility has not expired, opportunities may be available to develop nearby reserves through the existing infrastructure. This has become increasingly common where the infrastructure already installed is being exploited to develop much smaller fields than would otherwise be possible. These fields are not necessarily owned by the company which operates the host facilities, in which case a service charge (*tariff*) will be negotiated for the use of third party facilities.

Ultimately, all economically recoverable reserves will be depleted and the field will be decommissioned. Much thought is now going into decommissioning planning to devise procedures which will minimise the environmental effects without incurring excessive cost. Steel platforms may be cut off to an agreed depth below sea level or toppled over in deep waters, whereas concrete structures may be refloated, towed away and sunk in the deep ocean. Pipelines may be flushed and left in place. In shallow tropical waters opportunities may exist to use decommissioned platforms and jackets as artificial reefs in a designated offshore area.

Management of decommissioning costs is an issue that most companies have to face at some time. On land sites, wells can often be plugged and processing facilities dismantled on a phased basis, thus avoiding high spending levels just as hydrocarbons run out. Offshore decommissioning costs can be very significant and less easily spread as platforms cannot be removed in a piecemeal fashion. The way in which provision is made for such costs depends partly on the size of the company involved and on the prevailing tax rules.

Usually a company will have a portfolio of assets which are at different stages of the described life cycle. Proper management of the asset base will allow optimisation of financial, technical and human resources.

CHAPTER 2

PETROLEUM AGREEMENTS AND BIDDING

Introduction and Commercial Application: When the host government notifies its intent to offer exploration acreage, the oil company has an opportunity to *gain access*. In this section, we will introduce the form of invitation to bid and the agreement under which the oil company may compete for and explore that acreage. Two broad types of Petroleum Agreement exist: *Licence Agreements* and *Contract Agreements*.

In a Licence Agreement the Government issues exclusive rights to an oil company to explore within a specific area. The operations are financed by the licence holder who also sells all production, often paying a royalty on production, and always paying taxes on profits. Such a fiscal regime is often called a *Tax and Royalty system*. The Government may insist upon an obligatory level of State participation.

In a Contract Agreement, the oil company obtains the rights to an area through a contract with the Government or its representative NOC. Essentially the company acts as a contractor to the Government, again funding all operations. However, in this case, title to the produced hydrocarbons is retained by the Government, and the oil company is remunerated for its costs and provided a share of the profits either in cash or in kind (i.e. a share of the produced hydrocarbons). The most common form of this type of agreement is a *production sharing contract* (PSC), also known as a *production sharing agreement* (PSA), and more detail of this is provided in Chapter 14.

2.1. THE INVITATION TO BID

As Chapter 1 pointed out, the majority of the remaining world hydrocarbon reserves lie under the control of NOCs, and usually this will be developed by the NOC. Exceptions to this may arise for a variety of reasons. The NOC may not have the local expertise required, the host Government may not have sufficient funds or manpower or an asset may be unattractive to the NOC. In cases such as these, the host Government may invite third parties to participate in the region. Such an opportunity may be posted in the international press, trade journals or by specific invitation. The following is a typical invitation to bid (Figure 2.1).

The geographic area of interest is divided up into a number of blocks by a grid, which is usually orthogonal. The size of these blocks varies from country to country and even from area to area in some cases. For example, UK North Sea licence blocks are 10×20 km, Norwegian blocks 20×20 km, GoM blocks 3×3 miles and deepwater Angola blocks approximately 100×50 km (and roughly follow the shape of the coastline as shown in Figure 2.2).

The Government will decide at its discretion what blocks it wishes to include in any *bidding round*, but there is often a geographic progression, from say shallow water areas into deeper water as time moves on.

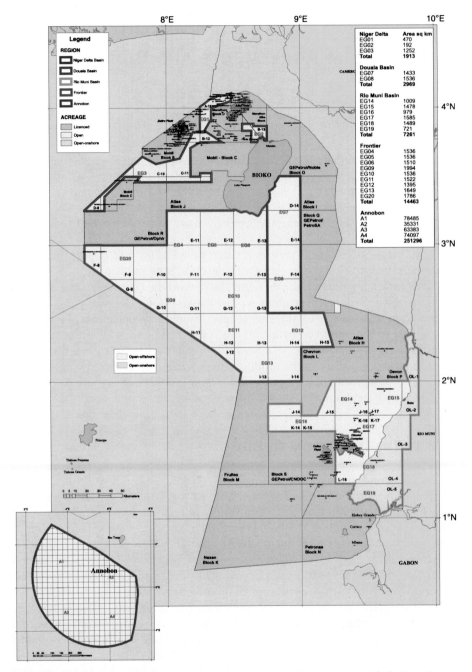

Figure 2.1 Licence map showing 2006 promotional blocks in Equatorial Guinea (source: www.equatorial.com).

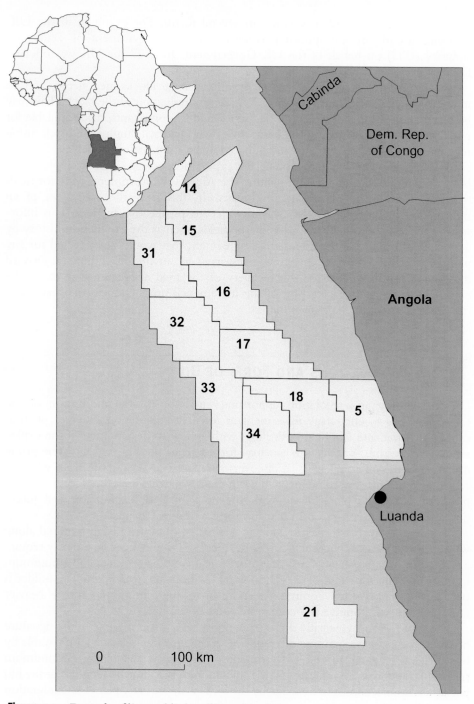

Figure 2.2 Example of licence blocks offshore Angola.

The invitation to bid may come in several forms. For example, in the UK, licensing rounds are announced periodically by the Department of Trade and Industry (DTI) on behalf of the UK Government. In 2007 the UK was offering licences in its *24th offshore licensing round.*

In any given UK licensing round, specific licence blocks are offered, and an interested bidder is left to his initiative to make an evaluation of the block. This may be based on speculative regional studies performed by consultants, made available for purchase by the author, or on the company's own understanding of the block, using regional data, analogue data and any public domain information available.

The invitation to bid may not be for exploration acreage. For example, some blocks offered by Sonatrach, representing the Algerian Government, were for fields that had many years of production history. In this case, the equivalent of an information memorandum (IM) was provided to prospective bidders. This information includes both technical data for the fields, such as the production history by well, and an outline of the commercial agreement that would be expected for any participation by a foreign investor. Investors were invited to submit a forward development plan to increase the recovery of the field above the base case. The commercial terms offer a fraction of the incremental production to the investor as the profit element of their investment.

2.2. MOTIVATIONS AND FORM OF BID

In offering an exploration opportunity in a block, the motivation of the Government is to encourage investment in form of exploration activities, such as shooting seismic and exploration drilling, with a view to development if the exploration is successful. A *signature bonus* may form part of the bid package. The prime objective of the oil company is to discover commercial hydrocarbons from which it can create profits by subsequent development, and it therefore considers the prospectivity of the block along with the costs of both exploration and future development. This risk–reward calculation is covered in Chapter 3.

The invitation to bid may include an outline of the form of bid required along with the fiscal terms applicable to any subsequent development. The bid may require a minimum *work programme* consisting of seismic data to be acquired and a minimum number of wells; for example 2000 km of 2-D seismic and four wells. The bidder is of course at liberty to commit to more than the minimum, and a heavier commitment will improve the competitiveness of the bid.

In many regions, especially those operating PSAs, it is normal to add a signature bonus to the work programme offered. This is the promise of a cash sum payable by the successful bidder to the Government on award of the block. A minimum signature bonus may be indicated in the invitation to bid, but this element of the bid package is again a choice to be made by the bidder. In the early phases of exploration in a basin, when the risks of exploration failure are high, signature bonuses are usually tens of millions of dollars. However, once the first discoveries have been made in the area, interest will be heightened and signature bonuses offered for subsequent nearby blocks can escalate to hundreds of millions of dollars. It is important to realise that this signature bonus, once paid, is a sunk cost and should be considered as part of the cost of exploration. It is not a tax-deductable cost against future revenues.

The offer will have a bid deadline, after which submitted bids will be opened by the Government, or its NOC representative. This may be done in public or more commonly behind the closed doors. The winning bids may be publicly announced, or kept confidential, depending on the country. The criterion by which the bids are then compared is normally the total value of the bid package – the combination of the work programme plus signature bonus. Of course, where the combined values of competitors are close, the Government will need to decide on the relative weighting it places on work programme versus cash offered in the signature bonus. The weighting is not always apparent to the bidders. Other considerations that the Government will take into account will be the bidders' technical competence, general reputation, any existing working relationships and any strategic reasons the Government may have to encourage particular entrants into the region.

The details of the winning bids may be publicly announced and published, which is both a useful piece of information for future bids and an interesting comparison for each bidder to make with their own offer. In some cases all bids are announced, in which case the margin by which the winner succeeded is clear – the winner of course hopes not to have outbid the next nearest competitor by an embarrassing sum, thereby 'leaving money on the table'.

2.3. BLOCK AWARD

The successful bid will result in award of the block, giving the rights to explore. Any signature bonus offered will be cashed by the Government. There is often a prescribed sequence of events that dictate the timing of carrying out the work programme and declaring a commercial interest in the block – meaning that the company intends to progress beyond the exploration stage and on to appraisal and possible development of a discovery in the block. In this case, the company will need to convert the exploration rights into development rights in the block.

Figure 2.3 shows an example of the provisions in a PSA for converting an exploration agreement into a production agreement.

The criteria for a commercial well would be based on production rate during testing of a discovery well, whereas the *declaration of a commercial discovery* (DCD) would depend on the oil company demonstrating that an economic development can be justified – this will need to pass internal economic screening criteria, further discussed in Chapter 14. In the example, Figure 2.3 below, the Government is due a bonus payable at DCD, and a further bonus when production from the development starts. Timeframes are typically imposed on the events, shown above for a PSA between the oil company and the Government.

In some cases there is a requirement to release only a fraction of the block if commerciality has not been declared after a specified period of time. Figure 2.4

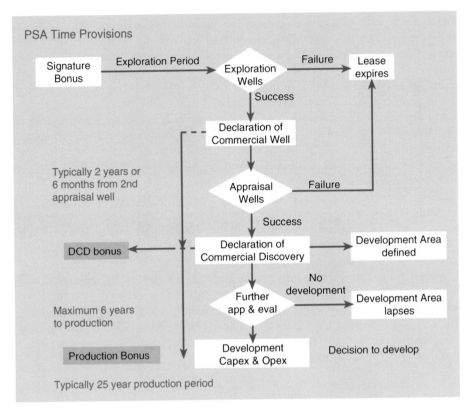

Figure 2.3 Example of sequence of events in a PSA.

shows an example of drilling up a commitment of three wells, and shooting 2-D seismic, whilst relinquishing fractions of the block during this time.

2.4. FISCAL SYSTEM

The Petroleum Agreement will also include a description of the *fiscal terms* by which the Government will claim its share of revenues during the production period. This will fall broadly into four categories, as shown in Table 2.1.

Within these broad categories, there are in excess of 120 different fiscal systems in place around the world. Some 50% of these are PSAs and 40% Tax and Royalty systems. More details of these two most common systems are covered in Chapter 14.

2.5. FARM-IN AND FARM-OUT

The participants in the block may change over time, for various reasons. Firstly, in a PSA the Government may choose to award the block to several companies, imposing a preferred split and a nominated operator. With the approval

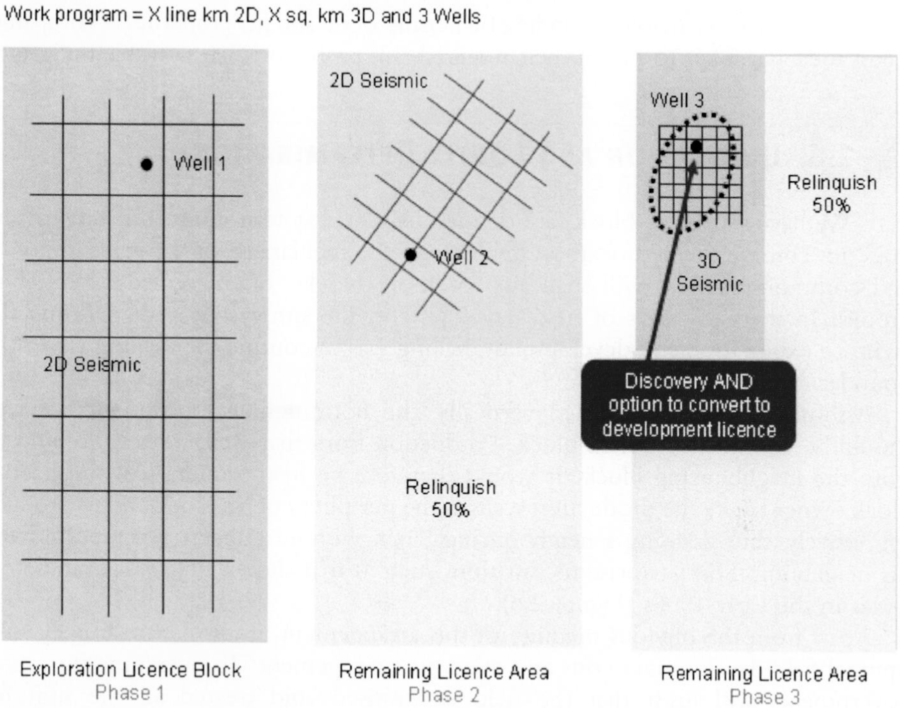

Work program = X line km 2D, X sq. km 3D and 3 Wells

Figure 2.4 Example of maturing of an exploration licence block.

Table 2.1 Broad categories of fiscal systems

Fiscal System	General Terms
Tax and royalty	Company pays royalty as a fraction of gross production and tax on net profits
Production sharing agreement	Company receives full cost recovery from production and a share of remaining profit oil
R-factor	Company pays a tax rate which is a function of the rate of return of the project (defined as cumulative revenues/cumulative expenditure)
Service agreement	Company receives remuneration for services or expertise provided

of the Government, the incumbents may choose to trade the initial splits. At any stage of the field life cycle, a company may choose to reduce its share in a block by selling a fraction to another company – this is known as 'farming out'. The company who accepts the share is said to have 'farmed in'. The farm-out may be for cash or for a trade in another interest.

A company may choose to farm out if it is unable to raise the capital required for development, or if it wishes to reduce its exposure in the project because it considers its position to be too risky.

There is an active market in trading ownership of oil and gas properties as companies adjust their portfolios to match their required risk profile or their available budgets.

2.6. UNITISATION AND EQUITY DETERMINATION

We have seen how blocks are defined by a grid system. Unfortunately, nature does not confine the hydrocarbon field size to the regularities of the grids imposed, and commonly a field will span two or more blocks, often owned by different groups. In the early days of field development, the simplest way of defining the rights to exploration and development drilling was to confine the drilling rig to the boundaries of the block.

Assuming wells were drilled vertically, the bottom hole location of the well should be within the owner's block. Production from that well, however, could be from the neighbouring block. It would therefore be in the interest of the licence block owner to site the production wells at the periphery of his block and to produce aggressively, thus draining a neighbouring block without concerns of reprisal from his neighbour. This gave rise to situations such as that shown below at Spindletop, Texas in the early 1900s (Figure 2.5).

Apart from the obvious inequity of this arrangement, it also led to hugely sub-optimal field development costs and reservoir management. To overcome this, most governments will insist that the field is 'unitised' and treated as one unit for development purposes. The owners of the field or the Government will nominate an operator, and the development will be planned based on the physical properties of

Figure 2.5   Field development at Spindletop, Texas, early 1900s.

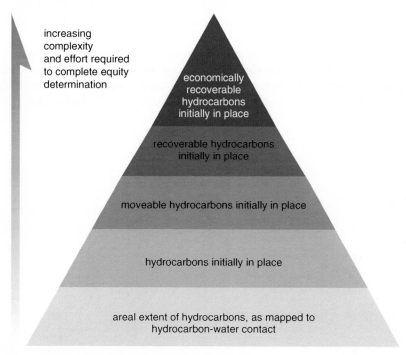

increasing
complexity
and effort required
to complete equity
determination

economically
recoverable
hydrocarbons
initially in place

recoverable hydrocarbons
initially in place

moveable hydrocarbons initially in place

hydrocarbons initially in place

areal extent of hydrocarbons, as mapped to
hydrocarbon-water contact

Figure 2.6 Options for the basis of equity.

the field, uninfluenced by ownership. The split of the costs of development and the resulting net cash flow will be determined by the 'equities' held by the owners of the licence blocks which the field straddles.

The basis for the equity determination is negotiated between the block owners (Figure 2.6). This basis could be

- areal extent of the accumulation, as mapped to the hydrocarbon–water contact
- hydrocarbons initially in place
- moveable hydrocarbons initially in place
- recoverable hydrocarbons initially in place
- economically recoverable hydrocarbons initially in place.

Moving toward the apex of Figure 2.6, the basis for equity becomes progressively more complex and lengthier to determine. The extreme case of economically recoverable reserves requires estimates of both the technical development plan and all of the economic assumptions such as costs and product prices, right through to the end of field life.

Prior to development, a 'deemed equity' may be agreed between the equity groups in order to set the proportional funding of the field development. This will usually be reviewed close to first production when more information is available from the development wells. Adjustments are then made to the initial funding to ensure that the correct contributions to the development costs have been made.

Once production has commenced and more information about the reservoir becomes available, it may become apparent that the initial equity is incorrect. If one of the equity groups feels that a revision to the equity is required, then a 're-determination' may be called, and new equities agreed. Again, this can be a costly exercise.

CHAPTER 3

EXPLORATION

Introduction and Commercial Application: This section will firstly examine the conditions necessary for the existence of a hydrocarbon accumulation. Secondly, we will see which techniques are employed by the industry to locate oil and gas deposits.

Exploration activities are aimed at finding new volumes of hydrocarbons, thus replacing the volumes being produced. The success of a company's exploration efforts determines its prospects of remaining in business in the long term.

3.1. HYDROCARBON ACCUMULATIONS

3.1.1. Overview

Several conditions need to be satisfied for the existence of a hydrocarbon accumulation, as indicated in Figure 3.1. The first of these is an area in which a suitable sequence of rocks has accumulated over geologic time, the *sedimentary basin*. Within that sequence there needs to be a high content of organic matter, the *source rock*. Through elevated temperatures and pressures these rocks must have reached *maturation*, the condition at which hydrocarbons are expelled from the source rock.

Migration describes the process which has transported the generated hydro-carbons into a porous type of sediment, the *reservoir rock*. Only if the reservoir is deformed in a favourable shape or if it is laterally grading into an impermeable formation does a *trap* for the migrating hydrocarbons exist.

3.1.2. Sedimentary basins

One of the geo-scientific breakthroughs of the last century was the acceptance of the concept of *plate tectonics*. It is beyond the scope of this book to explore the underlying theories in any detail. In summary, the plate tectonic model postulates that the positions of the oceans and continents are gradually changing through geologic times. Like giant rafts, the continents drift over the underlying mantle. Figure 3.2 shows the global configuration of major plate boundaries.

The features created by crustal movements may be mountain chains, like the Himalayas, where collision of continents causes *compression*. Conversely, the depressions of the Red Sea and East African Rift Basin are formed by *extensional plate movements*. Both type of movements form large-scale depressions into which sediments from the surrounding elevated areas (highs) are transported. These depressions are termed *sedimentary basins* (Figure 3.3). The basin fill can attain a thickness of several kilometres.

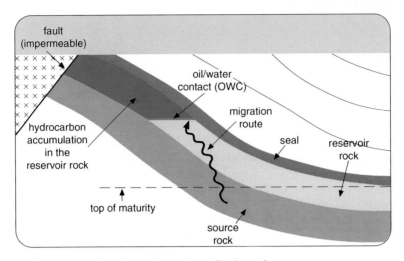

Figure 3.1 Generation, migration and trapping of hydrocarbons.

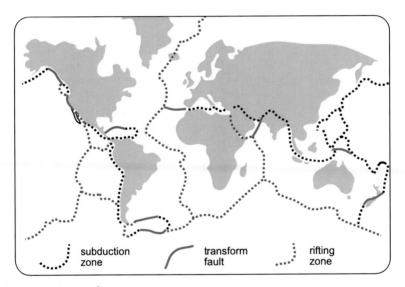

Figure 3.2 Global plate configuration.

3.1.3. Source rocks

About 90% of all the organic matter found in sediments is contained in *shales*. For the deposition of these *source rocks* several conditions have to be met: organic material must be abundant and a lack of oxygen must prevent the decomposition of the organic remains. Continuous sedimentation over a long period of time causes burial of the organic matter. Depending on the area of deposition, organic matter may consist predominantly of plant remnants or of *phytoplankton*. These are marine

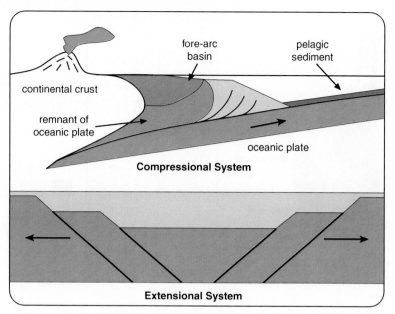

Figure 3.3   Sedimentary basins.

algae which live in the upper layers of the oceans, and upon death sink in vast quantities onto the seabed. Plant-derived source rocks often lead to 'waxy' crudes. An example of a marine source rock is the Kimmeridge clay which has sourced the large fields in the Northern North Sea. The coals of the carboniferous age have sourced the gas fields of the Southern North Sea.

3.1.4. Maturation

The conversion of sedimentary organic matter into petroleum is termed *maturation*. The resulting products are largely controlled by the composition of the original matter. Figure 3.4 shows the maturation process, which starts with the conversion of mainly *kerogen* into petroleum; but in very small amounts below a temperature of 50°C (kerogen: organic rich material which will produce hydrocarbon on heating). When kerogens are present in high concentrations in shale, and have not been heated to a sufficient temperature to release their hydrocarbons, they may form *oil shale deposits*.

The temperature rises as the sediment package subsides within the basinal framework. The peak conversion of kerogen occurs at a temperature of about 100°C. If the temperature is raised above 130°C for even a short period of time, crude oil itself will begin to 'crack' and gas will start to be produced. Initially the composition of the gas will show a high content of C4–C10 components (wet gas and condensate), but with further increases in temperature the mixture will tend towards the light hydrocarbons (C1–C3, dry gas). For more detail on the composition of hydrocarbons, refer to Section 6.2, Chapter 6.

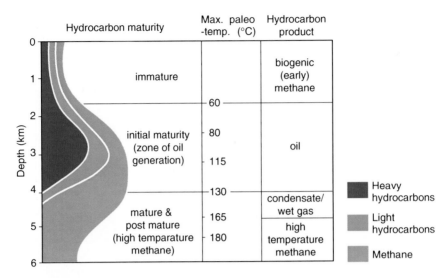

Figure 3.4 Hydrocarbon maturation.

The most important factor for maturation and hydrocarbon type is therefore heat. The increase of temperature with depth is dependent on the *geothermal gradient* which varies from basin to basin. An average value is about 3°C per 100 m of depth.

3.1.5. Migration

The maturation of source rocks is followed by the *migration* of the produced hydrocarbons from the deeper, hotter parts of the basin into suitable structures. Hydrocarbons are lighter than water and will therefore tend to move upwards through permeable strata.

Two stages have been recognised in the migration process. During *primary migration* the very process of kerogen transformation causes micro-fracturing of the impermeable and low porosity source rock which allows hydrocarbons to move into more permeable strata. In the second stage of migration the generated fluids move more freely along bedding planes and faults into a suitable reservoir structure. Migration can occur over considerable lateral distances of several tens of kilometres.

3.1.6. Reservoir rock

Reservoir rocks are either of *clastic* or *carbonate* composition. The former are composed of silicates, usually sandstone, the latter of biogenetically derived detritus, such as coral or shell fragments. There are some important differences between the two rock types which affect the quality of the reservoir and its interaction with fluids which flow through them.

The main component of sandstone reservoirs (siliciclastic reservoirs) is quartz (SiO_2). Chemically it is a fairly stable mineral which is not easily altered by changes in pressure, temperature or acidity of pore fluids. Sandstone reservoirs form after the sand grains have been transported over large distances and have deposited in particular *environments of deposition.*

Carbonate reservoir rock is usually found at the place of formation (in situ). Carbonate rocks are susceptible to alteration by the processes of *diagenesis.*

The pores between the rock components, for example the sand grains in a sandstone reservoir, will initially be filled with the *pore water.* The migrating hydrocarbons will displace the water and thus gradually fill the reservoir. For a reservoir to be effective, the pores need to be in communication to allow migration, and also need to allow flow towards the borehole once a well is drilled into the structure. The pore space is referred to as *porosity* in oil field terms. *Permeability* measures the ability of a rock to allow fluid flow through its pore system. A reservoir rock which has some porosity but too low a permeability to allow fluid flow is termed 'tight'. In Section 6.1, Chapter 6, we will examine the properties and lateral distribution of reservoir rocks in detail.

3.1.7. Traps

Hydrocarbons are normally of a lower density than formation water. Thus, if no mechanism is in place to stop their upward migration they will eventually seep to the surface. On seabed surveys in some offshore areas we can detect crater-like features (pock marks) which also bear witness to the escape of oil and gas to the surface. It is assumed that throughout the geologic past, vast quantities of hydrocarbons have been lost in this manner from sedimentary basins.

There are three basic forms of trap as shown in Figure 3.5. These are

- *Anticlinal traps* which are the result of ductile crustal deformations.
- *Fault traps* which are the result of brittle crustal deformations.
- *Stratigraphic traps* where impermeable strata seal the reservoir.

In many oil and gas fields throughout the world hydrocarbons are found in fault bound anticlinal structures. This type of trapping mechanism is called a *combination trap.*

Even if all of the elements described so far have been present within a sedimentary basin an accumulation will not necessarily be formed. One of the

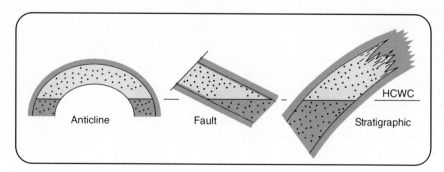

Figure 3.5   Main trapping mechanisms.

crucial questions in prospect evaluation is about the *timing* of events. The deformation of strata into a suitable trap has to precede the maturation and migration of petroleum. The reservoir seal must have been intact throughout geologic time. If a 'leak' occurred sometime in the past, the exploration well will only encounter small amounts of residual hydrocarbons. Conversely, a seal such as a fault may have developed early on in the field's history and prevented the migration of hydrocarbons into the structure.

In some cases bacteria may have *biodegraded* the oil, that is destroyed the light fraction. Many shallow accumulations have been altered by this process. An example would be the large heavy oil accumulations in Venezuela.

Given the costs of exploration ventures it is clear that much effort will be expended to avoid failure. A variety of disciplines are drawn in such as geology, geophysics, mathematics and geochemistry to analyse a prospective area. However, on average, even in very mature areas where exploration has been ongoing for years, only every third exploration well will encounter substantial amounts of hydrocarbons. In real 'wildcat' areas, basins which have not been drilled previously, only every tenth well is, on average, successful.

3.2. EXPLORATION METHODS AND TECHNIQUES

The objective of any exploration venture is to find new volumes of hydrocarbons at a low cost and in a short period of time. Exploration budgets are in direct competition with acquisition opportunities. If a company spends more money finding oil than it would do to buy the equivalent amount 'in the market place' there is little incentive to continue exploration. Conversely, a company which manages to find new reserves at low cost has a significant competitive edge since it can afford more exploration, find and develop reservoirs more profitably and can target and develop smaller prospects.

Once an area has been selected for exploration, the usual sequence of technical activities starts with the definition of a basin. The mapping of *gravity anomalies* and *magnetic anomalies* will be the first two methods applied. In many cases this data will be available in the public domain or can be bought as a 'non exclusive' survey. Next, a coarse two-dimensional (2D) seismic grid, covering a wide area, will be acquired in order to define *leads*, areas which show for instance a structure which potentially contains an accumulation (seismic methods will be discussed in more detail in the next section). Recently *electro-magnetic techniques* have also been deployed at this stage to assist in the delineation of basins and the identification of potential hydrocarbon accumulations. A particular exploration concept, often the idea of an individual or a team will emerge next. Since at this point very few hard facts are available to judge the merit of these ideas they are often referred to as 'play'. More detailed investigations will be integrated to define a 'prospect', a subsurface structure with a reasonable probability of containing all the elements of a petroleum accumulation, namely source rock, maturation, migration, reservoir rock and trap.

Eventually, only the drilling of an exploration well will prove the validity of the concept. A 'wildcat' well is drilled in a region with no prior well control. Wells either result in discoveries of oil and gas, or they find the objective zone to be water-bearing in which case they are termed 'dry'.

Exploration activities are potentially damaging to the environment. The cutting down of trees in preparation for an onshore seismic survey may result in severe soil erosion in years to come. Offshore, fragile ecological systems such as reefs can be permanently damaged by spills of crude or mud chemicals. Responsible companies will therefore carry out an *environmental impact assessment* (EIA) prior to activity planning and draw up contingency plans should an accident occur. In Chapter 5, a more detailed description of health, safety and environmental considerations will be provided.

3.2.1. Introduction to geophysical methods

There are various geophysical surveying methods that are routinely applied in the search for potential hydrocarbon accumulations. *Geophysical methods* respond to variations in physical properties of the earth's subsurface including its rocks, fluids and voids. They locate boundaries across which changes in properties occur. These changes give rise to an anomaly relative to a background value; this anomaly is the target which the methods are trying to detect.

The measurement of changes in signal strength along lines of a grid or network, 'profiling', allows anomalies to be mapped out spatially. Care should be taken to avoid *spatial* 'aliasing', the loss of fine detail information as a result of gathering data at only a small number of measuring stations (Figure 3.6). Time and budget often come into play at this stage.

It is important to remember that the mere acquisition and processing of data do not guarantee success of a survey: information is not equal to knowledge. Interpretation of geophysical data should always be carried out within a sound geological framework. Often several methods are used to complement one another or they are used in conjunction with other disciplines to develop a geologically

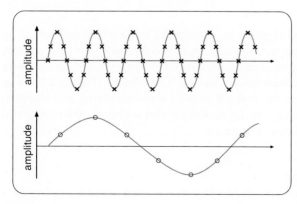

If the sample frequency is high e.g. measuring at times marked by a cross, then the wave is sampled adequantely with faithful representation of the input data.

If the sample frequency is low e.g. measuring at times marked by a circle, then the wave is sampled inadequantely with loss of the high frequency information and distortion of the input data.

Figure 3.6   Loss of information due to limited number of measurement points.

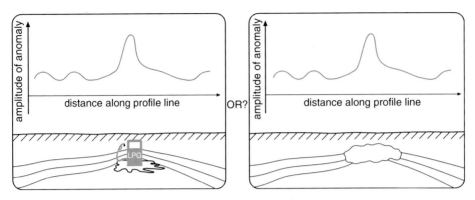

Figure 3.7   Alternative interpretations of the same anomaly response.

meaningful model that can explain the observed anomalies. This helps to reduce uncertainties and to address the principle of *equivalence* or 'non-uniqueness' where one anomaly can be modelled in a variety of ways (Figure 3.7).

3.2.1.1. Gravity surveys

The gravity method measures small variations of the earth's gravity field caused by density variations in geological structures. The measuring tool is a sophisticated form of spring balance designed to be responsive over a wide range of values. Fluctuations in the gravity field give rise to changes in the spring length which are measured (relative to a base station value) at various stations along the profile of a 2D network. The measurements are corrected for latitudinal position and elevation of the recording station to define the 'Bouguer' anomaly (Figure 3.8).

   The development of airborne gravity technology has allowed the surveying of previously inaccessible areas and of much larger basins than is currently practical with land-based measuring tools.

3.2.1.2. Magnetic surveys

The magnetic method detects changes in the earth's magnetic field caused by variations in the magnetic properties of rocks. In particular, basement and igneous rocks are relatively highly magnetic. If they are located close to the surface they give rise to anomalies with a short wavelength and high amplitude (Figure 3.9). The method is airborne (plane or satellite) which permits rapid surveying and mapping with good areal coverage. Like the gravity technique this survey is often employed at the beginning of an exploration venture.

3.2.1.3. CSEM seabed logging

Controlled source electro-magnetic (CSEM) surveying or seabed logging is a remote sensing technique which uses very low frequency electro-magnetic signals emitted from a source near the seabed (Figure 3.10). Receivers are placed on the

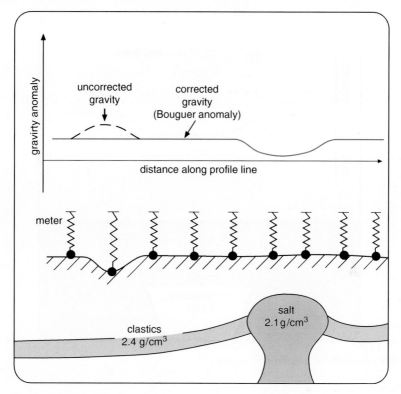

Figure 3.8 Principle of gravity surveying.

seabed at regular intervals and register anomalies and distortions in the electro-magnetic signal generated by resistive bodies, such as reservoirs saturated with hydrocarbons.

CSEM works best in deep water ($> 500 \, \text{m}$) in areas characterised by relatively simple sand-shale sequences (clastic reservoirs); it is particularly useful for surveying large traps (prospects) where other marine methods are less practical or economical. It is being increasingly used in conjunction with seismic data to verify likely fluid fill within the reservoir rocks of a prospect, thus helping to reduce risk and to improve the chance of success by allowing wells to be targeted in a more sophisticated way.

3.2.2. Seismic acquisition and processing

3.2.2.1. Introduction

Advances in seismic surveying techniques and the development of more sophisticated seismic processing algorithms over the last few decades have changed the way fields are developed and managed. From being a predominantly exploration focused tool, seismic surveying has progressed to become one of the most cost effective methods for optimising field production. In many cases, seismic data have allowed operators to extend the life of 'mature' fields by many years.

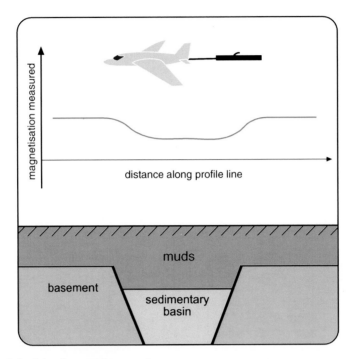

Figure 3.9 Principle of magnetic surveying.

Seismic surveys involve generating sound waves which propagate through the earth's rocks down to reservoir targets. The waves are reflected to the surface, where they are registered in receivers, recorded and stored for processing. The resulting data make up an acoustic image of the subsurface which is interpreted by geophysicists and geologists.

Seismic surveying is used in

- *exploration* for delineating structural and stratigraphic traps
- *field appraisal* and *development* for estimating reserves and drawing up FDPs
- *production* for reservoir surveillance such as observing the movement of reservoir fluids in response to production.

Seismic acquisition techniques vary depending on the environment (onshore or offshore) and the purpose of the survey. In an exploration area a seismic survey may consist of a loose grid of 2D lines. In contrast, in an area undergoing appraisal, a 3D seismic survey will be shot. In some mature fields a permanent 3D acquisition network might be installed on the seabed for regular (6–12 months) reservoir surveillance, called *ocean bottom stations* (OBS) or *ocean bottom cables* (OBC).

3.2.2.2. Principles of seismic surveying
The principles of seismic reflection surveying are set out below with the help of Figure 3.11.

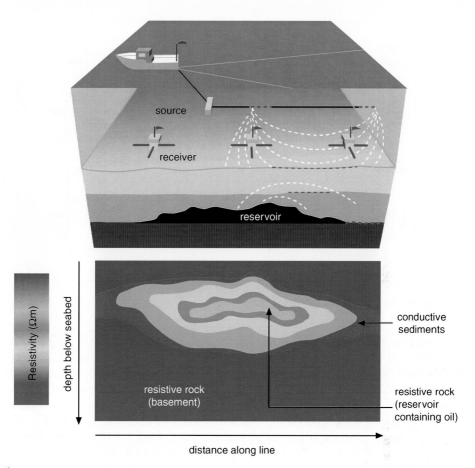

Figure 3.10   Principle of CSEM seabed logging.

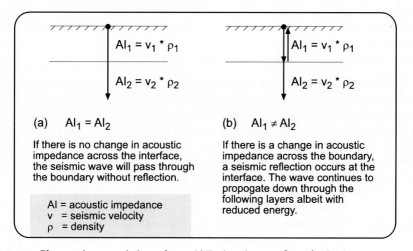

Figure 3.11   Changes in acoustic impedance (AI) give rise to reflected seismic waves.

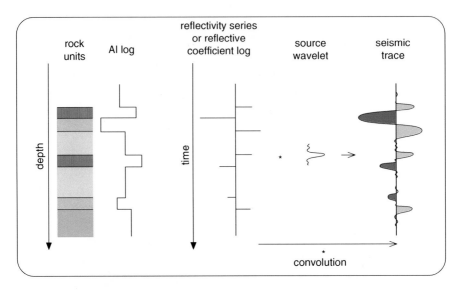

Figure 3.12 Convolution of a reflected seismic wave.

Sound waves are generated at the surface (onshore) or under water (offshore) and travel through the earth's subsurface. The waves are reflected back to the surface at the interface between two rock units where there is an appreciable change in 'acoustic impedance' (AI) across that interface. AI is the product of the density of the rock formation and the velocity of the wave through that particular rock (seismic velocity).

'Convolution' is the process by which a wave is modified as a result of passing through a filter. The earth can be thought of as a filter which acts to alter the waveform characteristics of the down-going wave (amplitude, phase, frequency). In schematic form (Figure 3.12) the earth can be represented either as an AI log in depth or as a series of spikes, called a *reflection coefficient log* or *reflectivity series* represented in the time domain. When the wave passes through the rocks its shape changes to produce a wiggle trace that is a function of the original source wavelet and the earth's properties.

Two attributes of the reflected signal are recorded.

- The *reflection time*, or travel time, is related to the depth of the interface or 'reflector' and the seismic velocity in the overburden.
- The *amplitude* is related to rock and fluid properties within the reflecting interval and various extraneous influences that need to be removed during processing.

When a seismic wave hits an interface at normal incidence (Figure 3.13a), part of the energy is reflected back to the surface and part of the energy is transmitted. In the case of oblique incidence the angle of the incident wave equals the angle of the reflected wave as shown in Figure 3.13b. Again part of the energy is transmitted to the following layer, but this time with a changed angle of propagation. A special case is shown in Figure 3.13c where an abrupt discontinuity, for example the edge of a tilted fault block, gives rise to 'diffractions', radial scattering of the incident seismic

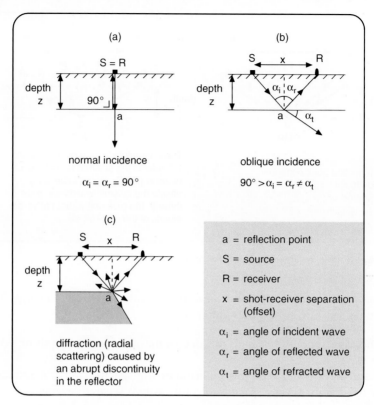

Figure 3.13   Reflection of waves at normal and oblique incidence.

energy. Such artefacts can impede interpretation of the seismic data but can be removed or suppressed during processing (as outlined later in this section).

3.2.2.3. Seismic data acquisition

The time it takes for the wave to travel from the source S to a reflection point a at depth z and up to a receiver R at an offset, or shot–receiver separation, x, is given by the ratio of the travel path and the velocity (Figure 3.14a). The acquisition system is arranged such that there are many shot–receiver pairs for each reflection point in the subsurface, also called 'common midpoint' or CMP.

Reflection times are measured at different offsets (x_1, x_2, x_3, ... x_n); the further away shot and receiver are for a particular reflection point in the subsurface, the longer the travel time. The difference in travel time between the zero offset case (normal incidence) and the non-zero offset case (oblique incidence) is called the *normal move out* (NMO) and is dependent on the offset, velocity and depth to the reflector. Collecting data from different offsets and also at different angles is important for imaging the subsurface properly, for instance where intermediate layers or structures impact on the amount of energy reaching the target (Figure 3.14b) or where they give rise to variations in seismic velocity.

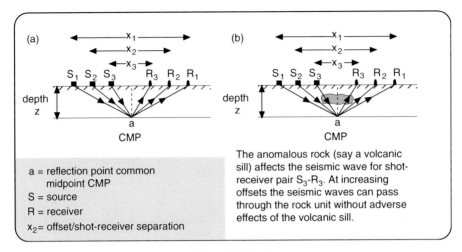

Figure 3.14   Source–receiver geometry for multiple offsets.

Seismic sources generate acoustic waves by the sudden release of energy. There are various types of sources and they differ in

- the amount of energy released: this determines the specific depth of penetration of the wave
- the frequencies generated: this determines the specific 'vertical resolution', or ability to identify closely spaced reflectors as two separate events.

There is usually a trade-off between the two depending on the objectives of the survey. Studies of deep crustal structures require low frequency signals capable of penetrating over 10 km into the earth, whereas a shallow geological survey requires a very high frequency signal which is allowed to die out after only a few hundred metres.

Typical sources for land surveys are truck-mounted *vibrating sources* or small *dynamite charge* sources detonated in a shallow hole. The most common marine sources are pneumatic sources such as *air guns* and *water guns* that expel air or water into the surrounding water column to create an acoustic pulse. There are also electrical devices such as sparkers, boomers and pingers that convert electrical energy into acoustic energy. Typically the latter produce less energy and have a higher frequency signal than pneumatic sources.

Seismic detectors are devices that register a mechanical input (seismic pulse) and transform it into an electrical output which is amplified before being recorded to tape. On land the receivers are called *geophones* and they are arranged in a spread on the ground or in shallow boreholes. At sea the receivers are called *hydrophones*, often clustered in arrays, and they are either towed in the water behind the boat or laid out on the sea floor in the case of OBC (Figure 3.15).

The *acquisition geometry*, or the configuration of source(s) and receivers depends on the objectives of the survey, characteristics of the subsurface geology and logistics. Seismic surveys can be acquired along straight lines, zig-zag lines, in a

Figure 3.15   Seismic data acquisition: survey vessels at sea.

square loop and even in a circular pattern. Over the last few years *multi-azimuth surveys* have become increasingly popular. Seismic data are acquired along different azimuths (Figure 3.16) to allow structures to be imaged at different angles thus enhancing the imaging of complex geology, such as radial fault patterns and areas affected by salt.

3.2.2.4. Borehole seismic surveying

In *vertical seismic profiling* (VSP) the seismic source is placed at the surface and the receiver array is lowered down a borehole. In the case of *borehole tomography* both source and receiver array are lowered into (different) boreholes and the source is fired at different depths (Figure 3.17). Typically the seismic sources use higher frequencies than in surface seismic surveys.

Advantages of borehole seismic techniques include improved resolution and the ability to predict or more accurately model the velocity variations between wells. Furthermore, the effects of the near-surface weathered layer are removed or suppressed. The result is that small-scale features and subtle variations in reservoir continuity can be imaged better than using conventional surface seismic data which has proved very powerful in field development and well planning. More recently it has also been used to help characterise tight gas sands and coal bed methane seams where very small features can have a dramatic impact on resource distribution and recovery.

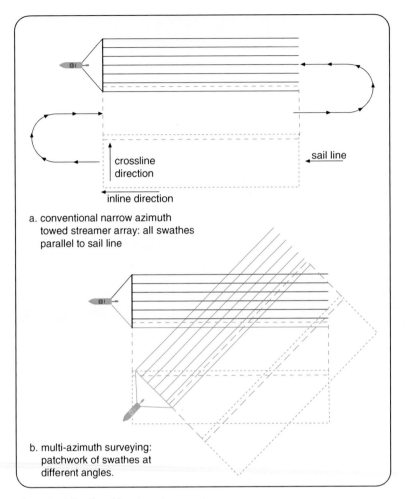

crossline
direction

sail line

inline direction

a. conventional narrow azimuth
   towed streamer array: all swathes
   parallel to sail line

b. multi-azimuth surveying:
   patchwork of swathes at
   different angles.

Figure 3.16   Principle of multi-azimuth surveying.

3.2.2.5. Seismic data processing

3.2.2.5.1. Introduction. The three main steps in *seismic data processing* are *deconvolution, stacking* and *migration*. Additional processes are required to prepare or enhance the seismic data before or after each of the main steps.

There are typically hundreds of traces in a 2D survey and thousands in a 3D survey. Once they have been sorted, static corrections must be applied to compensate for variations in topography, for example when seismic data are acquired in an area covered by sand dunes. 'Statics' also correct for variations in seismic velocity in the near-surface, for example when a seismic survey is acquired in a swampy area.

3.2.2.5.2. Deconvolution. The next stage in processing is deconvolution. In essence this is an inverse filtering procedure which removes or suppresses unwanted signals. It aims to collapse the wavelet and make it as sharp as possible so that it

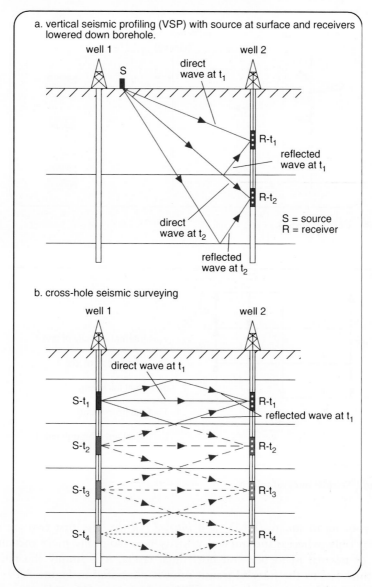

a. vertical seismic profiling (VSP) with source at surface and receivers lowered down borehole.

b. cross-hole seismic surveying

Figure 3.17 Principles of borehole seismic surveying.

resembles a spike (Figure 3.18). In effect deconvolution tries to remove the effects of the earth's filter by reproducing the geological boundaries as a reflectivity series.

3.2.2.5.3. Velocity analysis and NMO correction. From previous sections it has become clear that seismic velocity plays an important role in seismic surveying and processing. It is the one parameter that allows the seismic image to be converted into a geological depth section. There are several types of seismic velocity, such as

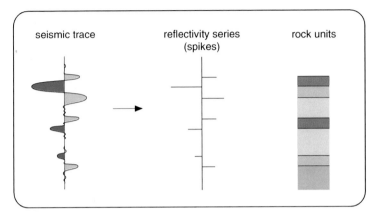

Figure 3.18   Deconvolution.

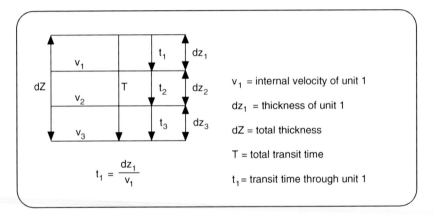

Figure 3.19   Seismic interval velocity.

average, root mean square (RMS) and *interval velocity*. The first two are statistical parameters only, whereas the interval velocity is geologically more meaningful. In the case of normal incidence and horizontal layers, it is simply the ratio of the interval thickness to the interval transit time as illustrated in Figure 3.19.

As mentioned previously, there is a difference in travel time between the zero offset case and the non-zero offset case for each CMP – this is known as NMO. Viewing the traces side by side (Figure 3.20a), it is clear that the NMO for each non-zero offset trace needs to be removed before the traces can be summed. The stacking velocity is the seismic velocity which results in the best correction for NMO (Figure 3.20b).

3.2.2.5.4. Stacking.   All the reflections from the various offsets associated with a CMP are summed, or 'stacked' to give one trace for each CMP; this leads to an improvement in the 'signal-to-noise ratio'. Signals from spurious noise tend to vary between the different traces and will, therefore, get cancelled out or at least suppressed. True geological signals from the different traces tend to be similar and are thus boosted during the stacking process.

3.2.2.5.5. Migration. Ideally, after stacking the seismic data are in the correct position and have the correct amplitudes. However, steeply dipping horizons cause reflections to be recorded at surface positions which are different to their actual subsurface position as shown in Figure 3.21. This also happens when large and sudden variations occur in seismic velocity.

The incident wave coming from the source at S_1 hits a point at position a and depth z and is reflected to the receiver at R_1. In the case of horizontal reflectors the

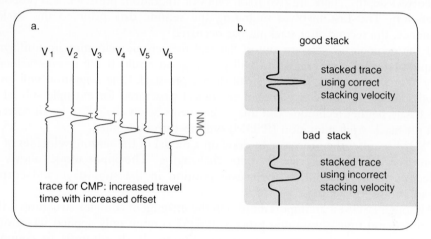

Figure 3.20 NMO correction.

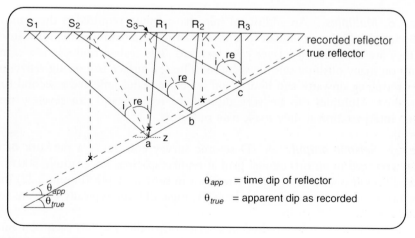

Figure 3.21 Migration.

travel time of the incident wave is the same as the travel time of the reflected wave. Point a at depth z is recorded at position a' at the surface and associated with depth z'; both the position and the depth are correct: $a = a'$ and $z = z'$.

In the case of steeply dipping reflectors the travel time of the incident wave is different to the travel time of the reflected wave. In the picture the travel time of the reflected wave is much smaller than the travel time of the incident wave. This leads to point a being recorded updip of its true position with a shift in surface position $(a \neq a')$ and a shift in depth $(z \neq z')$; the same occurs at point b and so on. The true dip (θ_{true}) of the reflector is imaged incorrectly and the apparent dip (θ_{app}) is shallower.

Migration is the process of repositioning reflected signals to show an event (geological boundary or other structure) at its true position in the subsurface and at its correct depth. There are two main types of migration: pre-stack and post-stack migration. The first involves migrating the seismic data prior to the stacking sequence, the second after stacking has occurred.

If the geological layers are almost flat and the seismic velocities are uniform, a simple post-stack time migration will give a good result. If the seismic velocities vary only a little or the dips are small then a pre-stack time migration will give a good solution. In areas of complex geological structures, for example sub-salt or sub-basalt, neither technique will image the events below the salt or basalt correctly and pre-stack *depth* migration (PSDM) will need to be applied.

PSDM requires the processor to draw up a model of the seismic velocities of the subsurface, this in itself can be quite challenging. The input model allows the reflectors to be restored to their true position in the subsurface and corrects apparent dips to true dips.

Although PSDM is an important tool in the imaging of complex structures it is an expensive and time-consuming process. PSDM is often only applied when other methods have failed to give a working solution. With advances in computer technology and processing capability, however, PSDM is likely to be become economic and more readily applied.

3.2.2.5.6. *Multiples.*

An additional step that is often required is the removal of 'multiples'. Multiples are signals that have been reflected at more than one interface and they are common in a layer cake scenario. The seabed multiple is a common feature on many offshore seismic sections. The sea surface is a strong reflector and waves travelling upwards can bounce off it before being reflected a second time at the seabed. Multiples can be very difficult to remove and can severely impede seismic interpretation if they mask true reflectors.

3.2.2.5.7. *Seismic output.*

A 2D seismic survey consists of a network of lines, usually arranged in an orthogonal grid at regular spacing, for example 500 m. The processed result is a series of seismic sections in time or depth (Figure 3.22) that tie at the nodes or intersections of the lines. A single 2D line typically contains several hundred traces.

A 3D seismic survey is acquired in a series of parallel swathes each containing a large number of inlines (sail lines) and crosslines (perpendicular to the sail lines)

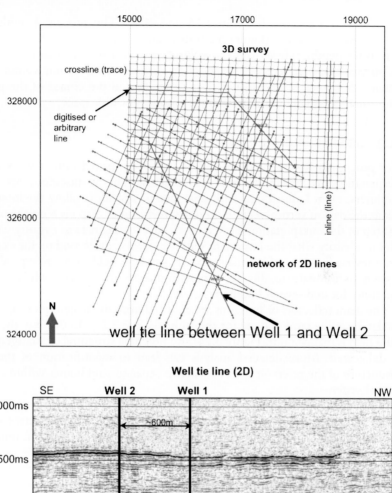

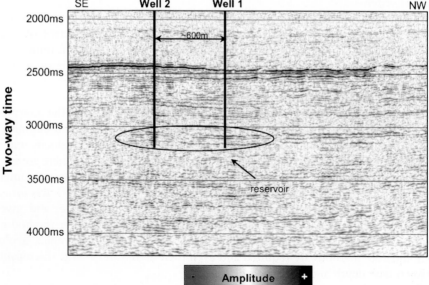

Figure 3.22 Basemap with network of 2D lines and 3D survey (top) and 2D well tie line in vertical section (bottom).

typically with a spacing between 12.5 and 50 m. The processed result is a 3D 'volume' or cube of data (Figure 3.22) that can be viewed along all three axes (line, trace, time/depth). These days the volumes can also be sliced along an 'arbitrary line' such as along the axis of a meandering channel. A 3D seismic volume typically contains thousands of traces. It is clear to see that in the course of the processing phase, such large volumes require huge amounts of disk space.

3.2.3. Seismic interpretation

After processing has been completed, the data are loaded onto a workstation for interpretation by geologists and geophysicists. The workstations are powerful computers, often Linux-based with dual screen capacity to allow the interpreter to look at the data in vertical section on one screen and in map view on the other. The first step in the interpretation cycle is to 'tie' the seismic data to existing well data in order to identify what the important reflector events correspond to, for example top of the reservoir or top of the main seal. In a mature field there are typically dozens of wells to calibrate to, but in exploration areas there may only be a couple, sometimes located several kilometres away.

The main reflectors or horizons are digitised from the screen (picked) and stored in a database; the same is done for the faults (Figure 3.23). In this way the structure of the field is mapped out (Figure 3.24) and potential structural or stratigraphic traps are delineated. More detailed analysis can lead to identification of the internal architecture of the reservoir interval, such as separate sand bodies within a complex channel system.

Nowadays geoscientists and engineers prefer to view seismic data not in terms of reflection data with the characteristic wavelet signature, but in terms of *acoustic impedance*. This is achieved by *seismic inversion*, a process which removes the influence of the wavelet and represents the data in a geologically meaningful way, namely as a function of rock properties. Inversion requires careful calibration to well data and knowledge of the broad geological model of the subsurface.

Once the interpretation has been completed in the time domain, the interpreted surfaces need to be converted to depth for use in the geological and engineering model. Depth conversion again requires knowledge of the seismic velocity and any significant variations, both lateral and vertical, that may be present. There are several methods of depth conversion. A simple method is to derive seismic interval velocities for a number of key intervals and then to calculate the thickness for each interval before summing them. This method is called 'isochoring' and gives a reasonable result in areas not affected by velocity variations. Another method is to build a velocity model based on stacking velocities. In areas of complex geology, more intricate methods are required and even then there can be large discrepancies between true depth and calculated depths.

3.2.4. Seismic attributes

The development of post-stack processing algorithms has allowed 3D seismic data to be interrogated in increasingly sophisticated ways. Structural attributes of the data

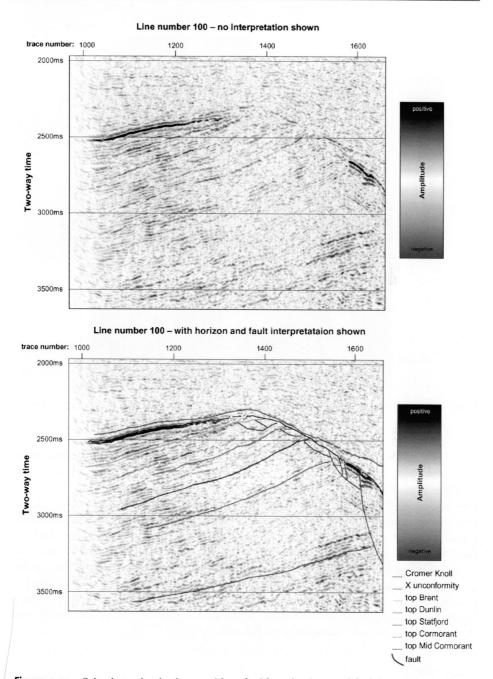

Figure 3.23 Seismic section in time – with and without horizon and fault interpretation.

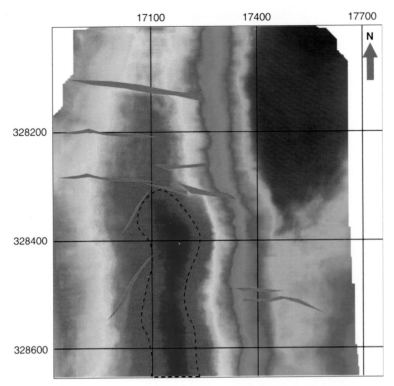

Figure 3.24 Top structure map with interpreted faults and hydrocarbon contact (dashed line).

such as dip, azimuth and degree of uniformity can help the interpreter to understand structural styles within a basin or to interpret intricate fault patterns (Figure 3.25).

Attributes derived from the amplitude characteristics of the data can give insights into rock properties, such as porosity and density, and in some cases fluid fill, for example hydrocarbon saturation (Figure 3.26). Detailed amplitude analysis work requires careful calibration and modelling before attempting to attribute rock or fluid properties to the amplitudes seen in the data. It is important to note that the results of the analysis are only as good as the input data and the quality of the model.

In recent years a technique called 'frequency decomposition', also called 'spectral decomposition', has been developed to analyse 3D seismic data in even more detail. It involves decomposing the amplitude signal into constituent frequency bands and studying the amplitude strength at each band.

3.2.5. AVO

Amplitude variation with offset (AVO) or angle (AVA) can be a powerful tool in the search for hydrocarbons. Instead of looking only at the stack of traces at all offsets, the traces at near, mid and far offsets can be stacked separately and their amplitudes compared. Given certain rock properties and hydrocarbon fill, changes occur in seismic amplitude and/or phase between the different offset stacks. For instance,

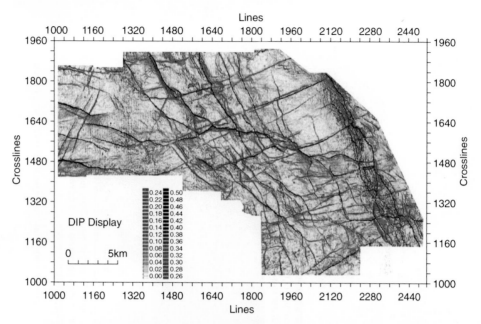

Figure 3.25 Structural attribute: dip map showing complex fault patterns induced by salt movement.

Figure 3.26 Stratigraphic attribute: the bright amplitudes (green, yellow, red) indicate the presence of turbidite channels; note the influence of faulting (red dashed line) on sediment transport (arrows).

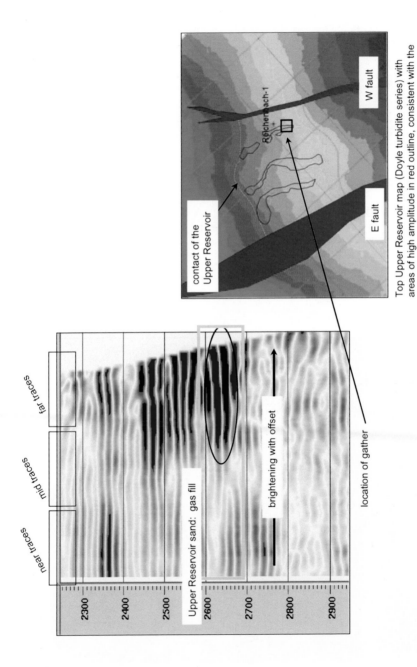

Figure 3.27   Gas filled sand of the Doyle turbidite series (Upper Reservoir) produces seismic amplitude increase with increasing offset.

with increasing angle there may be an increase in amplitude in the case of a gas-filled sand (Figure 3.27).

In order to achieve success with AVO analysis, careful modelling of rock properties and fluid fill is required to understand the variations that occur relative to the background trend. Furthermore, AVO is not suitable for all reservoir types and works better in young, poorly consolidated rocks, for example West African turbidites, than in some of the older, more cemented reservoirs encountered in the North Sea.

3.2.6. Time-lapse seismic surveys (4D seismic)

Seismic surveys can be repeated at difference times over the course of field life, for example at regular intervals after production has started. Changes in seismic amplitude and other attributes may occur on the post-production seismic data (monitor survey) when compared to the original pre-production seismic data (baseline survey). These changes are usually related to fluid movement and changes in fluid content as a result of depleting a reservoir (Figure 3.28).

Time-lapse seismic data can include repeat VSP surveys, 2D surveys or 3D surveys, the latter are termed '4D' data. Time-lapse surveys are becoming increasingly popular especially in mature fields where 4D data can highlight the presence of unswept compartments (Figure 3.29) or track the movement of flood

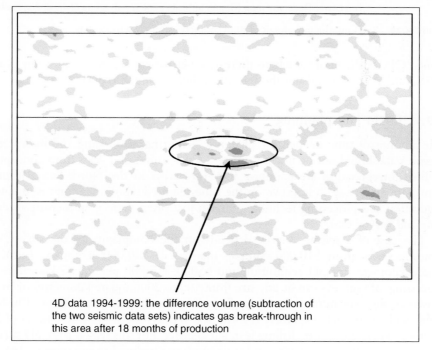

4D data 1994-1999: the difference volume (subtraction of the two seismic data sets) indicates gas break-through in this area after 18 months of production

Figure 3.28 4D seismic data: difference between 1994 and 1999 seismic data showing changes in amplitude as a result of production.

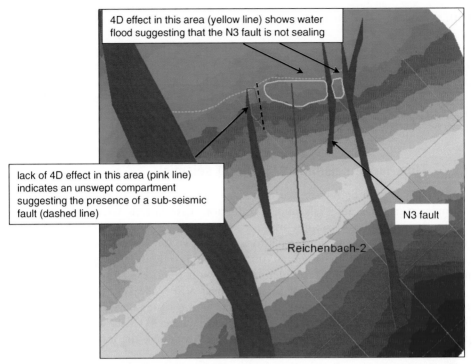

4D effect in this area (yellow line) shows water flood suggesting that the N3 fault is not sealing

lack of 4D effect in this area (pink line) indicates an unswept compartment suggesting the presence of a sub-seismic fault (dashed line)

N3 fault

Reichenbach-2

Top Lower Reservoir map (Conan turbidite series) with areas of 4D changes shown

Figure 3.29   Tracking the movement of fluid contacts using 4D seismic data.

fronts. Obviously, in areas where there is a permanent seismic acquisition system (OBC) the cost of acquiring the repeat survey(s) is much reduced.

3.2.7. Costs and planning

The amount of time needed for planning, acquiring, processing and interpreting seismic data should not be underestimated. Cycle times of 2 years from conception to final interpretation are common for 3D surveys in the North Sea. Although efforts are underway to improve on the time required, continued improvements in acquisition and processing technology mean that often there is an increase in cycle time and survey cost.

The cost of a seismic survey depends on the complexity of the survey, but typically varies from $10,000 (simple, marine) to $40,000 (complex, land) per square kilometre for 3D acquisition and $5000–$15,000 per square kilometre for processing. 3D surveys can be any size from 100 to 2000 square kilometres or more. However, the determining economic factor is often the ratio to well cost. Offshore wells can be extremely expensive (North Sea wells typically cost in the order of $20 million), whereas onshore drilling is much cheaper. For this reason large 3D surveys are often used offshore where companies are more inclined to use seismic data as a substitute for drilling at the appraisal stage.

CHAPTER 4

DRILLING ENGINEERING

Introduction and Commercial Application: Drilling operations are carried out during all stages of the project life cycle and in all types of environments. The main objectives are the acquisition of information and the safeguarding of production. Expenditure for drilling represents a large fraction of the total project's capital expenditure (CAPEX) (typically 20–60%), therefore an understanding of the techniques, equipment and cost of drilling is important.

An initial successful exploration well will establish the presence of a working petroleum system. In the following months, the data gathered in the first well will be evaluated and the results documented. The next step will be the appraisal of the accumulation requiring more wells. If the project is subsequently moved forward, development wells will have to be engineered. The following section will focus on these drilling activities, and will also investigate the interactions between the drilling team and the other E&P functions.

4.1. WELL PLANNING

The drilling of a well involves a major investment, ranging from a few million US$ for an onshore well to 100 million US$ plus for a deepwater exploration well. *Well engineering* is aimed at maximising the value of this investment by employing the most appropriate technology and business processes, to drill a 'fit for purpose' well, at the minimum cost, without compromising safety or environmental standards. Successful drilling engineering requires the integration of many disciplines and skills.

Successful drilling projects will require extensive planning. Usually, wells are drilled with one, or a combination, of the following objectives:

- to gather information
- to produce hydrocarbons
- to inject gas or water to maintain reservoir pressure or sweep out oil
- to dispose of water, drill cuttings or CO_2 (sequestration).

To optimise the design of a well it is desirable to have as accurate a picture as possible of the subsurface. Therefore, a number of disciplines will have to provide information prior to the design of the well trajectory and before a drilling rig and specific equipment can be selected.

The subsurface team will define optimum locations for the planned wells to penetrate the reservoir and in consultation with the well engineer agree on the desired trajectory through the objective sequence. In discussions with production and well engineers maximum hole inclination and required wellbore diameter will be determined. Wellhead locations, well design and trajectory are aimed at minimising the combined costs of well construction and seabed/surface facilities, whilst maximising production.

The accuracy of the parameters used in the well planning process will depend on the knowledge of the field or the region. Particularly during exploration drilling and the early stages of field development considerable uncertainty in subsurface data will prevail. It is important that the uncertainties are clearly spelled out and preferably quantified. Potential risks and problems expected or already encountered in *offset wells* (earlier wells drilled in the area) should be incorporated into the design of the planned well. This is often achieved by using a decision tree approach in the well planning phase. The optimum well design balances risk, uncertainty and cost with overall project value.

The basis for the well design is captured in a comprehensive document. This is then 'translated' into a drilling programme.

In summary, the well engineer will be able to design and cost the well in detail using the information obtained from the petroleum engineers, geoscientists and production engineers. In particular, he will plan the setting depth and ratings for the various casing strings, cementing programme, mud weights and mud types required during drilling, and select an appropriate rig and related hardware, for example drill bits.

The following sections will explain, in more detail, the terms introduced so far.

4.2. RIG TYPES AND RIG SELECTION

The type of rig which will be selected depends upon a number of parameters, in particular:

- cost and availability
- water depth of location (offshore)
- mobility/transportability (onshore)
- depth of target zone and expected formation pressures
- prevailing weather/metocean conditions in the area of operation
- experience of the drilling crew (in particular the safety record!).

The following types of rig can be contracted for offshore drilling (Figure 4.1).

Swamp barges operate in very shallow water (less than 20 ft). They can be towed onto location and are then ballasted so that they 'sit on bottom'. The drilling unit is mounted onto the barge. This type of unit is used in the swamp areas of, for example Nigeria, Venezuela and US Gulf Coast.

Drilling jackets are small steel platform structures which are used in areas of shallow and calm water. A number of wells may be drilled from one jacket. If a jacket is too small to accommodate a drilling operation, a jack-up rig (see Figure 4.2) is usually cantilevered over the jacket and the operation carried out from there. Once a viable development has been proven, it is extremely cost-effective to build and operate jackets in a shallow sea environment. In particular, they allow a flexible and step-wise

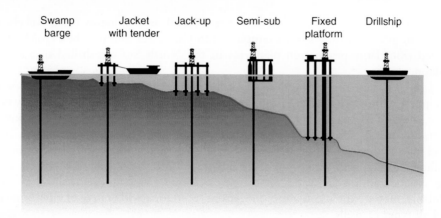

Figure 4.1   Offshore rig types.

Figure 4.2   Jack-up rig (courtesy of Reading & Bates, UK).

progression of field development activities. Phased developments using jackets are common in coastal waters, for example South China Sea and the shelf GoM. Wells drilled from large production platforms in the North Sea are drilled in a similar fashion.

Jack-up rigs are either towed to the drilling location (or alongside a jacket) or are equipped with a propulsion system. The three or four legs of the rig are lowered onto the seabed. After some penetration the rig will lift itself to a determined operating height above the sea level. If soft sediment is suspected at seabed, large mud mats will be placed on the seabed to allow a better distribution of weight. All drilling and supporting equipment are integrated into the overall structure. Jack-up rigs are operational in water depths up to about 450 ft and as shallow as 15 ft. Globally, they are the most common rig type, used for a wide range of environments and all types of wells.

Semi-submersibles are used for exploration and appraisal in water depths too great for a jack-up. A semi-submersible rig is a movable offshore vessel consisting of a large deck area built on columns of steel. Attached to these heavy-duty columns are at least two barge-shaped hulls called pontoons. Before operation commences on a specified location, these pontoons are partially filled with water and submersed in approximately 50 ft of water to give stability. A large-diameter steel pipe (riser) is connected to the seabed and serves as a conduit for the drill string. The *blowout preventer* (BOP) is also located at the seabed (subsea stack).

A combination of several anchors and dynamic positioning (DP) equipment assists in maintaining position. Relocation of the semi-submersible vessel is made possible by the utilisation of tugboats and/or propulsion machinery.

Heavy-duty semi-submersibles, for example Deepwater Horizon (rated 15,000 psi), can handle high reservoir pressures and operate in the most severe metocean conditions in water depths down to 3000 m (Figure 4.3).

Drill ships are used for deep and very deep water work. They can be less stable in rough seas than semi-submersibles. However, modern high-specification drill ships such as Discoverer Enterprise can remain stable, and on target during 100 knot winds using powerful thrusters controlled by a DP system. The thrusters counter the forces of currents, wind and waves to keep the vessel exactly on target, averaging less than 2 m off her mark, without an anchor.

Heavy-duty drill ships are capable of operating in water depths up to 3000 m (Figure 4.4).

In some cases, oil and gas fields are developed from a number of platforms. Some platforms will accommodate production and processing facilities as well as living quarters. Alternatively, these functions may be performed on separate platforms, typically in shallow and calm water. On all offshore structures, however, the installation of additional weight or space is costly. Drilling is only carried out during short periods of time if compared to the overall field life span and it is desirable to have a rig installed only when needed. This is the concept of tender-assisted drilling operations.

In *tender-assisted drilling*, a derrick is assembled from a number of segments transported to the platform by a barge. All the supporting functions such as storage, mud tanks and living quarters are located on the tender, which is a specially built

Figure 4.3   Semi-submersible rig (courtesy of Stena Drilling).

Figure 4.4   Drill ship 'Transocean Enterprise'.

Figure 4.5   Tender–assisted drilling.

spacious barge anchored alongside (Figure 4.5). It is thus possible to service a whole field or even several fields using only one or two tender–assisted derrick sets. In rough weather, barge type tenders quickly become inoperable and unsafe since the platform is fixed whereas the barge moves up and down with the waves. In these cases and in the hostile environment of the North Sea, a modified semi-submersible vessel may serve as a tender. Currently, purpose-built semi-submersible tenders are being introduced for some future field developments.

4.3. DRILLING SYSTEMS AND EQUIPMENT

Whether onshore or offshore drilling is carried out, the basic drilling system employed in both the cases will be the *rotary rig* (Figure 4.6). The parts of such a unit and the three basic functions carried out during rotary drilling operations are as follows:

- Torque is transmitted from a power source at the surface through a drill string to the drill bit.
- A drilling fluid is pumped from a storage unit down the drill string and up through the annulus. This fluid will bring the cuttings created by the bit action to the surface, hence clean the hole, cool the bit and lubricate the drill string.
- The subsurface pressures above and within the hydrocarbon-bearing strata are controlled by the weight of the drilling fluid and by large seal assemblies at the surface (BOPs).

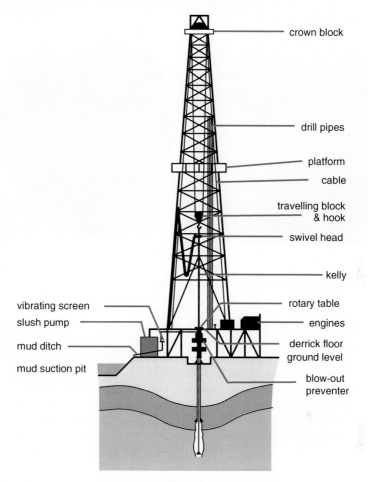

Figure 4.6   The basic rotary rig.

However, in practice, onshore and offshore drilling units are often quite different in terms of technology and degree of automatisation. This is largely driven by rig availability, costs and safety considerations, and will be explained in more detail in the following text.

We will now consider the rotary rig in operation, visiting all elements of the system. The type of rig operation described first is now found mainly in low-cost onshore areas. For complicated, more expensive wells, older rigs have usually been upgraded to include a *top drive system* and *automated pipe handling* as described later in this section. New rigs are usually built with this equipment as standard.

4.3.1. Drill bits

The most frequently used bit types are the roller cone or rock bit and the polycrystalline diamond compact bit or PDC bit (Figure 4.7).

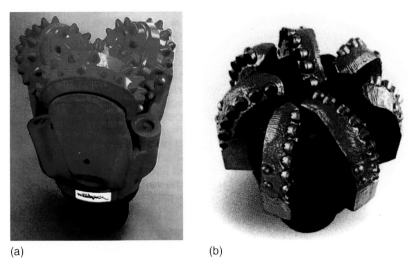

(a)   (b)

Figure 4.7   Roller cone bit (left) and PDC bit.

On a rock bit, the three cones are rotated and the attached teeth break or crush the rock underneath into small chips (cuttings). The cutting action is supported by powerful jets of drilling fluid which are discharged under high pressure through nozzles located at the side of the bit. After some hours of drilling (between 5 and 25 h depending on the formation and bit type), the teeth will become dull and the bearings wear out. Later on we will see how a new bit can be fitted to the drill string. The PDC bit is fitted with industrial diamond cutters instead of hardened metal teeth. This type of bit is popular because of its much better rate of penetration (ROP), longer lifetime and suitability for drilling with high revolutions per minute (rpm), which makes it the preferred choice for turbine drilling. The selection of bit type depends on the composition and hardness of the formation to be drilled and the planned drilling parameters.

Between the bit and the surface, where the torque is generated, we find the *drill string* (Figure 4.8). Whilst primarily being a means for power transmission, the drill string fulfils several other functions, and if we move up from the bit we can see what those are.

The drill collars (DCs) are thick-walled, heavy lengths of pipe. They keep the drill string in tension (avoiding buckling) and provide weight onto the bit. Stabilisers are added to the drill string at intervals to hold, increase or decrease the hole angle. The function of stabilisers will be explained in more detail in Section 4.5. The *bottom hole assembly* (BHA) described so far is suspended from the drill pipe, made up of 30 ft long sections of steel pipe (joints) screwed together. The drill string is connected to the kelly saver sub. A saver sub is basically a short piece of connecting pipe with threads on both ends. In cases where connections have to be made up and broken frequently, the sub 'saves' the threads of the more expensive equipment. The kelly is a six-sided piece of pipe that fits tightly into the kelly bushing which is fitted into the rotary table. By turning the latter, torque is transmitted from the kelly down

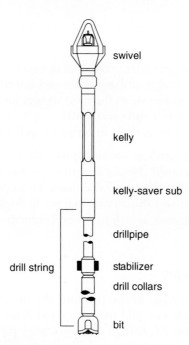

Figure 4.8 The drill string (schematic).

the hole to the bit. It may take a number of turns of the rotary table to initially turn the bit thousands of meters down the hole.

The kelly is hung from the travelling block. Since the latter does not rotate, a bearing is required between the block and the kelly. This bearing is called a swivel. Turning the drill string in a deep reservoir would be the dimensional equivalent to transmitting torque through an everyday drinking straw dangling from the edge of a 75-storey high-rise building! As a result, all components of the drill string are made of high-quality steels.

After the drilling has progressed for some time, a new piece of drill pipe will have to be added to the drill string (see below). Alternatively, the bit may need to be replaced or the drill string has to be removed for logging. In order to 'pull out of hole', hoisting equipment is required. On a basic rotary rig this consists of the hook which is connected to the travelling block. The latter is moved up and down via a steel cable (block line) which is spooled through the crown block onto a drum (draw works). The draw works, fitted with large brakes, move the whole drill string up and down as needed. The derrick or mast provides the overall structural support to the operations described.

For various reasons, such as to change the bit or drilling assembly, the drill string may have to be brought to surface. It is normal practice to pull 'stands' consisting of 90 ft sections of drill string and rack them in the mast rather than disconnecting all the segments. The procedure of pulling out of hole (POOH) and running in again is called a *round trip*.

The system described so far has been in operation for many decades. Several disadvantages are apparent:

- The requirement to add pipe after every 30 ft drilled is time consuming, results in longer openhole times, longer drilling times and lower quality wellbores.
- The rig floor is known to be one of the most dangerous areas of a rig, accounting for a large percentage of lost time accidents.
- The technology imposes limitations in terms of well trajectory and complexity.

For a modern mobile offshore drilling unit (MODU) day rates in excess of 500,000 US$ may be incurred. Hence any time savings constitute a considerable potential for cost reduction. This, and the desire to improve the safety record of drilling operations, has driven the automatisation of high-specification rigs, both offshore and onshore. The areas which have been dramatically changed over the last decade are as follows.

4.3.2. Top drive systems

Instead of having a rotary table in the rig floor the drive mechanism for the drill string is mounted on guide rails and moves up and down inside the derrick. This allows drilling in segments of 90 ft of pre-assembled pipe, significantly reducing connection time. The more continuous drilling process also results in better hole conditions and faster penetration rates. The latest rigs allow running of 120 ft sections and are equipped with two derricks – one drilling the well and the other used to concurrently pre-assemble drill strings (Figure 4.9).

4.3.3. Automated pipe handling

The manual labour on the rig floor has largely been replaced by a hydraulic system which picks up pipe from the rack, transports it up to the rig floor and inserts it into the drill string. The process is controlled by the rig crew manning a workstation housed beside the rig floor (Figure 4.10).

Earlier on when we described the cutting action of the drill bit, we learned about the drilling fluid or *mud*. The mud cools the bit and also removes the cuttings by carrying them up the annulus, outside the drill pipe. At the surface, the mud runs over a number of moving screens, the *shale shakers* (Figure 4.11), which remove the cutting for disposal. The fine particles which pass through the screens are then removed by *desanders* and *desilters*, the latter usually hydrocyclones.

Having been cleaned, the mud is transferred into the *mud tanks*, which are large treatment and storage units. From there a powerful pump brings the mud up through a pipe (stand pipe) and through a hose connected to the swivel (rotary hose) forcing it down the hole inside the drill string. Eventually the cleaned mud will exit again through the bit nozzles.

Originally, 'mud' was made from clay mixed with water – a simple system. Today the preparation and treatment of drilling fluid has reached a sophistication which requires specialist knowledge. The reason for this becomes clear if we consider the properties expected.

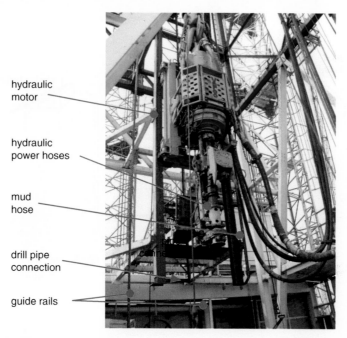

hydraulic
motor

hydraulic
power hoses

mud
hose

drill pipe
connection

guide rails

Figure 4.9 Top drive system.

Figure 4.10 The rig crew controlling the automated pipe handling system visible outside the control room.

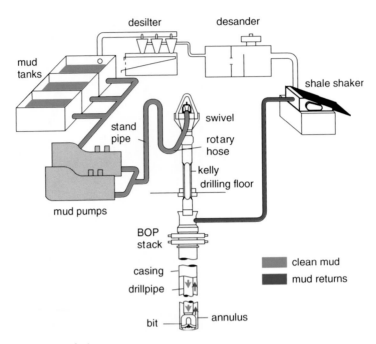

Figure 4.11 Mud circulation system.

In order to effectively lift the cuttings out of the hole a certain fluid *viscosity* needs to be achieved; yet the mud must remain pumpable. If the mud circulation stops, for instance to change the bit, the mud must *gel* and any material suspended in it must remain in suspension to avoid settling out at the bottom of the hole. It has to be stable under high temperatures and pressures as well as at surface conditions. Mud chemicals should not be removed by the mud–cleaning process. Drilling fluids have to be capable of carrying weighing material such as barites in order to control the formation pressures. They have to be compatible with the formations being drilled, for example they should prevent the swelling of formation clay and not permanently damage the reservoir zone. Last but not least, since these fluids are pumped, transported and disposed in large quantities, they should be environmentally friendly and cheap!

Often drilling fluids are made up using water and are called *water-based muds* (WBM). Another frequently employed system is based on oil, *oil-based mud* (OBM). The advantage of OBM is better lubrication of the drill string, compatibility with clay or salt formations and they give a much higher ROP. Diesel fuel was usually used for the preparation of OBM. During operations, large quantities of contaminated cuttings were formerly disposed of onto the seabed. This practice is no longer considered environmentally acceptable. If OBM or any other hazardous fluid is contained in the cuttings, a *closed-loop mud system* will be required. The cuttings will be either decontaminated in a dedicated onshore plant or re-injected as a slurry into a suitable formation. New mud compositions and systems are continuously being developed, for instance *synthetic drilling fluids* which rival the performance of OBM but are environmentally benign (e.g. synthetic oil–based mud [SOBM]).

The choice of drilling fluid has a major impact on the evaluation and production of a well. Later in this section, we will investigate the interaction between drilling fluids, logging operations and the potential damage to well productivity caused by *mud invasion* into the formation.

An important safety feature on every modern rig is the BOP. As discussed earlier on, one of the purposes of the drilling mud is to provide a hydrostatic head of fluid to counterbalance the pore pressure of fluids in permeable formations. However, for a variety of reasons (see Section 4.7) the well may 'kick', that is formation fluids may enter the wellbore, upsetting the balance of the system, pushing mud out of the hole, and exposing the upper part of the hole and equipment to the higher pressures of the deep subsurface. If left uncontrolled, this can lead to a *blowout*, a situation where formation fluids flow to the surface in an uncontrolled manner.

The BOPs are a series of powerful sealing elements designed to close off the annular space between the pipe and the hole through which the mud normally returns to the surface. By closing off this route, the well can be 'shut in' and the mud and/or formation fluids forced to flow through a controllable choke, or adjustable valve. This choke allows the drilling crew to control the pressure that reaches the surface and to follow the necessary steps for *'killing' the well*, that is restoring a balanced system. Figure 4.12 shows a typical set of BOPs. The annular

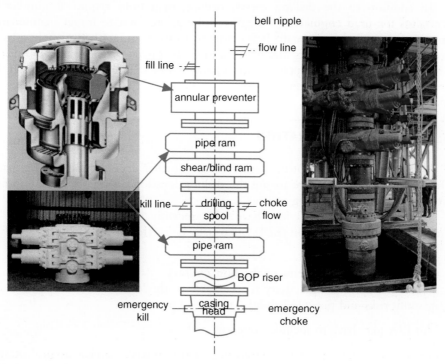

Figure 4.12   Blowout preventer.

preventer has a rubber sealing element that is hydraulically inflated to fit tightly around any size of pipe in the hole. Ram type preventers either grip the pipe with rubber-lined steel pipe rams, block the hole with blind rams when no pipe is in place or cut the pipe with powerful hydraulic shear rams to seal off the hole.

BOPs are opened and closed by hydraulic fluid stored under a pressure of 3000 psi in an accumulator, often referred to as a 'Koomy' unit.

All drilling activity will be carried out by the drill crew which usually works in 8- or 12-hour shifts. The driller and assistant driller will man the drilling controls on the rig floor from where instrumentation will enable them to monitor and control the drilling parameters, specifically

- hookload
- torque in drill string
- weight on bit (WOB)
- rotary speed (rpm)
- pump pressure and rate
- ROP (in min/ft)
- mud weight in and out of the hole
- volume of mud in the tanks.

Drilling operations which employ a measurement while drilling (MWD) system (see Section 4.5) will also provide the drilling engineers with formation parameters, downhole pressure and directional data in real time.

In addition to the drilling crews, drilling operations require a number of specialists for mud engineering, logging, fishing, etc., not to forget maintenance crews, cooks and cleaning staff. It is common to have some 40–90 people on site, depending on the type of rig and the location. The operation is managed on site by a 'Company Man' who represents the operator and a 'Rig Manager' who is the drilling contractor's representative.

4.4. SITE PREPARATION

Once the objectives of the well are clear, further decisions have to be made. One decision will be where to site the drilling location relative to the subsurface target and which type of rig to use.

If no prior drilling activities have been recently carried out in the area, an *environmental impact assessment (EIA)* will usually be carried out as a first step. An EIA is usually undertaken to

- meet the legal requirements of the host country
- ensure that the drilling activity is acceptable to the local environment
- quantify risks and possible liabilities in case of accidents.

An EIA may have to include concerns such as

- protection of sites of special interest (e.g. nature reserves, archaeological sites)
- noise control in built-up areas

- air emission
- effluent and waste disposal
- pollution control
- visual impact
- traffic (rig transport and supply)
- emergency response (e.g. fire, spills).

The EIA is an important document, often on the drilling project's critical path. In new areas, the required environmental data may not be available. Data collection may stretch over several seasons to capture such parameters as currents, migration paths, breeding habitats or weather patterns.

4.4.1. Onshore sites

A *site survey* will be carried out, from which a number of geotechnical parameters can be established, for example carrying capacity of the soil at the planned location, possible access routes, surface restrictions like built-up areas, lakes, nature reserves, the general topography and possible water supplies. The survey will allow the adequate preparation of the future location. For instance, onshore in a swamp area the soil needs to be covered with support mats.

The size of the rig site will depend on operational requirements and possible constraints imposed by the particular location. It will be determined by

- the type of derrick or mast (which will depend on the required loads); it must be possible to rig this up on site
- the layout of the drilling equipment
- the size of the waste pit
- the amount of storage space required for consumables and equipment
- the number of wells to be drilled
- whether the site will be permanent (in case of development drilling).

A land rig can weigh over 200 tons and is transported in smaller loads to be assembled on site.

Prior to moving the rig and all auxiliary equipment, the site will have to be cleared of vegetation and levelled. To protect against possible spills of hydrocarbons or chemicals, the surface area of a location should be coated with plastic lining and a closed draining system installed. Site management should ensure that any pollutant is trapped and properly disposed of.

If drilling and service personnel require accommodation at the well site, a camp will need to be constructed. For safety reasons, the camp will be located at a distance from the drilling rig and consist of various types of portacabins. For the camp, waste pits, access roads, parking space and drinking water supplies will be required.

4.4.2. Offshore sites

The survey requirements will depend on rig type and the extent of the planned development, for example single exploration well or drilling jacket installation.

A typical survey area is some 4 by 4 km, centred on the planned location. Surveys may include

- *Seabed survey*: employing high-resolution echo-sounding and side scan sonar imaging, an accurate picture of the sea bottom is created. The technique allows the interpreter to recognise features such as pipelines, reefs and wreckage. Particularly if a jack-up rig is considered, an accurate map of these obstructions is required to position the jack-up legs safely. Such a survey will sometimes reveal crater-like structures (pockmarks), which are quite common in many areas. These are the result of gas escape from deeper strata to the surface and could indicate danger from shallow gas accumulations.
- *Shallow seismic*: unlike 'deep' seismic surveys aimed at the reservoir section, the acquisition parameters of shallow surveys are selected to provide maximum resolution within the near-surface sedimentary layers (i.e. the top 800 m). The objective is to detect indications of shallow gas pockets or water zones. The gas may be trapped within sand lenses close to the surface and may enter the borehole if penetrated by the drill bit, resulting in a potential blowout situation. Gas chimneys are large-scale escape structures where leakage from a reservoir has created a gas-charged zone in the overburden. If shallow water zones are penetrated, they may flow to the surface of the seabed and reduce the load-bearing capacity of the conductor pile.
- *Soil boring*: where planned structures require soil support, for example drilling jackets or jack-up rigs, the load-bearing capacity has to be evaluated (just like on a land location). Usually a series of shallow cores are taken to obtain a sample of the sediment layers for investigation in a laboratory.

Particularly for jack-up rigs, site surveys may have to be carried out prior to each re-employment to ensure that the rig is positioned away from the previously formed 'footprints' (depressions on the seabed left by the jack-up legs on a previous job).

4.5. DRILLING TECHNIQUES

If we consider a well trajectory from surface to *total depth* (TD), it is helpful to look at the shallow section and the intermediate and reservoir intervals separately. The shallow section, usually referred to as *top hole*, consists of rather unconsolidated sediments, hence the formation strength is low and drilling parameters and equipment have to be selected accordingly.

The reservoir section is more consolidated and is the main objective to which the well is being drilled, hence the drilling process has to ensure that any productive interval is not damaged.

4.5.1. Top hole drilling

For the very first section of the borehole, a base from which to commence drilling is required. In a land location, this will be a cemented 'cellar' in which a conductor or

Figure 4.13   Cellar with Christmas tree on a land location.

stove pipe will be piled prior to the rig moving in. The cellar will later accommodate the 'Christmas tree' (an arrangement of seals and valves to control production), once the well has been completed and the rig has moved off location (Figure 4.13).

As in the construction industry, piling of the conductor is done by dropping weights onto the pipe or using a hydraulic hammer until no further penetration occurs. In an offshore environment, the conductor is either piled (e.g. on a platform) or a large-diameter hole is actually drilled, into which the conductor is lowered and cemented. Once the drill bit has drilled below the conductor the well is said to have been *spudded*.

The top hole will usually be drilled with a large-diameter bit (between 22 and 27 in. diameter). The drill bit (roller cone type) will be designed to drill predominantly soft formations. As a result of the hole diameter and the rapid penetration rate, vast quantities of drilled formation will have to be treated and removed from the mud circulation system. Often the ROP will be reduced to allow adequate removal of cuttings and conditioning of mud. In some cases, the problem is alleviated by first drilling a pilot hole with a smaller diameter bit ($12\frac{1}{4}$ in.) and later redrilling the section to the required size using a hole opener. This is essentially a larger diameter drill bit above the smaller diameter bit. Hole openers are also run if the hole has to be logged (most logging tools are not designed for diameters above $17\frac{1}{2}$ in.) and if accurate directional drilling is required.

A surface casing is finally cemented to prevent hole collapse and protect shallow aquifers.

4.5.2. Intermediate and reservoir section

Between the top hole and the reservoir section, in most cases, an *intermediate section* will need to be drilled. This section consists of more consolidated rocks than the top hole. The deviation angle is often increased significantly in this interval to reach the subsurface target, and lateral departures from the surface co-ordinates may reach several kilometres. Based on pore pressure prediction (from seismic or measured data from offset wells) the mud weight has to be determined. The pressure exerted by the mud column has to exceed the formation pressure in order to maintain overbalance and prevent the hole from collapsing but has to be lower than the fracture pressure of the formation (Figure 4.14). If the formation strength is exceeded, fracturing may occur, resulting in mud losses and formation damage. Borehole/formation stability is the realm of *geomechanics*. Challenges in well planning arise when rock strength and thus borehole stability show considerable variations depending on hole angle and direction, as shown in Figure 4.15. In this example, the small difference between fracture gradient and collapse gradient at high deviation may require a revision of the initially planned well trajectory through the intermediate and/or reservoir section.

An intermediate casing is usually set above the reservoir in order to protect the water-bearing, hydrostatically pressured zones from influx of possibly overpressured

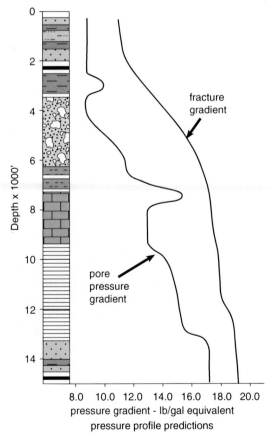

Figure 4.14 Mud weight envelope has to be between pore pressure gradient and fracture gradient.

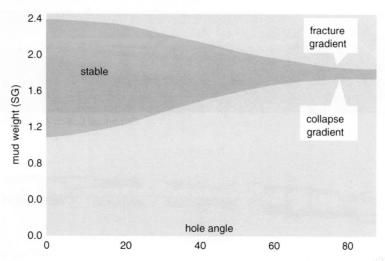

Figure 4.15   Example of relationship between mud weight and borehole stability.

hydrocarbons and to guarantee the integrity of the wellbore above the objective zone. In mature fields where production has been ongoing for many years, the reservoir may show depletion pressures considerably lower than the hydrostatically pressured zones above. Casing and cementing operations are covered in Section 4.6.

Before continuing to examine the aspects of drilling through the reservoir, remember that the reservoir is the prime objective of the well and a very significant future asset to the company. If the drilling process has impaired the formation, production may be deferred or totally lost. In exploration wells, the information from logging and testing may not be sufficient to fully evaluate the prospect if the hole is not on gauge, necessitating sidetracking or even an additional well. On the other hand, there is considerable scope to improve productivity and information value of the well by carefully selecting the appropriate technology and practices.

In some areas such as the Central North Sea, offshore Canada, and onshore California, *high pressure high temperature* (HPHT) accumulations are present. Wells may encounter reservoir temperatures in excess of 370°F (190°C) and pressures above 15,000 psi. These are challenging conditions for drilling fluids, mud motors, gauges and logging tools. In particular, components such as batteries, sensors, electronics and seal elastomers had to be specifically developed for these extreme conditions.

4.5.3. Directional drilling

Directional drilling is usually done with a *rotary steerable system* (Figure 4.16). A downhole steering and control unit is located in the near-bit assembly. A set of small electronically controlled rotating stabiliser pads (actuators) exert a continuous directional force onto a drive shaft which orients the drill bit into the desired

Cambridge Drilling Automation

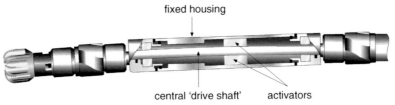

fixed housing

central 'drive shaft'   activators

Gyro/Data *Inc.*

Figure 4.16   Rotary steerable system.

direction. The drill string is rotated at the same time, allowing hole cleaning. A control unit near the bit ensures that the hole angle is not increased or decreased rapidly creating 'dog legs' which will result in excessive torque and drag. The rotary steerable system is combined with logging tools in the drill string close to the bit, allowing a continuous optimisation of the well trajectory.

Mud turbines and mud motors are also used for directional drilling. Rotational movement of the drill string is restricted to the motor or turbine section, whilst the rest of the drill string moves by 'sliding' or being rotated at a lower speed to ensure hole cleaning. In the example of the turbine shown in Figure 4.17, the mud is pumped between the rotor and the stator section, inducing a rotational movement which is transmitted onto the drill bit. Motors and turbines are being replaced by the rotary steerable system for cost and operational reasons. Their use is increasingly limited to such applications as kicking off a sidetrack or where a sharp change in angle is required in a short-radius horizontal well.

Advances in drilling and completion technology today allow us to construct complicated wells along 3D trajectories. In addition to vertical wells, directional drilling allows us to build, maintain or drop hole angle and to turn the drill bit into different directions. Thus, we are able to optimise the wellpath in terms of reservoir quality, production or injection requirements. Sometimes constraints at the surface (e.g. built-up areas) or subsurface (e.g. shallow gas, faults, lenticular reservoirs) may require a particular well trajectory to be followed.

The steering of the well is supported by the stabilisers which form part of the drill string. The blades can be activated and deactivated from the surface depending on whether the angle is to be maintained, increased or decreased (Figure 4.18).

High deviation angles (above $60°$) may cause excessive drag or torque whilst drilling, and will also make it difficult to later service the well with standard wireline tools.

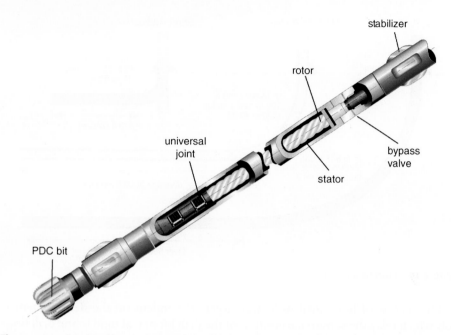

Figure 4.17   Mud turbine.

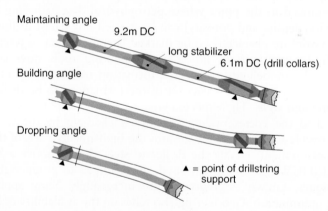

Figure 4.18   Types of assemblies for directional drilling.

4.5.4. Horizontal drilling

Given the lateral distribution of reservoir rock or reservoir fluids, a horizontal well may provide the optimum trajectory. Figure 4.19 shows the types of horizontal wells being drilled. The build-up rate of angle is the main distinction from a drilling point of view. Medium radius wells are preferred since they can be drilled, logged and completed with fairly standard equipment. The horizontal drilling target can be controlled within a vertical window of less than 2 m.

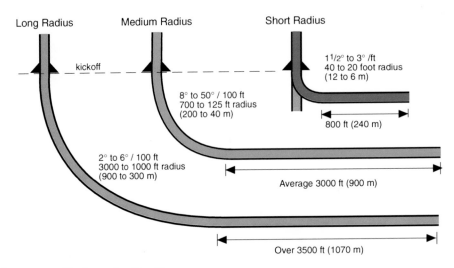

Figure 4.19 Horizontal well radii.

The success of horizontal wells was largely dependent on the development of tools which relay the subsurface position of the drill bit in real time to the drill floor. Improvements in this technology have greatly improved the accuracy with which well trajectories can be targeted. *MWD* is achieved by the insertion of a sonde into the drill string close to the bit. Initially providing only directional data, the tools have been improved to the point where petrophysical data gathering (gamma ray [GR], resistivity, density and porosity) can be carried out whilst drilling.

Most reservoirs are characterised by marked lateral changes in reservoir quality corresponding to variations in lithology. Computing tools now commercially available allow the modelling of expected formation responses 'ahead of the bit'. This is possible in areas where a data set of the formations to be drilled has been acquired in previous wells. The expected GR and density response is then simulated and compared to the corresponding signature picked up by the tool. Thus, in theory, it is possible to direct the bit towards the high-quality parts of the reservoir. Resistivity measurements enable the driller to steer the bit above a hydrocarbon water contact (HCWC), a technique used, for example, to produce thin oil rims. These techniques, known as *geosteering*, are increasingly being applied to field development optimisation. Geosteering also relies on the availability of high-quality seismic and possibly detailed palaeontological sampling.

4.5.5. Multilateral wells

Drilling a number of holes, branching from a central borehole, is an attractive option in the following cases:

- where reservoir productivity is low but can be significantly improved by increasing the reservoir surface area exposed to the well (Figure 4.20)

Figure 4.20   Model of multilateral well to increase reservoir contact.

- in reservoirs which are lenticular
- where reservoir layers are vertically segregated by permeability barriers.

Whilst drilling and, in particular, completing (see Chapter 10) is more complex, multilateral wells have the advantage that they only require one borehole from the surface. If there are no spare conductors on the platform, this is an attractive option. In mature fields, multilaterals are often best suited to drain remaining pockets of hydrocarbons. In subsea developments, multilaterals can offer a substantial cost advantage over conventional wells.

To commence drilling of each branch, either a rotary steerable system or a *whipstock* are used. The latter is a curved steel wedge which is inserted into the borehole, forcing the drilling assembly in the planned azimuth (Figure 4.21).

4.5.6. Extended reach drilling

An extended reach well is loosely defined as having a horizontal displacement of at least twice the vertical depth. With current technology, a ratio of over 4 (horizontal displacement/vertical depth) can be achieved.

Extended reach drilling (ERD) wells are technically more difficult to drill and because of the degree of engineering required for each well, the term 'designer well' is frequently used.

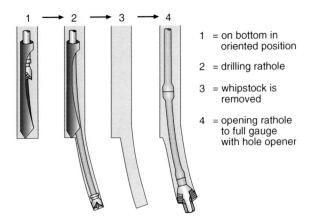

1 → 2 → 3 → 4

1 = on bottom in oriented position

2 = drilling rathole

3 = whipstock is removed

4 = opening rathole to full gauge with hole opener

Figure 4.21   Kicking off with a whipstock.

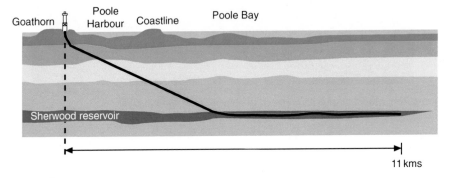

Figure 4.22   Extended reach drilling (BP, Wytch Farm).

ERD will be considered

- where surface restrictions exist
- where marginal accumulations are located several miles from existing platforms/ clusters
- where ERD allows a reduction in the number of platforms required.

The high deviation (often up to 85°) and the long horizontal displacement expose the drill string to extreme drag and torque. Hole cleaning (cutting removal) and cementing of casing is more difficult due to the increased effect of gravity forces compared to low-angle wells. Thus, ERD wells usually require heavier and better-equipped rigs compared to standard wells, and take longer to drill. Top drive systems are routinely employed in combination with rotary steerable systems.

Not surprisingly, costs are several times higher than conventional wells. Nevertheless, overall project economics may favour ERD over other development options. For example, BP developed the offshore part of the Wytch Farm Oilfield (which is located under Poole Harbour in Dorset, UK) from an onshore location. The wells targeted the reservoir at a vertical depth of 1500 m with a lateral displacement of over 11,000 m (Figure 4.22). The alternative was to build a drilling location on an artificial island in Poole Bay. ERD probably saved a considerable amount of money and advanced first oil by several years.

Offshore, subsea satellite development may be a viable alternative to ERD wells.

4.5.7. Slim hole drilling

Slim hole drilling has been used by the mining industry for a number of years. Recently, the oil industry has been developing rigs, drill string components and logging tools that will allow smaller diameter holes and completions. One definition used for 'slim holes' is 'a well in which 90% or more of the length has a diameter of 7 in. or less'. In principle, slim hole drilling has the potential to drill wells at greatly reduced cost (estimates range from 40 to 60%). The cost reductions accrue from several sources:

- less site preparation
- easier equipment mobilisation
- reduction in the amount of consumables (drill bits, cement, muds, fuel)
- less cuttings to dispose of
- smaller equipment.

A slim hole rig weighs about one-fifth of a conventional rig and its small size can open new frontiers by making exploration economic in environmentally sensitive or inaccessible areas.

The following table highlights the potential of slim hole wells:

Type of Rig	Conventional	Slim Hole
Hole diameter (in.)	8.5	3–6
Drill string weight (tons)	40	5–7
Rig weight (tons)	80	10
Drill site area (%)	100	25
Installed power (kW)	350	70–100
Mud tank capacity (bbl)	500	30
Hole volume (bbl/1000 ft)	60	6–12
Crew size	25–30	12–15

The greatly reduced hole volume of slim hole wells can lead to problems if an influx is experienced (see Section 4.7). The maximum depth drillable with slim hole configurations is another current limitation of this technology.

Some slim hole rigs were adapted from units employed by mining exploration companies and are designed to allow continuous coring rather than breaking the formation into cuttings. These rigs are sometimes employed for data gathering wells in exploration ventures. They are ideally suited for remote locations since they can be transported in segments by helicopter.

Figure 4.23   Coiled tubing drilling unit.

4.5.8. Coiled tubing drilling

A special version of slim hole drilling has emerged as a viable alternative: *coiled tubing drilling* (CTD) (Figure 4.23). Whilst standard drilling operations are carried out using joints of drill pipe, CTD employs a seamless tubular made of high-grade steel. The diameter varies between $1\frac{3}{4}$ and $3\frac{1}{2}$ in. Rather than being segmented, the drill string is reeled onto a large-diameter drum.

The advantages of CTD are several:

- nearly no pipe handling
- better well control allows at-balance or even underbalanced drilling, resulting in higher penetration rates and reduced potential for formation damage
- less environmental impact
- lower cost for site preparation, lower day rates, lower mobilisation and demobilisation costs
- easier completion by using the computerised tomography (CT) as a production string.

However, CTD is limited to slim holes, and the reliability of some of the drill string components such as downhole motors is still improving. Presently, the cost of building a new customised CTD rig limits the wider application of this technology.

4.6. CASING AND CEMENTING

Imagine that a reservoir exists at a depth of 2500 m. We could attempt to drill one straight hole all the way down to that depth. That attempt would end either

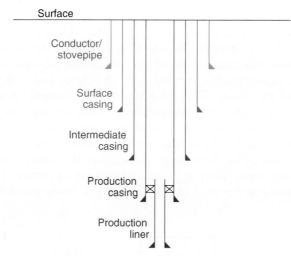

Figure 4.24 Casing scheme.

with the hole collapsing around the drill bit, with the loss of drilling fluid into formations with low pressure or in the worst case with the uncontrolled flow of gas or oil from the reservoir into unprotected shallow formations or to the surface (blowout). Hence, from time to time, the borehole needs to be stabilised and the drilling progress safeguarded.

The casing design will usually start with a 23 in. conductor, then $18\frac{5}{8}$ in. surface casing, $13\frac{3}{8}$ in. intermediate casing above reservoir, $9\frac{5}{8}$ in. production casing across reservoir section and possibly 7 in. production 'liner' over a deeper reservoir section (Figure 4.24). A liner is a casing string which is clamped with a packer into the bottom part of the previous casing; it does not extend all the way to the surface, and thus saves cost.

Casing joints are available in different grades, depending on the expected loads to which the string will be exposed during running, and the lifetime of the well. The main criteria for casing selection are

- *Collapse load*: originates from the hydrostatic pressure of drilling fluid, cement slurry outside the casing and later on by 'moving formations', for example salt
- *Burst load*: this is the internal pressure the casing will be exposed to during operations
- *Tension load*: caused by the string weight during running in; it will be highest at the top joints
- *Corrosion service*: carbon dioxide (CO_2) or hydrogen sulphide (H_2S) in formation fluids will cause rapid corrosion of standard carbon steel and therefore special steel may be required
- *Buckling resistance*: the load exerted on the casing if under compression.

The casing will also carry the BOPs described earlier.

Running casing is the process by which 40 ft sections of steel pipe are screwed together on the rig floor and lowered into the hole. The bottom two joints will contain a *guide shoe*, a protective cap which facilitates the downward entry of the casing string through the borehole. Inside the guide shoe is a one-way valve which will open when cement/mud is pumped down the casing and is displaced upwards on the outside of the string. The valve is necessary because at the end of the cementing process the column of cement slurry filling the annulus will be heavier than the mud inside the casing and 'U tubing' would occur without it.

To provide a second barrier in the string, a *float collar* is inserted in the joint above the guide shoe. The float collar also catches the *bottom plug* and *top plug*, between which the *cement slurry* is placed. The slurry of cement (Figure 4.25) is pumped down between the two rubber seals (plugs). Their function is to prevent

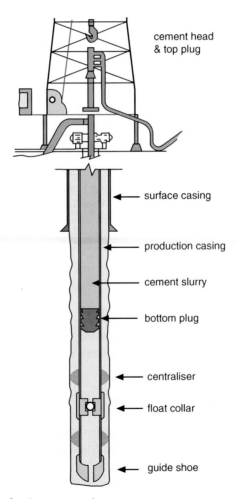

cement head
& top plug

surface casing

production casing

cement slurry

bottom plug

centraliser

float collar

guide shoe

Figure 4.25 Principle of casing cementation.

contamination of the cement with drilling fluid which would cause a bad cement bond between borehole wall and casing. Once the bottom plug bumps into the float collar, it ruptures and the cement slurry is pushed down through the guide shoe and upwards outside the casing. Thus, the annulus between casing and borehole wall is filled with cement.

The success of a cement job depends partly on the velocities of the cement slurry in the annulus. A high pump rate will result in *turbulent* flow which results in a better bond than the slower, *laminar* flow. The cement has to be placed evenly around each casing joint. This becomes more difficult with increasing deviation angle since the casing joints will tend to lie on the lower side of the borehole preventing cement slurry entering between casing and borehole wall. To avoid this happening, steel springs or *centralisers* are placed at intervals outside the string to centralise the casing in the borehole.

Once the cementation has been completed, the rig will 'wait on cement' (WOC), that is wait until the cement hardens prior to running in with a new assembly to drill out the plugs, float collar and shoe, all of which are made of easily drillable materials.

The process described so far is called *primary cementation*, the main purpose of which is to

- bond the casing to the formation and thereby support the borehole wall
- prevent the casing from buckling in critical sections
- separate the different zones behind the casing and thereby prevent fluid movement between permeable formations
- seal off troublesome horizons such as lost circulation zones.

Sometimes primary cementations are not successful, for instance if the cement volume has been wrongly calculated, if cement is lost into the formation or if the cement has been contaminated with drilling fluids. In this case, a remedial or *secondary cementation* is required. This may necessitate perforating the casing at a given depth and then pumping cement through the perforations.

A similar technique may also be applied later in the well's life to seal off perforations through which communication with the formation has become undesirable, for instance if water breakthrough has occurred (squeeze cementation).

Plug back cementations, that is cement placement inside the casing and across the perforations may be required prior to sidetracking a well or in the course of decommissioning.

The chemistry of cement slurries is complex. Additives will be used to ensure the slurry remains pumpable long enough at the prevailing downhole pressures and temperatures but sets (hardens) quickly enough to avoid unnecessary delays in the drilling of the next hole section. The cement also has to attain sufficient compressive strength to withstand the forces exerted by the formation over time. A *spacer* fluid is often pumped ahead of the slurry to clean the borehole of mudcake and thereby achieve a better *cement bond* between formation and cement.

4.7. DRILLING PROBLEMS

Drilling equipment and drilling activities have to be carried out in complex and often hostile environments. Surface and subsurface conditions may force the drilling rig and crew to operate close to their limits. Sometimes non-routine or unexpected operating conditions will reach the rating of equipment and normal drilling practices may not be adequate for a given situation. Thus, drilling problems can and do occur.

4.7.1. Stuck pipe

This term describes a situation whereby the drill string cannot be moved up or down or rotated. The pipe can become stuck as a result of mechanical problems during the drilling process itself or because of the physical and chemical parameters of the formation being drilled. Most common reasons for stuck pipe are as follows:

- Excessive *pressure differentials* between the borehole and the formation. For instance, if the pressure of the mud column is much higher than the formation pressure, the drill pipe may become 'sucked' against the borehole wall (*differential sticking*). This often happens when the pipe is stationary for some time, for example whilst taking a deviation survey. Prevention methods include reduced mud weights, addition of friction reducing components to the mud, continuous rotation/moving of string, addition of centralisers or use of spiral DCs to minimise contact area between string and formation, or a low fluid loss mud system.
- Some *clay minerals* may absorb some of the water contained in the drilling mud. This will cause the clays to *swell* and eventually reduce the borehole size to the point where the drill pipe becomes stuck. Prevention is through mud additives, for example potassium salt, which prevent clay swelling.
- Unstable formations or a badly worn drill bit may result in *undergauged* holes. An example of an unstable formation is salt which can 'flow' whilst the drilling is in progress, closing around the drill pipe. Prevention is by addition of stabilisers and string reamers to the drilling assembly.
- Residual stresses in the formation, resulting from regional tectonic forces may cause the borehole to collapse or deform, resulting in stuck pipe. Sometimes high mud weights may help delay deformation of the borehole.
- If the well trajectory shows a severe dogleg (sudden change in angle or direction), the movement of the string may result in a groove cut into the borehole wall by the drill pipe. Eventually the pipe will become stuck, a process termed *key seating* (Figure 4.26). The best prevention is the avoidance of doglegs, and frequent reaming, insertion of stabilisers on top of DCs or insertion of key seat wipers in the string (string reamers).

In many cases, the point at which the pipe is stuck can be determined by means of a 'free point indicator tool', a special electrical strain gauge device run on wireline inside the drill pipe which will measure axial and angular deformation.

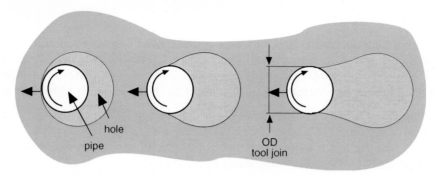

Figure 4.26 Development of key seating (plan view of hole).

An initial estimate of where the string is stuck can be calculated by applying a pull on the drill string in excess of the drill string weight and measuring the observed stretch in the pipe. This information may be used to decide where to 'back off' the string if the deeper part cannot be recovered.

If the string indeed cannot be retrieved by overpull, an explosive or chemical charge is lowered inside the pipe to the top of the stuck interval and the pipe above the stuck point is recovered after severing the string. Since drilling assemblies and redrilling of the borehole in a sidetrack are expensive, a further attempt to retrieve the tubulars (often called a 'fish') left in the hole will then be made. This is one application of fishing operations as described below.

4.7.2. Fishing

Fishing refers to the retrieval of a foreign object from the borehole. Fishing operations will be required if the object is expected to hamper further drilling progress either by jamming the string or damaging the drill bit. This 'junk' often consists of small non-drillable objects, for example bit nozzles, rock bit cones or broken off parts of equipment. Other common causes for fishing are

- drill pipe left in the hole (either as a result of twist off, string back off or cementing operations)
- items that have been dropped into the hole which can cause major drilling problems (e.g. rig floor tools, parts of the drill string).

Bottom hole assemblies and certain types of downhole equipment (e.g. logging tools, MWD tools) cost several hundred thousand US$. Some logging tools will have radioactive sources which may need to be recovered or isolated for safety and legal reasons. However, prior to commencing fishing operations, a cost–benefit assessment will have to be made to establish that the time and equipment attributable to the fishing job is justified by the value of the fish or the cost of sidetracking the hole.

Due to the different nature of 'junk', a wide variety of fishing tools are employed.

4.7.3. Lost circulation

During drilling operations large volumes of drilling mud are sometimes lost into a formation. In this case, normal mud circulation is no longer possible and the fluid level inside the borehole will drop, creating a potentially dangerous situation as described below. The formations in which lost circulation can be a problem are

- a highly *porous, coarse* or *vuggy formation* which does not allow the build-up of an effective mudcake
- a *karst structure*, that is a limestone formation which has been eroded resulting in a large-scale, open system comparable to a cave
- a densely *fractured* interval
- a *low-strength formation* in which open fractures are initiated by too high a mud pressure in the borehole.

The consequences of lost circulation are dependent on the severity of the losses, that is how quickly mud is lost and whether the formation pressures in the openhole section are hydrostatic or above hydrostatic, that is overpressured (see below). Mud is expensive and losses are undesirable but they can also lead to a potentially hazardous situation. Moderate losses may be controlled by adding 'lost circulation material' (LCM) to the mud system, such as mica flakes or coconut chippings. The LCM will plug the porous interval by forming a sealing layer around the borehole preventing further mud invasion. However, LCM may also plug elements of the mud circulation system, for example bit nozzles and shale shaker screens, and may later impair productivity or injectivity of the objective intervals. In severe cases, the losses can be controlled by *squeezing cement* slurry into the trouble horizon. This is obviously not a solution if the formation is the reservoir section!

If sudden total losses occur in a hydrostatically pressured interval, for example in a karstified limestone, the decision may actually be taken to drill ahead without drilling mud but using large quantities of surface water to cool the bit. The fluid level in the annulus will usually stabilise at a certain depth; this type of operation is also referred to as 'drilling blind with a floating mud cap'. Since no cuttings are returned to surface, mudlogging is no longer possible, therefore preventing early reservoir evaluation.

In the event of a sudden loss of mud in an interval containing overpressures, the mud column in the annulus will drop, thereby reducing the hydrostatic head acting on the formation to the point where formation pressure exceeds mud pressure. Formation fluids (oil, gas or water) can now enter the borehole and travel upwards. In this process, the gas will expand considerably as it loses its initial pressure due to the reduction of hydrostatic head above the gas bubble. The last line of defence left is the BOP. However, although the BOP will prevent fluid or gas escape to the surface, 'closing in the well' may lead to two potentially disastrous situations:

1. Formation breakdown (fracture development) in a shallower, weaker formation and subsequent uncontrolled flow from the deeper to the shallower formation (*internal blowout*).
2. Formation breakdown and subsequent liquefaction of the near-surface strata and the initiation of *cratering* below the rig. This will result in a *surface blowout*.

When drilling through *normally pressured* formations, the mud weight in the well is controlled to maintain a pressure greater than the formation pressure to prevent the influx of formation fluid. A typical overbalance would be in the order of 200 psi. A larger overbalance would encourage excessive loss of mud into the formation, slow down drilling and potentially cause differential sticking. If an influx of formation fluid into the borehole did occur due to insufficient overbalance, the lighter formation fluid would reduce the pressure of the mud column, thus encouraging further influx, and an unstable situation would occur, possibly leading to a blowout. Hence, it is important to avoid the influx of formation fluid by using the correct mud weight in the borehole all the time. This is the 'first line of defence'.

When drilling into an *overpressured formation*, the mud weight must be increased to prevent influx. If this increased mud weight could cause large losses in shallower, normally pressured formations, it is necessary to isolate the normally pressured formation behind casing before drilling into the overpressured formation. The prediction of overpressures is therefore important in well design.

Similarly, when drilling into an *underpressured formation*, the mud weight must be reduced to avoid excessive losses into the formation. Again, it may be necessary to set a casing before drilling into underpressures.

Considerable effort will be made to predict the onset of overpressures ahead of the drill bit. The most reliable indications are gas readings, porosity-depth trends, ROP and shale density measurements.

If a situation arises whereby formation fluid or gas enters the borehole, the driller will notice an increase in the total volume of mud. Other indications such as a sudden increase in penetration rate and a decrease in pump pressure may also indicate an influx. Much depends on a quick response of the driller to close in the well before substantial volumes of formation fluid have entered the borehole. Once the BOP is closed, the new mud gradient required to restore balance to the system can be calculated. The heavier mud is then circulated in through the drill string and the lighter mud and influx is circulated out through the choke line. Once overbalance is restored, the BOP can be opened again and drilling operations continue.

4.8. Costs and Contracts

The actual *well costs* are divided into

- *Fixed costs*: casing and tubulars, logging, cementing, drill bits, mobilisation charges, rig move
- *Daily costs*: contractor services, rig time, consumables
- *Overheads*: offices, salaries, pensions, health care, travel.

A fairly significant charge is usually made by the drilling contractor to modify and prepare the rig for a specific drilling campaign. This is known as a *mobilisation cost*. A similar charge will cover 'once off' expenses related to terminating the operations for a particular client, and is called a *demobilisation cost*. These costs can be significant, say 5–10 million US$.

The actual costs of a well show considerable variations and are dependent on a number of factors, for example

- type of well (exploration, appraisal, development)
- well trajectory (vertical, deviated, horizontal, multilateral)
- total depth
- subsurface environment (temperatures, pressures, corrosiveness of fluids)
- type and rating of rig
- type of operation (land, marine)
- infrastructure available, transport and logistics
- climate and geography (tropical, arctic, remoteness of location).

4.8.1. Contracts

Most companies hire a drilling contractor to supply equipment and manpower rather than having their own rigs and crews. The reasons for this are threefold:

- a considerable investment is required to build/buy a rig
- rig and crew need to be maintained and paid regardless of the operational requirements and activities of the company
- drilling contractors can usually operate more cheaply and efficiently than a company which carries out drilling operations as a non-core activity.

Before a contract is awarded a tender procedure is usually carried out (very different from the tender described earlier!). Thus, a number of suitable companies are invited to bid for a specified amount of work. Bids will be evaluated based on price, rig specifications and the past performance of the contractor, with particular attention to their safety record. Several types of contract are used.

4.8.1.1. Turnkey contract

This type of contract requires the operator to pay a fixed amount to the contractor upon completion of the well, whilst the contractor furnishes all the material and labour and handles the drilling operations independently. The difficulty with this approach is to ensure that a 'quality well' is delivered to the company since the drilling contractor will want to drill as quickly and cheaply as possible. The contractor therefore should guarantee an agreed measurable quality standard for each well. The guarantee should specify remedial actions which will be implemented should a substandard well be delivered.

4.8.1.2. Footage contract

The contractor is paid per foot drilled. Whilst this will provide an incentive to 'make hole' quickly, the same risks are involved as in the turnkey contract. Footage contracts are often used for the section above the prospective reservoir where hole conditions are less crucial from an evaluation or production point of view.

4.8.1.3. Incentive contract

This method of running drilling operations has been very successfully applied in recent years and has resulted in considerable cost savings. Various systems are in operation, usually providing a bonus for better than average performance. The contractor agrees with the company on the specifications for the well. Then the 'historic' cost of similar wells which have been drilled in the past is established. This allows estimation of the costs expected for the new well. The contractor will be entirely in charge of drilling the well, and cost savings achieved will be split between company and contractor.

4.8.1.4. Day rate contract

As the name implies the company basically rents the rig and crew on a per day basis. Usually the oil company also manages the drilling operation and has full control over the drilling process. This type of contract actually encourages the contractor to spend as much time as acceptable 'on location'. With increased cost consciousness, day rate contracts have become less favoured by most oil companies.

Actual contracts often involve a combination of the above. For instance, an operator may agree to pay footage rates to a certain depth, day rates below that depth, and standby rates for days when the rig is on site, but not drilling.

4.8.1.5. Partnering and alliances

In recent years, a new approach to contracting has evolved and is gaining rapid acceptance in the industry. The concept has become known as *partnering* and can be seen as a progression of the incentive contract. Whilst the previously described contractual arrangements are restricted to a single well project or a small number of wells in which a contractor is paid by a client for the work performed, partnering describes the initiation of a long-term relationship between the asset holder (e.g. an oil company) and the service companies (e.g. drilling contractor and equipment suppliers). It includes the definition and merging of *joint business objectives*, the *sharing* of financial risks and rewards and is aimed at an improvement in efficiency and reduction of operating costs. Therefore, a partnering contract will not only address *technical* issues but also include *business process quality management*. The latter has proven to result in more efficient and economic use of resources, for instance the setting up of 'joint implementation teams' has replaced the practice of having separate teams in contractor and operator offices, essentially performing the same tasks.

The industry is increasingly acknowledging the value of contractors and service companies in improving their individual core capabilities through *alliances*, that is a joint venture for a particular project or a number of projects. A *lead contractor*, for example a drilling company, may form alliances with a number of *subcontractors to be able to cover a wider spectrum of activities*, for example *completions, workovers and well interventions*.

CHAPTER 5

SAFETY AND THE ENVIRONMENT

Introduction and Commercial Application: Safety and the environment have become important elements of all parts of the field life cycle, and involve all of the technical and support functions in an oil company. The Piper Alpha disaster in the North Sea in 1988 triggered a major change in the approach to management of safety within the industry. Companies recognise that good safety and environmental management make economic sense and are essential to guaranteeing long-term presence in the market. Stakeholders, be they governments, non–government organisations (NGOs) or financing entities will scrutinise the *HSE* (health, safety and environment) performance of an operator on a continuous basis.

Many techniques have been developed for management of the safety and environmental impact of operations, and much science is applied to these areas. The objective of this section is to demonstrate how HSE concerns can have a significant impact on all aspects of a field development and subsequent production operations, and that safety and the environment must be the concern of all employees.

5.1. SAFETY CULTURE

One of the leaders in industrial safety management is the chemicals company, Dupont which has a history of top safety performance going back to the early 20th century. Initially, the company was a manufacturer of dynamite, hence, safety had a high priority! The company recognised that good safety performance must start with *management commitment* to safety, but that the level of *employee commitment* ultimately determines the safety performance. The following diagram expressed their findings (Figure 5.1).

At point A, despite full management commitment to safety performance, with low employee commitment to safety, the number of accidents remains high; employees only follow procedures laid out because they feel they have to. At the other extreme, point B, when employee commitment is high, the number of accidents reduces dramatically; employees feel responsible for their own safety as well as that of their colleagues. Employee commitment to safety is an *attitude* of mind rather than a taught discipline, and can be enhanced by training and (less effectively) incentive schemes.

Safety performance is measured by companies in many different ways. To benchmark safety performance on an industry wide scale, globally recognised standards are required. A commonly used method is the recording of the number of accidents, or *lost time incidents* (LTI). An LTI is an incident which causes a person to stay away from work for one or more days. *Recordable injury frequency* (RIF) is the number of injuries that require medical treatment per 100 employees.

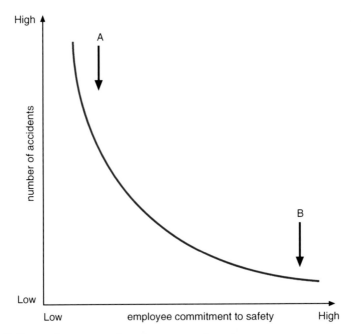

Figure 5.1 Safety performance and employee commitment.

Another measure might be the monetary cost of a safety incident. Many techniques are applied to improve the company's safety performance, such as writing work procedures and equipment standards, training staff, performing safety audits and using hazard studies in the design of plant and equipment. These are all very valid and important techniques, but one of the most effective methods of influencing safety performance is to create a *safety culture* within the company.

The practising engineer has an excellent opportunity to influence the safety of operations by applying techniques such as *hazard and operability studies* (HAZOP) to the design of plant layout and equipment. This technique involves determining the potential hazard of an operation under normal and abnormal operating conditions, and considering the probability and consequences of an accident. This type of study is now commonly applied to new platform design and to the evaluation of refurbishment on existing platforms. Some examples of innovations in platform design which have resulted from this type of study are

- *freefall lifeboats*, launched from heat shielded slipways on offshore platforms
- *emergency shutdown valves* installed on the seabed and topsides in incoming and outgoing pipelines, designed to isolate the platform from all sources of oil and gas in an emergency
- *protected emergency escape routes* with heat shielded stairways, to provide at least two escape routes from any point on the platform
- *physical separation of accommodation modules* from the drilling/process/compression modules (creating a pressurised 'safe haven'). On an integrated platform the areas are at opposite ends of the platform, and are separated by fire and blast walls

- *fire resistant coatings* on structural members
- *computerised control* and shutdown of process equipment.

In both safety and environment issues, the engineer should try to eliminate the hazard *at source*. For example, one of the most hazardous operations performed in both the offshore and onshore environments is *transport*, amongst which helicopter flying has the most incidents per hour of exposure. At feasibility study stage in, say, an offshore development, the engineer should be considering alternatives for reducing the flying exposure of personnel. Options to consider might include

- boat transport (catamaran, fast crew boat)
- longer shifts (2 weeks instead of one)
- minimum manned operation
- unmanned operation.

Working down this list, we see more innovative approaches. The unmanned option using computer-assisted operation (CAO) (discussed in Section 12.2, Chapter 12) would improve safety of personnel and reduce operating cost. This is an example of innovation and the use of technology by the engineer, and is driven by an awareness of safety.

Accident investigation indicates that there are often many individual causes to an accident, and that a series of incidents occur simultaneously to 'cause' the accident. The following figure is called the 'safety triangle', and shows the approximate ratios of occurrence of accidents with different severities. This is based on industrial statistics (Figure 5.2).

An LTI is a *lost time incident*, mentioned earlier as an accident which causes one or more days away from work. A non-LTI injury does not result in time away from

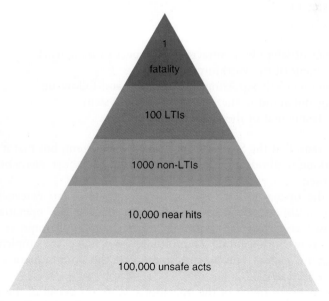

Figure 5.2   The safety triangle.

work. A near hit (often called a near miss) is an incident which causes no injury, but had the potential to do so (e.g. a falling object hitting the ground, but missing personnel). An example of an unsafe act would be a poorly secured ladder, where no incident occurs, but which potentially could have been the cause of an incident.

The safety triangle shows that there are many orders of magnitude more unsafe acts than LTIs and fatalities. A combination of unsafe acts often results in a fatality. Addressing safety in industry should begin with the base of the triangle; trying to eliminate the unsafe acts. This is simple to do, in theory, since most of the unsafe acts arise from carelessness or failure to follow procedures. In practice, reducing the number of unsafe acts requires *personal commitment* and a *safety culture*.

5.2. SAFETY MANAGEMENT SYSTEMS

The UK Government enquiry into the *Piper Alpha disaster* in the North Sea in 1988 had a significant impact on working practices and equipment and has helped to improve offshore safety around the world. One result has been the development of a *safety management system* (SMS) which is a method of integrating work practices, and is a form of quality management system. Major oil companies have each developed their own specific SMS, to suit local environments and modes of operation, but the SMS typically addresses the following areas (recommended by the Cullen Enquiry into the Piper Alpha disaster)

- organisational structure
- management personnel standards
- safety assessment
- design procedures
- procedures for operations, maintenance, modifications and emergencies
- management of safety by contractors in respect of their work
- the involvement of the workforce in safety
- accident and incident reporting, investigation and follow-up
- monitoring and auditing the operation of the system
- systematic reappraisal of the system.

It is important that the SMS is not a stand-alone system, but that it is integrated into the working methods of a company. Some of the above elements of an SMS will be discussed.

Auditing the operation of a system may be done by an external audit team composed of qualified people from within or outside the operating company. However, involvement of the workforce in the audit will improve the level of information, assist with gaining commitment and make the implementation of recommendations easier. This is consistent with the commitment of employees mentioned in Section 5.1.

Contractors perform much of the operational work on behalf of the oil company, because they can supply the specialist skills required. Contractor teams may range from

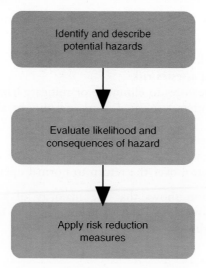

Figure 5.3   General approach to quantitative risk analysis (QRA).

individuals to large groups, and their tasks may take days or months. The contractors are therefore the group with the highest exposure to the operations, and often the least familiar with the particular practices on an installation, since they move between oil companies, and between installations. Special attention must be given to incorporating the contractors into the prevailing SMS by familiarising them with a new location and work practices. This may be achieved through a safety induction-training course.

Design procedures are developed with the intention of improving the safety of equipment. Tools used in this step are hazard and operability studies and *quantitative risk analysis* (QRA). The following scheme may be used (Figure 5.3).

In the first step, a screening process will be applied to separate the major potential hazards, and a *risk register* or *risk matrix* is established; major hazards will be addressed in more detail (see also Section 15.3.2, Chapter 15). QRA techniques are used to evaluate the extent of the risk arising from hazards with the potential to cause major accidents, based on the prediction of the likelihood and consequence of the event. This assessment will be based on engineering judgement and statistics of previous performance. Where necessary, risk reduction measures will be applied until the level of risk is acceptable or 'ALARP'. The acronym stands for as low as reasonably practicable, and is a term often used in the context of safety-critical and high-integrity systems. The ALARP principle is that the residual risk shall be as low as reasonably practicable. For the risk to be ALARP it must be demonstrated that the cost involved in reducing the risk further would be grossly disproportionate to the benefit gained. In other words it would be possible to spend infinite time, effort and money attempting to reduce a risk to zero.

'Permit to Work' procedures are written to ensure that activities are performed in a systematic way. Before conducting work that involves confined space entry, work on energy systems, ground disturbance in locations where buried hazards may exist, or hot work in potentially explosive environments, a permit must be obtained that

- defines scope of work
- identifies hazards and assesses risk
- establishes control measures to eliminate or mitigate hazards
- links the work to other associated work permits or simultaneous operations (SIMOPS)
- is authorised by the responsible person(s)
- communicates above information to all involved in the work
- ensures adequate control over the return to normal operations.

Accident investigation shows that the majority of accidents occur because procedures are not followed, and this contributes mostly to the base of the safety triangle introduced at the end of Section 5.1.

 ## 5.3. ENVIRONMENT

Environmental standards have become a critical part of any business. Many companies account for their performance as part of an annual 'sustainability report' which is audited independently and scrutinised by stakeholders. Whilst individual companies tend to have their own specific *environmental management system* (EMS), global standards have been established, such as ISO 14001. This is an EMS that helps an organisation to identify environmental risks and impacts that may occur as a result of its activities and ensure they are routinely managed. ISO 14001 is designed to support environmental protection and the prevention of pollution in balance with socio-economic needs. Since its principles are generic they can be applied to almost any type of organisation and many large oil and gas companies have adopted its framework.

Adherence to environmental standards is not only required to meet the legislative requirements in host countries, but is also viewed as good business because it is

- cost effective
- providing a competitive edge
- essential to ensuring continued operations in an area
- helpful in gaining future operations in an area.

The approval of loans from major banks for project finance is usually conditional on acceptable environmental management.

Environmental vulnerability varies considerably from area to area. For example, the North Sea, whose waters are displaced into the Atlantic over a 2 year period, is a much more robust area than the Caspian Sea which is enclosed. The EMS of an operating company should reflect those differences.

5.3.1. Environmental impact assessment (EIA)

EIAs have already been introduced in Section 4.4, Chapter 4. The objective of an EIA is to document the potential physical, biological, social and health effects of a

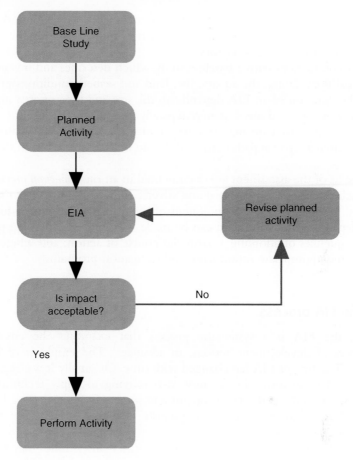

Figure 5.4 Application of an EIA.

planned activity (Figure 5.4). This will enable decision makers to determine whether an activity is acceptable and if not, identify possible alternatives. Typically, EIAs will be carried out for

- seismic
- exploration and appraisal drilling
- development drilling and facilities installation
- production operations
- decommissioning and abandonment.

To allow objectivity of the findings, EIAs are usually carried out by independent specialists or organisations. It will involve not only scientific experts, but also require consultation with official and representative bodies such as the government ministries for the environment, fisheries, food, agriculture and local water. In activities which may impact on local population (terminals, refineries, access roads, land developments) agreement with local NGOs may be critical. The EIA process and its outcome will have a significant impact on the overall project schedule. Hence, early consultation, stakeholder participation and maximising the use of local knowledge are critical success factors.

An EIA commences with a baseline study which describes and inventorises the natural initial flora, fauna, the aquatic life, land and seabed conditions prior to any activity. The duration of an EIA depends on the size and type of area under study, and the previous work done, but may typically take at least 6 months. However, to establish a valid database may require monitoring over several seasons (years). The EIA is often a 'critical path item' and should not be omitted from the planning schedule.

The results of the assessment are documented in an *environmental impact statement* (EIS), which discusses the beneficial and adverse impacts considered to result from the activity. The report is one component of the information upon which project approval depends. A final decision can be made with due regard being paid to the likely consequences of adopting a particular course of action, and where necessary by introducing appropriate monitoring and mitigation programs.

5.3.2. The EIA process

Essentially, the EIA is a systematic process that examines the environmental consequences of development actions, in advance.[1] The emphasis of EIA is on prevention. The role of EIA has changed with time. Originally it was regarded very much as a defensive tool whereas now it is moving to apply techniques which can add positive value to the environment and society in general. The EIA should be regarded as a process which is constantly changing in response to shifting environmental pressures.

The EIA is a 'before project' audit. However, baseline studies may be required again later on in the project, for instance to help refine impact predictions. Baseline studies can account for a significant part of the overall EIA cost since they require extensive field studies.

Certain key stages in the EIA process have been adopted by many countries. These broad stages reflect what is considered to be good practice within environmental assessment and include

- *Screening*: undertaken to decide which projects should be subject to environmental assessment. Screening may be partly determined by local EIA regulations. Criteria used include threshold, size of project and sensitivity of the environment.
- *Scoping*: identifies, at an early stage, the most significant issues to be included in the EIA. Many early EIAs were criticised because they were encyclopaedic and included irrelevant information.
- *Consideration of alternatives*: seeks to ensure that the proposer has considered other feasible options including location, scales, processes, layouts, operating conditions and the 'no action' option.

[1]This section: courtesy of CORDAH.

- *Project description*: includes a clarification of the purpose and rationale of the project.
- *EIA preparation*: is the scientific and objective analysis of the scale, significance and importance of impacts identified. Various methods have been developed, in relation to baseline studies; impact identification; prediction; evaluation and mitigation, to execute this task.
- *Public consultation and participation*: aims to assure the quality, comprehensiveness and effectiveness of the EIA, as well as to ensure that the public's views are adequately taken into consideration in the decision-making process.
- *EIS presentation*: a vital step in the process, the documentation serves to communicate the findings of the EIA process to interested parties.
- *Review*: involves a systematic appraisal by a government agency or independent review panel.
- *Decision-making* on the project involves a consideration by the relevant authority of the EIS (including consultation responses) together with any material considerations.
- *Monitoring*: is normally adopted as a mechanism to check that any conditions imposed on the project are being enforced or to check the quality of the affected environment.
- *Auditing*: follows on from monitoring. Auditing is being developed to test the scientific accuracy of impact predictions and as a check on environmental management practices. It can involve comparing actual outcomes with predicted outcomes, and can be used to assess the quality of predictions and the effectiveness of mitigation. It provides vital feedback into the EIA process.

5.4. CURRENT ENVIRONMENTAL CONCERNS

The following section outlines some of the current environmental concerns the oil and gas industry is facing.

5.4.1. Greenhouse emissions

Methane and carbon dioxide are two significant contributors to greenhouse gases released into the atmosphere, in particular through the venting and burning of fossil fuels. The CO_2 level in the atmosphere has increased from 280 ppm in the late 19th century to 375 ppm today. This increase correlates with a rise in global temperature. Concerns about this trend have resulted in a number of new realities such as

- the Kyoto Protocol
- creation of a 'carbon credit' trading scheme
- the strong public and governmental scrutiny of oil and gas operations worldwide, mentioned earlier.

These realities have to be accounted for when designing or operating oil and gas assets, and are concerns of both petroleum and surface engineers.

5.4.2. Gas venting and flaring

Gas venting has historically been used in many operations as a means of disposal for excess associated gas. Alternatively, gas was flared, a process that emits carbon dioxide and water vapour into the air, another contributor to global warming. Much effort is now being spent on gathering excess gas and to make commercial use of it where possible, or otherwise to re-inject it into reservoirs. Some countries, such as Norway, have introduced a carbon tax, which penalises companies for venting or flaring gas.

5.4.3. CO_2 sequestration

New large-scale gas projects, such as the Gorgon development on the Australian North West shelf and the Sleipner field in Norway have dedicated CO_2 sequestration schemes whereby carbon dioxide separated from natural gas or flue gases from combustion are injected into suitable formations in the subsurface. The benefits are increased oil recovery and the ultimate capture of CO_2 known as sequestration. Seismic surveys and pressure measurements are employed to monitor the integrity of the storage.

5.4.4. Oil-in-water emissions

When water is produced along with oil, the separation of water from oil invariably leaves some water in the oil. The current oil-in-water emission limit into the sea is commonly 40 ppm. Oily water disposal occurs on processing platforms, some drilling platforms and at oil terminals. The quality of water disposed from terminals remains an area of scrutiny, especially since the terminals are often near to local habitation and leisure resorts. If the engineer can find a means of reducing the produced water at source (e.g. water shut-off or re-injection of produced water into reservoirs) then the surface handling problem is much reduced. Recent trends are for governments to disallow oil-in-water emissions entirely, forcing operators to install *produced water re-injection schemes* (PWRI).

5.4.5. Ozone-depleting substances

The 'Montreal Protocol on Substances that Deplete the Ozone Layer' is an international agreement designed to protect the stratospheric ozone layer. It stipulates that the production and consumption of compounds that deplete ozone in the stratosphere, including chlorofluorocarbons (CFCs), halons, carbon tetrachloride and methyl chloroform have to be phased out. Scientific theory and evidence suggest that, once emitted to the atmosphere, these compounds could significantly deplete the stratospheric ozone layer that shields the planet from damaging UV radiation. Some of these substances are used in fire-suppression equipment or in gas refrigeration processes.

5.4.6. Waste management

An oil and gas operation produces much waste material, such as contaminated drill cuttings, completion and workover fluids, chemical waste, radioactive scale, oil sludge and spent catalysts. For all those compounds, systems have to be put in place which effectively deal with containment, transport and disposal. Whilst this will often involve contractor companies, the operating oil or gas company will remain responsible for the waste management process.

Unfortunately, our industry carries a legacy of many cases where oil spills and inadequate waste management have lead to pollution or harm to humans and the environment. Subsequent litigation and persecution have resulted in hefty fines, compensation claims and the loss of reputation.

CHAPTER 6

RESERVOIR DESCRIPTION

Introduction and Commercial Application: The success of oil and gas field development is largely determined by the reservoir: its size, complexity, productivity and the type and quantity of fluids it contains. To optimise a development plan, the characteristics of the reservoir must be well defined. Often the level of information available is significantly less than that required for an accurate description of the reservoir, and estimates of the real situation need to be made. It is often difficult for surface engineers to understand the origin of the uncertainty with which the subsurface engineer must work, and the ranges of possible outcomes provided by the subsurface engineer can be frustrating. This section will describe what controls the uncertainties, and how data are gathered and interpreted to try to form a model of the subsurface reservoir.

The section is divided into four parts, which discuss the common reservoir types from a geological viewpoint, the fluids which are contained within the reservoir, the principal methods of data gathering and the ways in which these data are interpreted. Each section is introduced by pointing out its commercial relevance.

6.1. RESERVOIR GEOLOGY

Introduction and Commercial Application: The objective of reservoir geology is the description and quantification of geologically controlled reservoir parameters and the prediction of their lateral variation. Three parameters broadly define the reservoir geology of a field:

- depositional environment
- structure
- diagenesis.

To a large extent the reservoir geology controls the producibility of a formation, that is to what degree transmissibility to fluid flow and pressure communication exists. Knowledge of the reservoir's geological processes has to be based on extrapolation of the very limited data available to the geologist, yet the *geological model* is the base on which the FDP will be built.

In the following section, we will examine the relevance of depositional environments, structures and diagenesis for field development purposes.

6.1.1. Depositional environment

With a few exceptions, reservoir rocks are *sediments*. The two main categories are *siliciclastic rocks*, usually referred to as 'clastics' or 'sandstones', and *carbonate rocks*.

Most reservoirs in the GoM and the North Sea are contained in a clastic depositional environment; many of the giant fields of the Middle East are contained in carbonate rocks. Before looking at the significance of depositional environments for the production process, let us investigate some of the main characteristics of both the categories.

6.1.1.1. Clastics

The deposition of a clastic rock is preceded by the *weathering* and *transport* of material. *Mechanical* weathering will be induced if a rock is exposed to severe temperature changes or freezing of water in pores and cracks (e.g. in some desert environments). The action of plant roots forcing their way into bedrock is another example of mechanical weathering. Substances (e.g. acid waters) contained in surface waters can cause *chemical weathering*. During this process minerals are dissolved and the less stable ones, like feldspars, are leached. Chemical weathering is particularly severe in tropical climates.

Weathering results in the breaking up of rock into smaller components which can then be transported by agents such as water (rivers, sea currents), wind (deserts) and ice (glaciers). There is an important relationship between the mode of transport and the energy available for the movement of components. *Transport energy* determines the *size*, *shape* and degree of *sorting* of sediment grains. Sorting is an important parameter controlling properties such as porosity. Figure 6.1 shows the impact of sorting on reservoir quality.

Poorly sorted sediments comprise very different particle sizes, resulting in a dense rock fabric with low porosity. As a result, the connate water saturation is high, leaving little space for the storage of hydrocarbons. Conversely, a very well sorted sediment will have a large volume of 'space' between the evenly sized components, a lower connate water saturation and hence a larger capacity to store hydrocarbons.

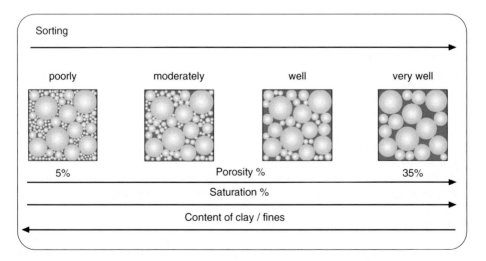

Figure 6.1   Impact of sorting on reservoir quality.

Connate water is the water which remains in the pore space after the entry of hydrocarbons.

Quartz (SiO_2) is one of the most stable minerals and is therefore the main constituent of sandstones which have undergone the most severe weathering and transportation over considerable distance. These sediments are called 'mature' and provide 'clean' high-quality reservoir sands. In theory, porosity is not affected by the size of the grains but is purely a percentage of the bulk rock volume. In nature, however, sands with large well-sorted components may have higher porosities than the equivalent sand comprising small components. This is simply the result of the higher transport energy required to move large components, hence a low probability of fine (light) particles such as clay being deposited.

Very clean sands are rare and normally variable amounts of *clay* will be contained in the reservoir pore system; the clays being the weathering products of rock constituents such as feldspars. The quantity of clay and its distribution within the reservoir exerts a major control on permeability and porosity. Figure 6.2 shows several types of *clay distribution*.

Laminae of clay and clay drapes act as vertical or horizontal baffles or barriers to fluid flow and pressure communication. *Dispersed clays* occupy pore space which in a clean sand would be available for hydrocarbons. They may also obstruct pore throats, thus impeding fluid flow. Reservoir evaluation is often complicated by the presence of clays. This is particularly true for the estimation of hydrocarbon saturation.

Bioturbation, due to the burrowing action of organisms, may connect sand layers otherwise separated by clay laminae, thus enhancing vertical permeability. On the

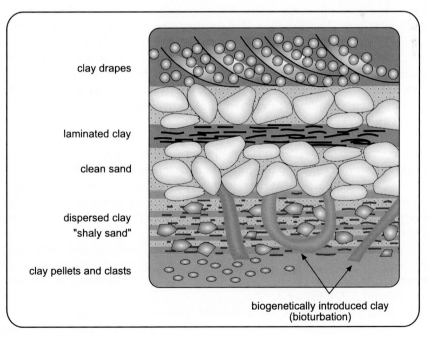

Figure 6.2 Types of clay distribution.

other hand, bioturbation may homogenise a layered reservoir resulting in an unproducible sandy shale.

6.1.1.2. Carbonate rocks

Carbonate rocks are not normally transported over long distances, and we find carbonate reservoir rocks mostly at the location of origin, 'in situ'. They are usually the product of marine organisms. However, carbonates are often severely affected by diagenetic processes. A more detailed description of altered carbonates and their reservoir properties is given below in the description of 'diagenesis'.

6.1.1.3. Depositional environment

Weathering and transportation is followed by the sedimentation of material. The *depositional environment* can be defined as an area with a typical set of physical, chemical and biological processes which result in a specific type of rock. The characteristics of the resulting sediment package are dependent on the intensity and duration of these processes. The physical, chemical, biological and geomorphic variables show considerable differences between and within particular environments. As a result, we have to expect very different behaviour of such reservoirs during hydrocarbon production. Depositional processes control porosity, permeability, net to gross ratio (N/G), extent and lateral variability of reservoir properties. Hence the production profile and ultimate recovery (UR) of individual wells and accumulations are heavily influenced by the environment of deposition.

For example, the many deepwater fields located in the GoM are of Tertiary age and are comprised of complex sand bodies which were deposited in a deepwater turbidite sequence. The Prudhoe Bay sandstone reservoir in Alaska is of Triassic/ Cretaceous age and was deposited by a large shallow water fluvial–alluvial fan delta system. The Saudi Arabian Ghawar limestone reservoir is of Jurassic age and was deposited in a warm, shallow marine sea. Although these reservoirs were deposited in very different depositional environments, they all contain producible accumulations of hydrocarbons, though the fraction of recoverable oil varies. In fact, Prudhoe Bay and Ghawar are amongst the largest in the world, each containing over 20 billion barrels of oil.

There exists an important relationship between the depositional environment, reservoir distribution and the production characteristics of a field (Table 6.1).

It is important to realise that knowledge of depositional processes and features in a given reservoir will be vital for the correct siting of the optimum number of appraisal and development wells, the sizing of facilities and the definition of a reservoir management policy.

To derive a reservoir geological model, various methods and techniques are employed: mainly the analysis of core material, wireline logs, high-resolution seismic and outcrop studies. These data gathering techniques are discussed in Sections 6.3, Chapter 6 and 3.2, Chapter 3.

The most valuable tools for a detailed environmental analysis are cores and wireline logs. In particular, the gamma ray (GR) response is useful since it captures the changes in energy during deposition. Figure 6.3 links depositional environments

Table 6.1 Characteristics of selected environments

Depositional Environment	Reservoir Distribution	Production Characteristic
Deltaic (distributary channel)	Isolated or stacked channels usually with fine-grained sands. May or may not be in communication	Good producers; permeabilities of 500–5000 mD. Insufficient communication between channels may require infill wells in late stage of development
Shallow marine/coastal (clastic)	Sand bars, tidal channels. Generally coarsening upwards. High subsidence rate results in 'stacked' reservoirs. Reservoir distribution dependent on wave and tide action	Prolific producers as a result of 'clean' and continuous sand bodies. Shale layers may cause vertical barriers to fluid flow
Shallow water carbonate (reefs and carbonate muds)	Reservoir quality governed by diagenetic processes and structural history (fracturing)	Prolific production from karstified carbonates. High and early water production possible. 'Dual porosity' systems in fractured carbonates. Dolomites may produce H_2S
Shelf (clastics)	Sheet-like sand bodies resulting from storms or transgression. Usually thin but very continuous sands, well sorted and coarse between marine clays	Very high productivity but high-quality sands may act as 'thief zones' during water or gas injection. Action of sediment burrowing organisms may impact on reservoir quality

to GR response. The GR response measures the level of natural GR activity in the rock formation. Shales have a high GR response, whilst sands have low responses.

A *funnel-shaped GR* log is often indicative of a *deltaic environment* whereby clastic, increasingly coarse sedimentation follows deposition of marine clays. *Bell-shaped GR* logs often represent a *channel environment* where a fining upwards sequence reflects decreasing energy across the vertical channel profile. A modern technique for sedimentological studies is the use of formation imaging tools which provide a very high quality picture of the formations forming the borehole wall. These are described in more detail in Section 6.4.8.

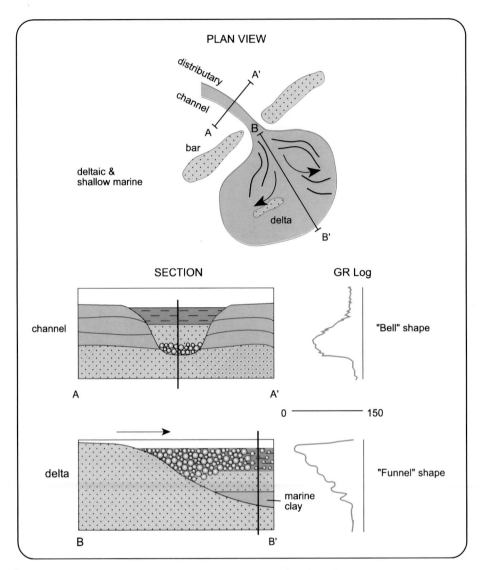

Figure 6.3 Depositional environments, sand distribution and GR log response.

6.1.2. Reservoir structures

As discussed in Chapter 3, the earth's crust is part of a dynamic system and movements within the crust are partly accommodated by rock deformation. Like any other material, rocks may react to stress with an elastic, ductile or brittle response, as described in the stress–strain diagram in Figure 6.4.

It is rare to be able to observe elastic deformations (which occur for instance during earthquakes) since by definition an elastic deformation does not leave any record. However, many subsurface or surface features are related to the other

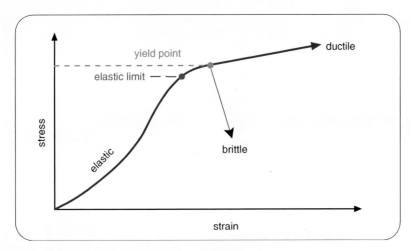

Figure 6.4 The stress–strain diagram for a reservoir rock.

two modes of deformation. The *composition* of the material, *confining pressure, rate of deformation* and *temperature* determine which type of deformation will be initiated.

If a rock is sufficiently stressed, the yield point will eventually be reached. If a brittle failure is initiated, a plane of failure will develop which we describe as a *fault*. Figure 6.5 shows the terminology used to describe *normal, reverse* and *wrench* faults.

Since faults are zones of inherent weakness they may be *reactivated* over geologic time. Usually, faulting occurs well after the sediments have been deposited. An exception to this is a *growth fault* (also termed a *syn-sedimentary fault*), shown in Figure 6.6. They are extensional structures and can frequently be observed on seismic sections through deltaic sequences. The fault plane is curved and in a three-dimensional view has the shape of a spoon. This type of plane is called listric. Growth faults can be visualised as submarine landslides caused by rapid deposition of large quantities of water-saturated sediments and subsequent slope failure. The process is continuous and concurrent with sediment supply, hence the sediment thickness on the downthrown (continuously downward moving) block is expanded compared to the upthrown block.

A secondary feature is the development of *rollover anticlines* which form as a result of the downward movement close to the fault plane which decreases with increasing distance from the plane. Rollover anticlines may trap considerable amounts of hydrocarbons.

Growth faulted deltaic areas are highly prospective since they comprise thick sections of good-quality reservoir sands. Deltas usually overlay organic-rich marine clays which can source the structures on maturation. Examples are the Niger, Baram or Mississippi Deltas. Clays, deposited within deltaic sequences may restrict the water expulsion during the rapid sedimentation/compaction. This can lead to the generation of *overpressures*.

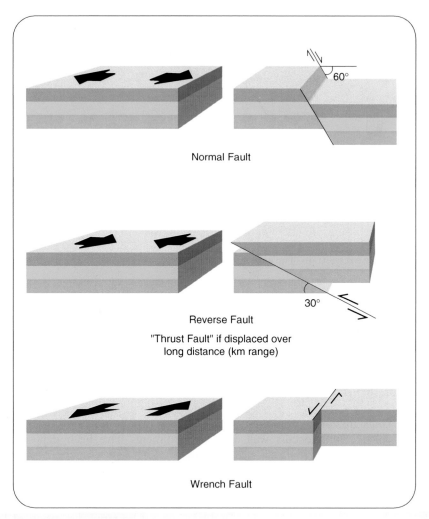

Figure 6.5 Types of faulting.

Faults may extend over several hundreds of kilometres or may be restricted to the deformation of individual grains. They create vast potential traps for the accumulation of oil and gas. However, they often dissect reservoirs and seal fluid and pressures in numerous individual compartments. Each of these isolated blocks may require individual dedicated wells for production and injection. Reservoir *compartmentalisation* through *small-scale faulting* can thus severely downgrade the profitability of a field under development. In the worst case, faulting is not detected until development is in an advanced stage. Early 3D seismic surveys will help to obtain a realistic assessment of fault density and possibly indicate the sealing potential of individual faults. However, small-scale faults with a displacement (*throw*) of less than some 5–10 m are not detectable using seismic alone. Geostatistical techniques can then be used to predict their frequency and direction.

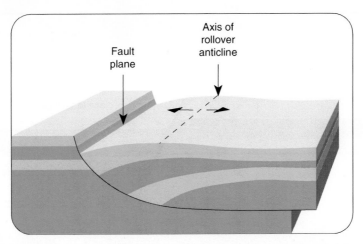

Figure 6.6 Geometry of growth faulting and resulting anticline (rollover) (after Petroleum Handbook, 1983).

Four mechanisms have been suggested to explain how faults provide seals. The most frequent case is that of *clay smear* and *juxtaposition* (Figure 6.7):

- *Clay smear*: soft clay, often of marine origin, is smeared into the fault plane during movement and provides an effective seal.
- *Juxtaposition*: faulting has resulted in an impermeable rock 'juxtaposed' against a reservoir rock.

Other, less frequent fault seals are created by

- *Diagenetic healing*: late precipitation of minerals on or near the fault plane has created a sealing surface (see Section 6.1.3 for more detail).
- *Cataclasis*: the fault movement has destroyed the rock matrix close to the fault plane. Individual quartz grains have been 'ground up' creating a seal comprising of 'rock flour'.

In many cases, faults will only restrict fluid flow, or they may be 'open', that is *non-sealing*. Despite considerable efforts to predict the probability of fault sealing potential, a wholly reliable method to do so has not yet emerged. Fault seal modelling is further complicated by the fact that some faults may leak fluids or pressures at a very small rate, thus effectively acting as *seal on a production time scale* of only a couple of years. As a result, the simulation of reservoir behaviour in densely faulted fields is difficult and predictions should be regarded as crude approximations only.

Fault seals are known to have been ruptured by excessive differential pressures created by production operations, for example if the hydrocarbons of one block are produced whilst the next block is kept at original pressure. Uncontrolled cross-flow and inter-reservoir communication may be the result.

Whereas faults displace formerly connected lithologic units, *fractures* do not show appreciable displacement. They also represent planes of brittle failure and affect hard

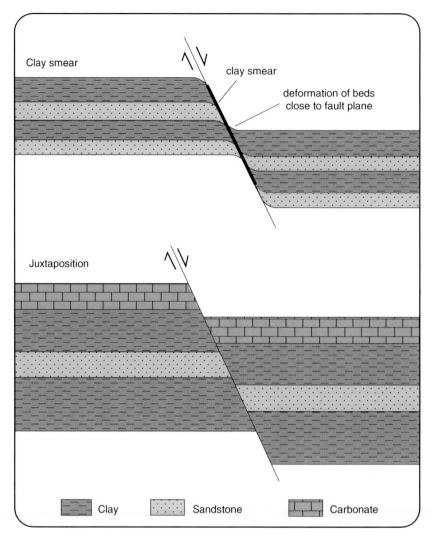

Figure 6.7   Fault seal as a result of clay smear and juxtaposition.

or *competent* lithologies rather than ductile or *incompetent* rocks such as claystone. Frequently fractures are oriented normal to bedding planes (Figure 6.8).

Carbonate rocks are more frequently fractured than sandstones. In many cases, open fractures in carbonate reservoirs provide high porosity/high permeability pathways for hydrocarbon production. The fractures will be continuously re-charged from the tight (less permeable) rock matrix. During field development, wells need to be planned to intersect as many natural fractures as possible, for example by drilling horizontal wells.

Folds are features related to compressional, ductile deformation (Figure 6.9). They form some of the largest reservoir structures known. A fold pair consists of *anticline* and *syncline*.

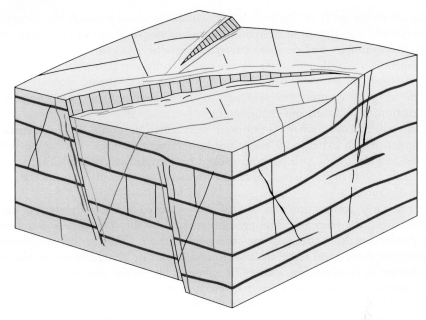

Figure 6.8   A fractured reservoir.

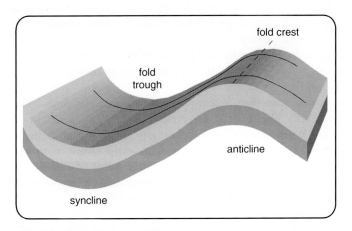

Figure 6.9   Fold terminology.

6.1.3. Diagenesis

The term *diagenesis* describes all chemical and physical processes affecting a sediment after deposition. Processes related to sub-aerial weathering and those which happen under very high pressures and temperatures are excluded from this category. The latter are grouped under the term *metamorphosis*. Diagenesis will alter the geometry and chemistry of the pore space as well as the composition of the rock. Many of these changes are controlled by the oxidising potential (eH) and the acidity/alkalinity (pH) of the pore water which circulates through the formation. Consequently, the migration of hydrocarbons and the displacement of water out of the pore system may end or at least retard diagenetic processes.

Diagenesis will either increase or decrease porosity and permeability and cause a marked change in reservoir behaviour compared to an unaltered sequence.

The diagenetic processes relevant to field development are compaction, cementation, dissolution and replacement.

Compaction occurs when continuous sedimentation results in an increase of overburden which expels pore water from a sediment package. Pore space will be reduced and the grains will become packed more tightly together. Compaction is particularly severe in clays which have an extremely high porosity of some 80% when freshly deposited.

In rare cases, compaction may be artificially initiated by the withdrawal of oil, gas or water from the reservoir. The pressure exerted by the overburden may actually help production by 'squeezing out' the hydrocarbons. This process is known as 'compaction drive', and some shallow accumulations in Venezuela are produced in this manner in combination with EOR schemes like steam injection.

If compaction occurs as a result of production, careful monitoring is required. The Ekofisk Field in the Norwegian North Sea made headlines when, as a result of hydrocarbon production, the pores of the fine-grained carbonate reservoir 'collapsed' and the platforms on the seabed started to sink. The situation was later remedied by inserting steel sections into the platform legs. Compaction effects are also an issue in the Groningen gas field in Holland where subsidence in the order of 1 m is expected at the surface.

Compaction reduces porosity and permeability. As mentioned earlier during the introduction of growth faults, if the expulsion of pore water is prevented, over-pressures may develop.

Cementation describes the 'glueing' together of components. The 'glue' often consists of materials like quartz or various carbonate minerals. They may be introduced to the system by either percolating pore water and/or by precipitation of minerals as a result of changes in pressure and temperature. Compaction may, for instance, lead to quartz dissolution at the contact point of individual grains where pressure is highest. In areas of slightly lower pressure, for example space between the pores, precipitation of quartz may result (Figure 6.10).

This kind of pressure solution/precipitation is active over prolonged periods of time and may almost totally destroy the original porosity. Precipitation of material may also occur in a similar way on the surface of fault planes, thus creating an effective seal via a process introduced earlier as *diagenetic healing*.

Dissolution and replacement. Some minerals, in particular carbonates, are not chemically stable over a range of pressures, temperatures and pH. Therefore, there will be a tendency over geologic time to change to a more stable variety as shown in Figure 6.11.

Rainwater, for instance, will pick up atmospheric CO_2 and react with calcium carbonate (limestone) to form a soluble substance, calcium bicarbonate. This reaction gives water its natural 'hardness'.

$$CaCO_3 + H_2O + CO_2 \rightarrow Ca(HCO_3)_2$$

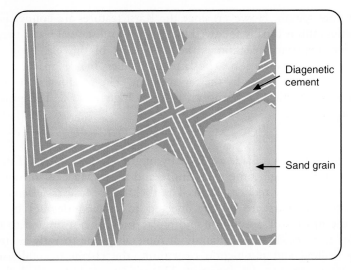

Figure 6.10   Destruction of porosity by cementation.

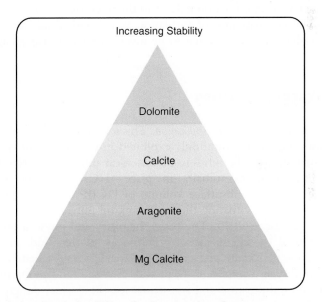

Figure 6.11   Relative chemical stability of carbonate minerals.

Surface water is usually undersaturated in calcium ions (Ca^{2+}). Where (even saturated) surface water mixes with seawater, *mixing zone corrosion* will dissolve calcium carbonate. Evidence of this occurring may be seen on islands.

The dissolution of carbonates can create spectacular features like those found in many caves. The process is termed *karstification*. Some reservoirs are related to *Karst*. Examples are the Bohai Bay Field in China or the Nang Nuan oil field in the Gulf of Thailand. These reservoirs are characterised by high initial production from the large open pore system. However, since the Karst features are connected downdip to the waterleg, this is usually followed by rapid and substantial water breakthrough.

A further important reaction is the *replacement* of the Ca^{2+} ion in calcium carbonate by a magnesium ion. The latter is smaller, hence 'space' or porosity is created in the mineral lattice by the replacement. The resulting mineral is *dolomite* and the increase in effective porosity can be as high as 13%. The process can be expressed as

$$2CaCO_3 + Mg^{2+} \rightarrow CaMg(CO_3)_2 + Ca^{2+}$$

The magnesium ion is made available by migrating pore waters. If the process is continuous on a geologic time scale, more and more Mg^{2+} is introduced to the system and the porosity reduces again. The rock has been *over-dolomitised*.

Carbonate reservoirs are usually affected to varying degrees by diagenesis. However, the process of dissolution and replacement is not limited to carbonates. Feldspar, for instance, is another family of minerals prone to early alterations.

During drilling and production operations, the chemical equilibrium in the reservoir pore system may be disturbed. This is particularly true if drilling mud or injection water enter the formation. The resulting reaction can lead to the precipitation of minerals around the borehole or in the reservoir, and may severely damage productivity. The compatibility of formation water with fluids introduced during drilling and production therefore has to be investigated at an early stage.

6.2. RESERVOIR FLUIDS

Introduction and Commercial Application: This section introduces the various types of hydrocarbons which are commonly exploited in oil and gas field developments. The initial distribution of the fluids in the reservoir must be described to be able to estimate the hydrocarbons initially in place (HCIIP) in the reservoir. The relationship between the subsurface volume of HCIIP and the equivalent surface volume is important in estimating the stock tank oil initially in place (STOIIP) and the gas initially in place (GIIP). The basic chemistry and physical properties of the fluid types are used to differentiate the behaviour of the fluids under producing conditions. For the petroleum and process engineers, a representative description of the reservoir fluid type is important to predict how the fluid properties will change with pressure and temperature and is essential for the correct design of the surface processing facilities. Looking further downstream, the chemical engineer would be concerned about the composition of the hydrocarbon fluids to determine the yields of various fractions which may be achieved.

6.2.1. Hydrocarbon chemistry

The fluids contained within petroleum accumulations are mixtures of *organic compounds*, which are mostly hydrocarbons (molecules composed of hydrogen and carbon atoms), but may also include sulphur, nitrogen, oxygen and metal compounds.

This section will concentrate on the hydrocarbons, but will also explain the significance of the other compounds in the processing of the fluids.

Petroleum fluids vary significantly in appearance, from gases, through clear liquids with the appearance of lighter fuel, to thick black, almost solid liquids. In terms of weight percent of crude oil, for example the carbon element represents 84–87%, the hydrogen element 11–14% and the other elements typically less than 1%. Despite this fairly narrow range of weight percent of the carbon and hydrogen elements, crude oil can vary from a light-brown liquid with a viscosity close to that of water, to a very high viscosity tar-like fluid.

The diversity of the appearance is due to the many ways in which the carbon atoms are able to bond to each other, from single carbon atoms to molecules containing hundreds of carbon atoms linked together in linear chains, to cyclic arrangements of carbon atoms. It is the ability of carbon molecules to combine together in long chains (catenate) which makes organic (i.e. carbon containing) compounds far more numerous than those of other elements, and the basis of living matter.

The various arrangements of carbon atoms can be categorised into 'series' which describe a common molecular structure. The series are based on four main categories which refer to

- the arrangement of the carbon molecules
    - open chain (which may be straight chain or branched)
    - cyclic (or ring)
- the bonds between the carbon molecules
    - saturated (or single) bond
    - unsaturated (or multiple) bond.

6.2.1.1. The alkanes

The largest series is that of the *alkanes* or *paraffins*, which are open chain molecules with saturated bonds, and have the general formula C_nH_{2n+2}.

Figure 6.12 shows the way in which the molecules are visualised, their chemical symbol and the names of the first three members of the series. The carbon atom has

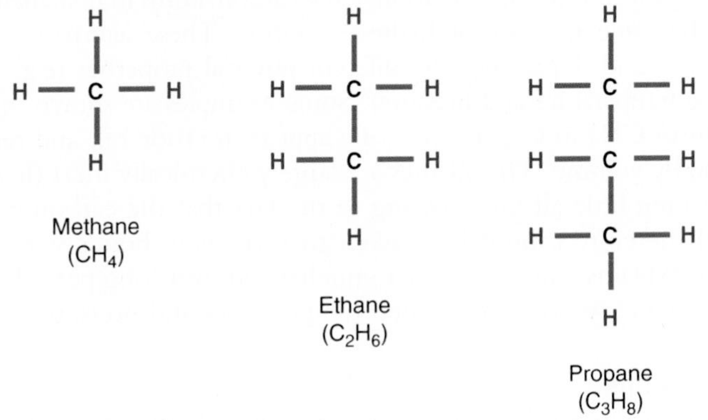

Figure 6.12 Examples from the alkane (paraffin) series.

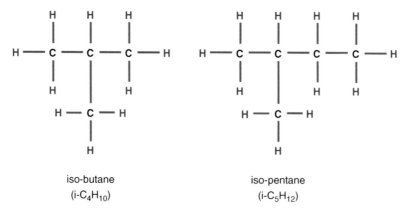

Figure 6.13 Isomers of the paraffin series.

four bonds that can join with either one or more carbon atoms (a unique property) or with atoms of other elements, such as hydrogen. Hydrogen has only one bond, and can therefore join with only one other atom.

Under standard conditions of temperature and pressure (STP), the first four members of the alkane series (methane, ethane, propane and butane) are gases. As the length of the carbon chain increases, the density of the compound increases: C_5H_{12} (pentane) to $C_{17}H_{36}$ are liquids, and from $C_{18}H_{36}$, the compounds exist as wax-like solids at STP.

The most common prefixes are written below using the alkane series as an example, and the prefixes are italicised:

C_1	*meth*ane
C_2	*eth*ane
C_3	*prop*ane
C_4	*but*ane
C_5	*pent*ane
C_6	*hex*ane

Beyond propane, it is possible to arrange the carbon atoms in branched chains whilst maintaining the same number of hydrogen atoms. These alternative arrangements are called *isomers*, and display slightly different physical properties (e.g. boiling point, density, critical temperature and pressure). Some examples are shown in Figure 6.13.

Alkanes from CH_4 to $C_{40}H_{82}$ typically appear in crude oil, and represent up to 20% of the oil by volume. The alkanes are largely chemically inert (hence the name paraffins, meaning little affinity), owing to the fact that the carbon bonds are fully saturated and therefore cannot be broken to form new bonds with other atoms. This probably explains why they remain unchanged over long periods of geological time, despite their exposure to elevated temperatures and pressures.

6.2.1.2. The olefins

Open chain hydrocarbons which are undersaturated, that is having at least one carbon–carbon double bond, are part of the olefin series, and have the ending '-ene'.

Those with one carbon–carbon double bond are called mono-olefins or *alkenes*, for example ethylene $CH_2 = CH_2$.

The double bond is not stronger than the single bond; on the contrary, it is more vulnerable, making unsaturated compounds more chemically reactive than the saturates.

In the longer carbon chains, two carbon–carbon double bonds may exist. Such molecules are called diolefins (or dienes), such as butadiene $CH_2 = CH–CH = CH_2$.

6.2.1.3. Acetylenes

Acetylenes are another series of unsaturated hydrocarbons which include compounds containing a carbon–carbon triple bond, for example acetylene itself:

$$CH \equiv CH$$

Olefins are uncommon in crude oils due to the high chemical activity of these compounds which causes them to become saturated with hydrogen. Similarly, acetylene is virtually absent from crude oil, which tends to contain a large proportion of the saturated hydrocarbons, such as the alkanes.

Whilst the long-chain hydrocarbons (above 18 carbon atoms) may exist in solution at reservoir temperature and pressure, they can solidify at the lower temperatures and pressures experienced in surface facilities, or even in the tubing. The fraction of the longer chain hydrocarbons in the crude oil is therefore of particular interest to process engineers, who will typically require a detailed laboratory analysis of the crude oil composition, extending to the measurement of the fraction of molecules as long as C_{30}.

6.2.1.4. Ring or cyclic structures

The *naphthenes* (C_nH_{2n}), or cycloalkanes, are ring or cyclic saturated structures, such as cyclohexane (C_6H_{12}), though rings of other sizes are also possible. An important series of cyclic structures is the *arenes* (or *aromatics*, so called because of their commonly fragrant odours), which contain carbon–carbon double bonds and are based on the benzene molecule (Figure 6.14).

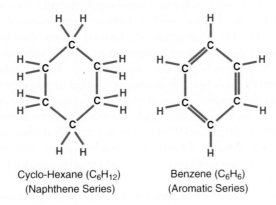

Cyclo-Hexane (C_6H_{12})
(Naphthene Series)

Benzene (C_6H_6)
(Aromatic Series)

Figure 6.14   Ring or cyclic structures.

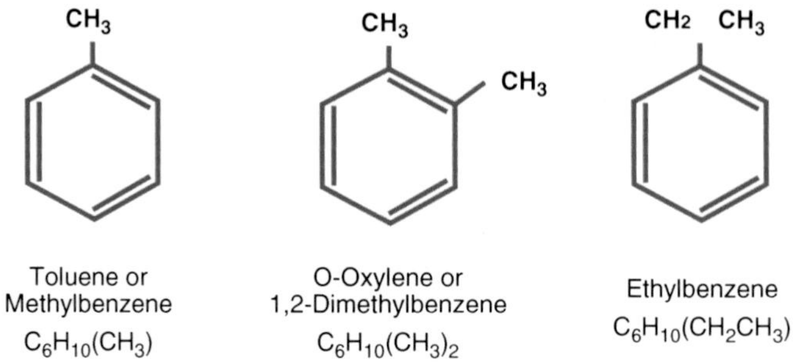

Figure 6.15 Derivatives of benzene.

Although benzene contains three carbon–carbon double bonds, it has a unique arrangement of its electrons (the extra pairs of electrons are part of the overall ring structure rather than being attached to a particular pair of carbon atoms) which allows benzene to be relatively unreactive. Benzene is, however, known to be a cancer-inducing compound.

Some of the common aromatics found in crude oil are the simple derivatives of benzene in which one or more alkyl groups (CH_3) are attached to the basic benzene molecule as a side chain which takes the place of a hydrogen atom. These arenes are either liquids or solids under standard conditions (Figure 6.15).

6.2.1.5. Non-hydrocarbon components of petroleum fluids

The non-hydrocarbon components of crude oil may be small in volume percent, typically less than 1%, but their influence on the product quality and the processing requirements can be considerable. It is therefore important to identify the presence of these components as early as possible, and certainly before the field development planning stage, to enable the appropriate choice of processing facilities and materials of construction to be made.

Sulphur and its products are the most common impurity in crude oil, ranging from 0.2 to over 6% in some Mexican and Middle Eastern crudes, with an average of 0.65% by weight. Corrosive sulphur compounds include free sulphur, *hydrogen sulphide* (H_2S, which is also highly toxic) and *mercaptans* of low molecular weight (e.g. ethyl mercaptan, C_2H_2SH). Mercaptans are formed during the distillation of crude oil, and require special alloys in plant equipment to avoid severe corrosion. The non-corrosive sulphur compounds are the *sulphides* (e.g. diethyl sulphide ($C_2H_5)_2S$), which are not directly corrosive, but require careful temperature control during processing to avoid decomposition to the corrosive products. Sulphur compounds have a characteristic bad smell, and both corrosive and non-corrosive forms are generally undesirable in crude oils. Corrosion due to H_2S is known as '*sour*' *corrosion*.

Some natural gases contain a high H_2S content; above 30% in some Canadian producing wells, where the sulphur is recovered from the product stream and is sold commercially.

Nitrogen content in crude oil is typically less than 0.1% by weight, but can be as high as 2%. The nitrogen compounds in crude oil are complex, and remain largely unidentified. Gaseous nitrogen reduces the calorific value and hence sales price of the hydrocarbon gas. Natural gas containing significant quantities of nitrogen must be blended with high calorific value gas to maintain a uniform product quality.

Oxygen compounds are present in some crude oils, and decompose to form naphthenic acids upon distillation. These may be highly corrosive.

Carbon dioxide (CO₂) is a very common contaminant in hydrocarbon fluids, especially in gases and gas condensate, and is a source of '*sweet*' *corrosion* problems. CO_2 in the gaseous phase dissolves in any water present to form carbonic acid (H_2CO_3) which is highly corrosive. Its reaction with iron creates iron carbonate ($FeCO_3$):

$$Fe + H_2CO_3 \Rightarrow FeCO_3 + H_2$$

The corrosion rate of steel in carbonic acid is faster than in hydrochloric acid! Correlations are available to predict the rate of steel corrosion for different partial pressures of CO_2 and different temperatures. At high temperatures the iron carbonate forms a film of protective scale on the steel's surface, but this is easily washed away at lower temperatures (again a corrosion nomogram is available to predict the impact of the scale on the corrosion rate at various CO_2 partial pressures and temperatures).

CO_2 corrosion often occurs at points where there is turbulent flow, such as in production tubing, piping and separators. The problem can be reduced if there is little or no water present. The initial rates of corrosion are generally independent of the type of carbon steel, and chrome alloy steels or duplex stainless steels (chrome and nickel alloy) are required to reduce the rate of corrosion.

Other compounds which may be found in crude oil are metals such as vanadium, nickel, copper, zinc and iron, but these are usually of little consequence. Vanadium, if present, is often distilled from the feedstock of catalytic cracking processes, since it may spoil catalysis. The treatment of emulsion sludges by biotreatment may lead to the concentration of metals and radioactive material, causing subsequent disposal problems.

Natural gas may contain helium, hydrogen and mercury, though the latter is rarely a significant contaminant in small quantities.

6.2.1.6. Classification of crude oils for refining

There are a total of 18 different hydrocarbon series, of which the most common constituents of crude oil have been presented – the alkanes, the cycloalkanes and the arenes. The more recent classifications of hydrocarbons are based on a division of the hydrocarbons into three main groups: alkanes, naphthenes and aromatics, along with the organic compounds containing the non-hydrocarbon atoms of sulphur, nitrogen and oxygen.

As a general guide, crude oil is commonly classified in the broad categories of paraffinic, naphthenic (meaning that on distillation the residue is asphalt rather than a wax) or intermediate. These classes act as a guide to the commercial value of the refined products of the crude oil, with the lighter ends (shorter carbon chains) commanding more value. Figure 6.16 indicates a first-stage *fractional distillation* of crude oil.

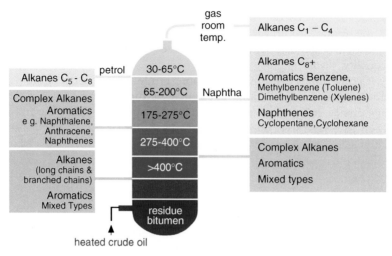

Figure 6.16   Fractional distillation of crude oil.

6.2.2. Types of reservoir fluid

Reservoir fluids are broadly categorised using those properties which are easy to measure in the field, namely oil and gas gravity, and the producing *gas:oil ratio* (GOR) which is the volumetric ratio of the gas produced at STP to the oil produced at STP. The commonly used units are shown in the following table:

	Volumes of Gas	Volumes of Oil
Oil field units	Standard cubic feet (scf)	Stock tank barrels (stb)
Metric units	Standard cubic metres (sm^3)	Stock tank cubic metres (stm^3)

STP are commonly defined as 60°F (298 K) and 1 atm (14.7 psia or 101.3 kPa).

Oil gravity is most commonly expressed in degrees API, a measure defined by the American Petroleum Institute as

$$\text{API} = \frac{141.15}{\gamma_o} - 131.5$$

where γ_o is the specific gravity of oil (relative to water = 1, measured at STP).

The *API gravity* of water is 10°. A light crude oil would have an API gravity of 40°, whilst a heavy crude would have an API gravity of less than 20°. In the field, the API gravity is readily measured using a calibrated hydrometer.

There are no definitions for categorising reservoir fluids, but the following table indicates typical GOR, API and gas and oil gravities for the five main types. The compositions show that the dry gases contain mostly paraffins, with the fraction of longer chain components increasing as the GOR and API gravity of the fluids decrease.

Type	Dry Gas	Wet Gas	Gas Condensate	Volatile Oil	Black Oil
Appearance at surface	Colourless gas	Colourless gas+some clear liquid	Colourless+significant clear/straw liquid	Brown liquid some red/green colour	Black viscous liquid
Initial GOR (scf/stb)	No liquids	>15000	3000–15000	2500–3000	100–2500
Degrees API	–	60–70	50–70	40–50	<40
Gas specific gravity (air=1)	0.60–0.65	0.65–0.85	0.65–0.85	0.65–0.85	0.65–0.8
Composition (mol%)					
C_1	96.3	88.7	72.7	66.7	52.6
C_2	3.0	6.0	10.0	9.0	5.0
C_3	0.4	3.0	6.0	6.0	3.5
C_4	0.17	1.3	2.5	3.3	1.8
C_5	0.04	0.6	1.8	2.0	0.8
C_6	0.02	0.2	2.0	2.0	0.9
C_{7+}	0.0	0.2	5.0	11.0	27.9

6.2.3. The physical properties of hydrocarbon fluids

6.2.3.1. General hydrocarbon phase behaviour

The strict definition of a phase is 'any homogeneous and physically distinct region that is separated from another such region by a distinct boundary'. For example, a glass of water with some ice in it contains one component (the water) exhibiting three phases: liquid, solid and gaseous (the water vapour). The most relevant phases in the oil industry are liquids (water and oil), gases (or vapours) and, to a lesser extent, solids.

As the conditions of pressure and temperature vary, the phases in which hydrocarbons exist and the composition of the phases may change. It is necessary to understand the initial condition of fluids to be able to calculate surface volumes represented by subsurface hydrocarbons. It is also necessary to be able to predict phase changes as the temperature and pressure vary both in the reservoir and as the fluids pass through the surface facilities, so that the appropriate subsurface and surface development plans can be made.

Phase behaviour describes the phase or phases in which a mass of fluid exists at given conditions of pressure, volume (the inverse of the density) and temperature (PVT). The simplest way to start to understand this relationship is by considering a single component, say water, and looking at just two of the variables, say pressure and temperature.

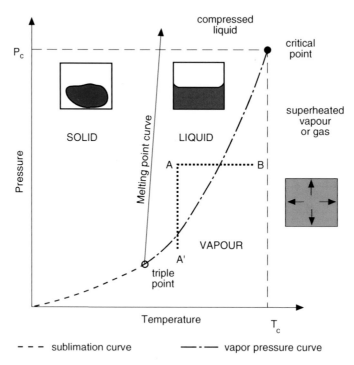

Figure 6.17   Pressure–temperature *phase diagram*.

Figure 6.17 shows the phase boundaries between the component in the solid, liquid and gas (vapour) states. Starting with the liquid (water) at point A, as the temperature is increased the *boiling point* is approached until the boiling point curve is reached, at which point the water boils and turns to steam (gas). Starting from the situation of the gaseous phase at point B, if the temperature is reduced the *dew point* curve is approached, and when the dew point is reached, the component changes from the gaseous phase to the liquid phase. For a single component, the boiling point curve and the dew point curve are coincident, and are known as the *vapour pressure curve*. Of course the phase boundary between the liquid and solid phases is the melting point curve.

At the *triple point* all three phases can co-exist, and this point is a unique property of pure substances. At the *critical point*, defined by the critical temperature (T_c) and pressure (P_c), it becomes impossible to distinguish between the gaseous and liquid phases; the highly compressed gas has the same density and appearance as a high-temperature liquid. The effect of the increased pressure and attractive forces between molecules is to move molecules together and increase the density (as when a gas becomes a liquid), but the increasing temperature increases the kinetic energy of the molecules and tends to drive them apart, thus reducing the density (as when a liquid becomes a gas). At the critical point, the phases become indistinguishable, and beyond the critical point just one state exists, and is usually referred to as a supercritical fluid.

In the production of hydrocarbon reservoirs, the process of *isothermal depletion* is normally assumed, that is reducing the pressure of the system whilst maintaining a constant temperature. Hence, a more realistic movement on the pressure–temperature plot is from point A to A′.

Now using a hydrocarbon component, say ethane, as an example, let us consider the other parameter, volume, using a plot of pressure vs. specific volume (i.e. volume per unit mass of the component, the inverse of the density). The process to be described could be performed physically by placing the liquid sample into a closed cell (PVT cell), and then reducing the pressure of the sample by withdrawing the piston of the cell and increasing the volume contained by the sample.

Starting at condition A with the ethane in the liquid phase, and assuming isothermal depletion, then as the pressure is reduced, the specific volume increases as the molecules move further apart. The relationship between pressure and volume is governed by the compressibility of the liquid ethane.

Once the *bubble point* is reached (at point B), the first bubble of ethane vapour is released. From point B to C, liquid and gas co-exist in the cell, and the pressure is maintained constant as more of the liquid changes to the gaseous state. The system exhibits infinite compressibility until the last drop of liquid is left in the cell (point C), which is the dew point. Below the dew point pressure only gas remains in the cell, and as pressure is reduced below the dew point, the volume increase is determined by the *compressibility* of the gas. The gas compressibility is much greater than the liquid compressibility, and hence the change of volume for a given reduction in pressure (the gradient of the curve on the pressure–volume plot) is much lower than for the liquid. Eventually the point A′ is reached (Figure 6.18).

If the experiment is now reversed, starting from A′ and increasing the pressure, the first drop of ethane liquid would appear at point C, the *dew point* of the gas. Remember that throughout this process, isothermal conditions are maintained.

The experiment could be repeated at a number of different temperatures and initial pressures to determine the shape of the *two-phase envelope* defined by the

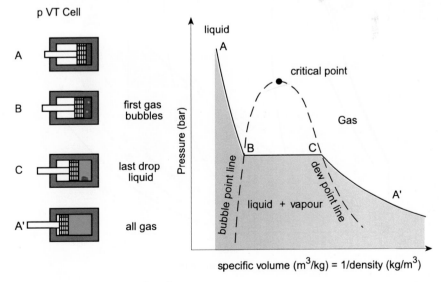

Figure 6.18   Pressure vs. specific volume.

bubble point line and the dew point line. These two lines meet at the critical point, where it is no longer possible to distinguish between a compressed gas and a liquid.

It is important to remember the significance of the bubble point line, the dew point line and the two-phase region, within which gas and liquid exist in equilibrium.

So far we have considered only a single component. However, reservoir fluids contain a mixture of hundreds of components, which adds to the complexity of the phase behaviour. Now consider the impact of adding one component to the ethane, say n-heptane (C_7H_{16}). We are now discussing a binary (two-component) mixture, and will concentrate on the pressure–temperature phase diagram.

Figure 6.19 shows that each component has its own vapour pressure curve and critical point when we consider the components in isolation. The n-heptane vapour pressure curve is shifted down and to the right on the diagram, indicating that it requires higher temperatures and lower pressure to move n-heptane from the liquid to the gaseous phase. This is generally true for longer chain hydrocarbon components.

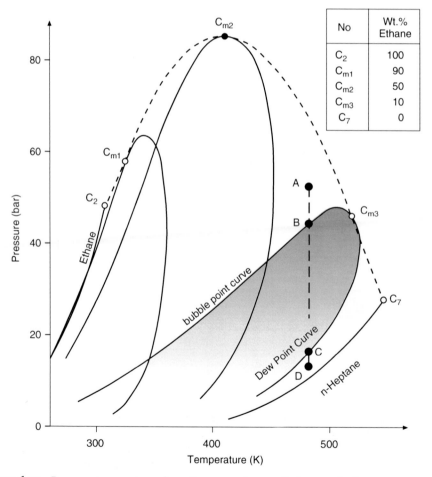

Figure 6.19   Pressure–temperature phase diagram; mixture of ethane and n-heptane.

When the two components are mixed together (say in a mixture of 10% ethane, 90% *n*-heptane), the bubble point curve and the dew point curve no longer coincide, and a *two-phase envelope* appears. Within this two-phase region, a mixture of liquid and gas exists, with both components being present in each phase in proportions dictated by the exact temperature and pressure, that is the composition of the liquid and gaseous phases within the two-phase envelope is not constant. The mixture has its own critical point C_{m3}.

Using this mixture as an example, consider starting at pressure A and isothermally reducing the pressure to point D on the diagram. At point A the mixture exists entirely in the liquid phase. When the pressure drops to point B, the first bubble of gas is evolved, and this will be a bubble of the lighter component, ethane. As the pressure continues to drop, the gaseous phase will acquire more of the heavier component and hence the liquid volume decreases. At point C, the last drop of liquid remaining will be composed of the heavier component, which itself will vaporise as the dew point is crossed, so that below the dew point the mixture exists entirely in the gaseous phase. Outside the two-phase envelope the composition is fixed, but varies with pressure inside the two-phase envelope.

Moving back to the overall picture, it can be seen that as the fraction of ethane in the mixture changes, the position of the two-phase region and the critical point change, moving to the left as the fraction of the lighter component (ethane) increases.

The example of a binary mixture is used to demonstrate the increased complexity of the phase diagram through the introduction of a second component in the system. Typical reservoir fluids contain hundreds of components, which makes the laboratory measurement or mathematical prediction of the phase behaviour more complex still. However, the principles established above will be useful in understanding the differences in phase behaviour for the main types of hydrocarbon identified.

6.2.3.2. Phase behaviour of reservoir fluid types

Figure 6.20 helps in explaining how the phase diagrams of the main types of reservoir fluid are used to predict fluid behaviour during production and how this influences field development planning. It should be noted that there are no values on the axes, since the scales will vary for each fluid type. Figure 6.20 shows the relative positions of the phase envelopes for each fluid type.

The four vertical lines on the diagram show the isothermal depletion loci for the main types of hydrocarbon: gas (incorporating dry gas and wet gas), gas condensate, volatile oil and black oil. The starting point, or initial conditions of temperature and pressure, relative to the two-phase envelope are different for each fluid type.

6.2.3.3. Dry gas

The initial condition for the dry gas is outside the two-phase envelope, and is to the right of the critical point, confirming that the fluid initially exists as a single-phase gas. As the reservoir is produced, the pressure drops under isothermal conditions, as indicated by the vertical line. Since the initial temperature is higher than the maximum temperature of the two-phase envelope (the *cricondotherm* – typically less

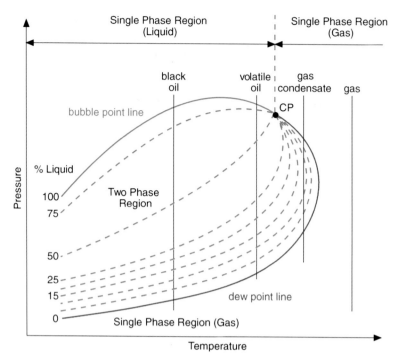

Figure 6.20 Pressure–temperature phase envelopes for main hydrocarbon types – showing initial conditions relative to the phase envelope only.

than 0°C for a dry gas), the reservoir conditions of temperature and pressure never fall inside the two-phase region, indicating that the composition and phase of the fluid in the reservoir remains constant.

In addition, the separator temperature and pressure of the surface facilities are typically outside the two-phase envelope, so no liquids form during separation. This makes the prediction of the produced fluids during development very simple, and gas sales contracts can be agreed with the confidence that the fluid composition will remain constant during field life in the case of a dry gas.

6.2.3.4. Wet gas

Compared to a dry gas, a wet gas contains a larger fraction of the C_2–C_6 components, and hence its phase envelope is moved down and to the right. Whilst the reservoir conditions remain outside the two-phase envelope, so that the reservoir fluid composition remains constant and the gaseous phase is maintained, the separator conditions are inside the two-phase envelope. As the dew point is crossed, the heavier components condense as liquids in the separator. The exact volume percent of liquids which condense depends upon the separator conditions and the spacing of the iso-vol lines for the mixture (the lines of constant liquid percentage shown in Figure 6.20). These heavier components are valuable as light ends of the fractionation range of petroleum, and sell at a premium price. It is usually worthwhile to recover these liquids, and to leave the sales gas as a dry gas (predominantly methane, CH_4). Note that the term wet gas does not refer to water content, but rather to the gas composition containing more of the heavier hydrocarbons than a dry gas.

6.2.3.5. Gas condensate

The initial temperature of a gas condensate lies between the critical temperature and the cricondotherm. The fluid therefore exists at initial conditions in the reservoir as a gas, but on pressure depletion the dew point line is reached, at which point liquids condense in the reservoir. As can be seen from Figure 6.20, the volume percent of liquids is low, typically insufficient for the saturation of the liquid in the pore space to reach the critical saturation beyond which the liquid phase becomes mobile. These liquids therefore remain trapped in the reservoir as an immobile phase. Since these liquids are valuable products, there is an incentive to avoid this condensation in the reservoir by maintaining the reservoir pressure above the dew point. This is the reason for considering *recycling of gas* in these types of reservoir (Figure 6.21).

Gas is produced to surface separators which are used to extract the heavier ends of the mixture (typically the C_{5+} components). The dry gas is then compressed and re-injected into the reservoir to maintain the pressure above the dew point. As the recycling progresses, the reservoir composition becomes leaner (less heavy components), until eventually it is not economic to separate and compress the dry

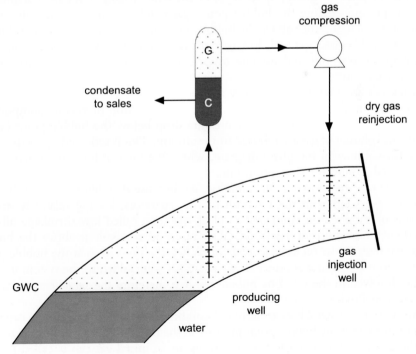

Figure 6.21   Gas recycling process.

gas, at which point the reservoir pressure is 'blown down' as for a wet gas reservoir. The sales profile for a recycling scheme consists of early sales of condensate liquids and delayed sale of gas. An alternative method of keeping the reservoir above the dew point but avoiding the deferred gas sales is by water injection, but this is rarely done as gas trapped behind an advancing gas–water contact represents a significant loss.

Figure 6.20 shows that as the pressure is reduced below the dew point, the volume of liquid in the two-phase mixture initially increases. This contradicts the common observation of the fraction of liquids in a volatile mixture reducing as the pressure is dropped (vaporisation), and explains why the fluids are sometimes referred to as *retrograde gas condensates*.

6.2.3.6. Volatile oil and black oil

For both volatile oil and black oil, the initial reservoir temperature is below the critical point, and the fluid is therefore a liquid in the reservoir. As the pressure drops, the bubble point is eventually reached, and the first bubble of gas is released from the liquid. The composition of this gas will be made up of the more volatile components of the mixture. Both volatile oils and black oils will liberate gas in the separators, whose conditions of pressure and temperature are well inside the two-phase envelope.

A volatile oil contains a relatively large fraction of lighter and intermediate components which vaporise easily. With a small drop in pressure below the bubble point, the relative amount of liquid to gas in the two-phase mixture drops rapidly, as shown in the phase diagram (Figure 6.20) by the wide spacing of the iso-vol lines. At reservoir pressures below the bubble point, gas is released in the reservoir, and is known as *solution gas*, since above the bubble point this gas was contained in solution. Some of this liberated gas will flow towards the producing wells, whilst some will remain in the reservoir and migrate towards the crest of the structure to form a secondary gas cap.

Black oils are a common category of reservoir fluids, and are similar to volatile oils in behaviour, except that they contain a lower fraction of volatile components and therefore require a much larger pressure drop below the bubble point before significant volumes of gas are released from solution. This is reflected by the position of the iso-vol lines in the phase diagram, where the lines of low liquid percentage are grouped around the dew point line.

Volatile oils are known as *high shrinkage oils* because they liberate relatively large amounts of gas either in the reservoir or the separators, leaving relatively smaller amounts of stabilised oil compared to black oils (also called low shrinkage oils).

When the pressure of a volatile oil or black oil reservoir is above the bubble point, we refer to the oil as *undersaturated*. When the pressure is at the bubble point we refer to it as *saturated oil*, since if any more gas were added to the system it could not be dissolved in the oil. The bubble point is therefore the *saturation pressure* for the reservoir fluid.

An oil reservoir which exists at initial conditions with an overlying gas cap must by definition be at the bubble point pressure at the interface between the gas and the oil, the gas–oil contact (GOC). Gas existing in an initial gas cap is called *free gas*, whilst the gas in solution in the oil is called dissolved or *solution gas*.

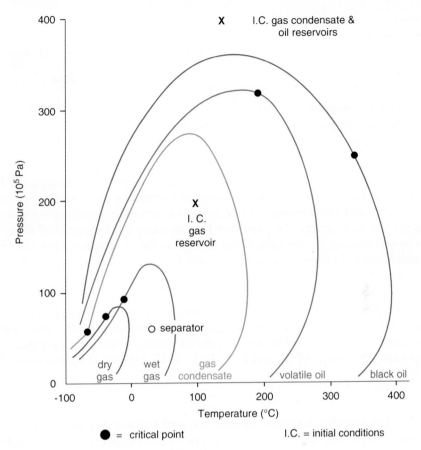

Figure 6.22 Relative positions of phase envelopes.

6.2.3.7. Comparison of the phase envelopes for different hydrocarbon types

Figure 6.22 shows the phase envelopes for the different types of hydrocarbons discussed, using the same scale on the axes. The higher the fraction of the heavy components in the mixture, the further to the right the two–phase envelope. Typical separator conditions would be around 50 bara and 15°C (note that 1 bar $= 10^5$ Pa).

6.2.4. Properties of hydrocarbon gases

The properties of hydrocarbon gases are relatively simple since the parameters of PVT can be related by a single equation. The basis for this equation is an adaptation of a combination of the classical laws of Boyle, Charles and Avogadro.

In the equation of state for an ideal gas, that is a gas in which the volume of the gas molecules is insignificant, attractive and repulsive forces between molecules are ignored, and molecules maintain their energy when they collide with each other.

$$PV = nRT \quad \text{The ideal gas law}$$

where

	Field Units	SI Units
P = absolute pressure	Psia	Bara
V = volume	ft^3	m^3
n = number of moles of gas	–	–
T = absolute temperature	°Rankine	°Rankine
R = universal gas constant	10.73 psia ft^3	8314.3 kJ/kmol K

The above equation is valid at low pressures where the assumptions hold. However, at typical reservoir temperatures and pressures, the assumptions are no longer valid, and the behaviour of hydrocarbon reservoir gases deviate from the ideal gas law. In practice, it is convenient to represent the behaviour of these 'real' gases by introducing a correction factor known as the *gas deviation factor* (also called the dimensionless compressibility factor, or *z*-factor) into the ideal gas law:

$$PV = znRT \quad \text{The real gas law}$$

The *z*-factor must be determined empirically (i.e. by experiment), but this has been done for many hydrocarbon gases, and correlation charts exist for the approximate determination of the *z*-factor at various conditions of pressure and temperature (Standing, M. B. and Katz, D. L. 1942. Density of natural gases. *Trans. AIME*).

6.2.4.1. Relationship between subsurface and surface gas volumes

The most important use of the real gas law is to calculate the volume which a subsurface quantity of gas will occupy at surface conditions, since when gas sales contracts are negotiated and gas is subsequently sold it is referred to in volumes at standard conditions of temperature (T_{sc}) and pressure (P_{sc}).

The relationship required is the gas expansion factor (E), and is defined for a given quantity (mass or number of moles) of gas as

$$E = \frac{\text{Volume of gas at standard conditions}}{\text{Volume of gas at reservoir conditions}} \quad \text{(scf/rcf) or (sm}^3/\text{rm}^3)$$

It can be shown using the real gas law, and the knowledge that at standard conditions $z = 1.0$, that for a reservoir pressure (P) and temperature (T):

$$E = \frac{1}{z} \frac{T_{sc}}{T} \frac{P}{P_{sc}} \quad \text{(vol/vol)}$$

The previous equation is only valid as long as there is no compositional change of the gas between the subsurface and the surface. The value of E is typically in the order of 200, in other words the gas expands by a factor of around 200 from subsurface to surface conditions. The actual value of course depends upon both the gas composition and the reservoir temperature and pressure. STP are commonly defined as 60°F (298 K) and 1 atm (14.7 psia or 101.3 kPa), but may vary from location to location, and between gas sales contracts.

In gas reservoir engineering, the *gas expansion factor, E*, is commonly used. However, in oil reservoir engineering it is often more convenient to refer to the gas *formation volume factor, B_g*, which is the reciprocal of E, and is expressed in units of rb/scf (using field units):

$$B_g \text{ (rb/scf)} = \frac{1}{5.615E}$$

The reason for this will become apparent in Chapter 9.

6.2.4.2. Gas density and viscosity

Density is the most commonly measured property of a gas, and is obtained experimentally by measuring the specific gravity of the gas (density of the gas relative to air = 1). As pressure increases, so does *gas density*, but the relationship is non-linear since the dimensionless gas compressibility (z-factor) also varies with pressure. The gas density (ρ_g) can be calculated at any pressure and temperature using the real gas law:

$$\rho_g = \frac{MP}{zRT}$$

where M is the molecular weight of the gas (lb/mol or kg/kmol).

Gas density at reservoir conditions is useful in calculating the pressure gradient of the gas when constructing pressure–depth relationships (see Section 6.2.8).

When fluid flow in the reservoir is considered, it is necessary to estimate the *viscosity* of the fluid, since viscosity represents an internal resistance force to flow given a pressure drop across the fluid. Unlike liquids, when the temperature and pressure of a gas is increased, the viscosity increases as the molecules move closer together and collide more frequently.

Viscosity is measured in poise. If a force of 1 dyn, acting on 1 cm^2, maintains a velocity of 1 cm/s over a distance of 1 cm, then the fluid viscosity is 1 P. For practical purposes, the centipoise (cP) is commonly used. The typical range of gas viscosity in the reservoir is 0.01–0.05 cP. By comparison, a typical water viscosity is 0.5–1.0 cP. Lower viscosities imply higher velocity for a given pressure drop, meaning that gas in the reservoir moves fast relative to oil and water, and is said to have a high mobility. This is further discussed in Chapter 9.

Measurement of gas viscosity at reservoir pressure and temperature is a complex procedure, and correlations are often used as an approximation.

6.2.4.3. Surface properties of hydrocarbon gases

Wobbe index.   The Wobbe index (WI) is a measurement of the quality of a gas and is defined as

$$WI = \frac{\text{Gross calorific value of the gas}}{(\text{Specific gravity of the gas})^{0.5}} \text{ or } \frac{\text{Energy density}}{(\text{Relative density of the gas})^{0.5}}$$

Measured in MJ/m^3 or Btu/ft^3, the WI has an advantage over the *calorific value* of a gas (the heating value per unit of weight, e.g. Btu/lb), which varies with the density of the gas. The WI is commonly specified in gas contracts as a guarantee of product quality. A customer usually requires a product whose WI lies within a narrow range, since a burner will need adjustment to a different fuel:air ratio if the fuel quality varies significantly. A sudden increase in heating value of the feed can cause a flame-out.

6.2.4.4. Hydrate formation

Under certain conditions of temperature and pressure, and in the presence of free water, hydrocarbon gases can form *hydrates*, which are a solid formed by the combination of water molecules and the methane, ethane, propane or butane. Hydrates look like compacted snow, and can form blockages in pipelines and other vessels. Process engineers use correlation techniques and process simulation to predict the possibility of hydrate formation, and prevent its formation by either drying the gas or adding a chemical (such as tri-ethylene glycol [TEG]), or a combination of both. This is further discussed in Section 11.1, Chapter 11.

6.2.5. Properties of oils

This section will firstly consider the properties of oils in the reservoir (compressibility, viscosity and density), and secondly, the relationship of subsurface to surface volume of oil during the production process (formation volume factor and GOR).

6.2.5.1. Compressibility of oil

Pressure depletion in the reservoir can normally be assumed to be isothermal, such that the isothermal *compressibility* is defined as the fractional change in volume per unit change in pressure, or

$$c = -\frac{1}{V}\frac{dV}{dP} \quad (\text{psi}^{-1}) \text{ or } (\text{bar}^{-1})$$

The value of the compressibility of oil is a function of the amount of dissolved gas, but is in the order of $10 \times 10^{-6}\,\text{psi}^{-1}$. By comparison, typical water and gas compressibilities are $4 \times 10^{-6}\,\text{psi}^{-1}$ and $500 \times 10^{-6}\,\text{psi}^{-1}$, respectively. Above the bubble point in an oil reservoir, the compressibility of the oil is a major determinant of how the pressure declines for a given change in volume (brought about by a withdrawal of reservoir fluid during production).

Reservoirs containing low-compressibility oil, having small amounts of dissolved gas, will suffer from large pressure drops after only limited production. If the expansion of oil is the only method of supporting the reservoir pressure, then abandonment conditions (when the reservoir pressure is no longer sufficient to produce economic quantities of oil to the surface) will be reached after production of probably less than 5% of the oil initially in place. Oil compressibility can be read from correlations.

6.2.5.2. Oil viscosity

Oil viscosity is an important parameter required in predicting the fluid flow, both in the reservoir and in surface facilities, since the viscosity is a determinant of the velocity with which the fluid will flow under a given pressure drop. Oil viscosity is significantly greater than that of gas (typically 0.2–50 cP compared to 0.01–0.05 cP under reservoir conditions).

Unlike gases, liquid viscosity decreases as temperature increases, as the molecules move further apart and decrease their internal friction. Like gases, oil viscosity increases as the pressure increases, at least above the bubble point. Below the bubble point, when the solution gas is liberated, oil viscosity increases because the lighter oil components of the oil (which lower the viscosity of oil) are the ones which transfer to the gaseous phase.

The same definition of viscosity applies to oil as gas, but sometimes the *kinematic viscosity* is quoted. This is the viscosity divided by the density ($u = \mu/\rho$), and has a straight-line relationship with temperature.

6.2.5.3. Oil density

Oil density at surface conditions is commonly quoted in °API, as discussed in Section 6.2.2. Recall,

$$\text{API} = \frac{141.5}{\gamma_o} - 131.5$$

where γ_o is the specific gravity of oil (relative to water $= 1$, measured at STP).

The oil density at surface is readily measured by placing a sample in a cylindrical flask and using a graduated hydrometer. The API gravity of a crude sample will be affected by temperature because the thermal expansion of hydrocarbon liquids is significant, especially for more volatile oils. It is therefore important to record the temperature at which the sample is measured (typically the flowline temperature or the temperature of the stock tank). When quoting the gravity of a crude, standard conditions should be used.

The downhole density of oil (at reservoir conditions) can be calculated from the surface density using the equation:

$$\rho_{orc} B_o = \rho_o + R_s \rho_g$$

where ρ_{orc} is the oil density at reservoir conditions (kg/m^3), B_o the oil formation volume factor (rm^3/stm^3), ρ_o the oil density at standard conditions (kg/m^3), R_s the solution GOR (sm^3/stm^3) and ρ_g the gas density at standard conditions (kg/m^3).

The density of the oil at reservoir conditions is useful in calculating the gradient of oil and constructing a pressure–depth relationship in the reservoir (see Section 6.2.8).

The previous equation introduces two new properties of the oil, the formation volume factor and the solution GOR, which will now be explained.

6.2.5.4. Oil formation volume factor and solution gas:oil ratio

Assuming an initial reservoir pressure above the bubble point (undersaturated reservoir oil), only one phase exists in the reservoir. The volume of oil (rm^3 or rb) at reservoir conditions of temperature and pressure is calculated from the mapping techniques discussed in Section 6.4.

As the reservoir pressure drops from the initial reservoir pressure towards the bubble point pressure (P_b), the oil expands slightly according to its compressibility. However, once the pressure of the oil drops below the bubble point, gas is liberated from the oil, and the remaining oil occupies a smaller volume. The gas dissolved in the oil is called the solution gas, and the ratio of the volume of gas dissolved per volume of oil is called the *solution* GOR (R_s, measured in scf/stb of sm^3/stm^3). Above the bubble point, R_s is constant and is known as the initial solution GOR (R_{si}), but as the pressure falls below the bubble point and solution gas is liberated, R_s decreases. The volume of gas liberated is ($R_{si}-R_s$) scf/stb.

As solution gas is liberated, the oil shrinks. A particularly important relationship exists between the volume of oil at a given pressure and temperature and the volume of the oil at stock tank conditions. This is the *oil formation volume factor* (B_o, measured in rb/stb or rm^3/stm^3).

The oil formation volume factor at initial reservoir conditions (B_{oi}, rb/stb) is used to convert the volumes of oil calculated from the mapping and volumetrics exercises to stock tank conditions. The value of B_{oi} depends upon the fluid type and the initial reservoir conditions, but may vary from 1.1 rb/stb for a black oil with a low GOR to 2.0 rb/stb for a volatile oil. Whenever volumes of oil are described, the volume quoted should be in stock tank barrels, or stock tank cubic metres, since these are the conditions at which the oil is sold. Quoting hydrocarbon volumes at reservoir conditions is of little commercial interest.

Figure 6.23 shows the change in oil volume as pressure decreases from the initial pressure, the amount of gas remaining dissolved in the oil and the volume of liberated gas.

If the reservoir pressure remains above the bubble point, then any gas liberated from the oil must be released in the tubing and the separators, and will therefore appear at the surface. In this case, the *producing* GOR (R_p) will be equal to R_s, that is every stock tank barrel of oil produced liberates R_s scf of gas at surface.

If, however, the reservoir pressure drops below the bubble point, then gas will be liberated in the reservoir. This liberated gas may flow either towards the producing wells under the hydrodynamic force imposed by the lower pressure at the well, or it may migrate upwards, under the influence of the buoyancy force, towards the crest of the reservoir to form a *secondary gas cap*. Consequently, the producing GOR (R_p) will differ from R_s. This is further discussed in Chapter 9.

In a saturated oil reservoir containing an initial gas cap, the producing GOR (R_p) may be significantly higher than the solution GOR (R_s) of the oil, as free gas in the gas cap is produced through the wells via a coning or cusping mechanism. *Free gas* is the gas existing in the gas cap as a separate phase, distinct from solution gas which is dissolved in the oil phase.

6.2.6. Fluid sampling and PVT analysis

The collection of representative reservoir fluid samples is important in order to establish the PVT properties – phase envelope, bubble point, R_s and B_o – and the

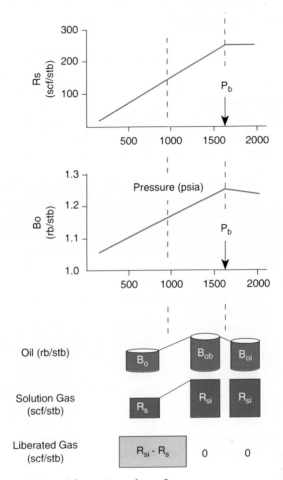

Figure 6.23   Solution GOR and formation volume factor vs. pressure.

physical properties – composition, density and viscosity. These values are used to determine the initial volumes of fluid in place in stock tank volumes and the flow properties of the fluid both in the reservoir and through the surface facilities, and to identify any components which may require special treatment, such as sulphur compounds.

Reservoir fluid sampling is usually done early in the field life in order to use the results in the evaluation of the field and in the process facilities design. Once the field has been produced and the reservoir pressure changes, the fluid properties will change as described in the previous section. Early sampling is therefore an opportunity to collect unaltered fluid samples.

Fluid samples may be collected downhole at near-reservoir conditions, or at surface. *Subsurface samples* are more expensive to collect, since they require downhole sampling tools, but are more likely to capture a representative sample, since they are targeted at collecting a single-phase fluid. A surface sample is inevitably a two-phase sample which requires recombining to recreate the reservoir fluid. Both *sampling techniques* face the same problem of trying to capture a representative sample (i.e. the correct proportion of gas to oil) when the pressure falls below the bubble point.

6.2.6.1. Subsurface samples

Subsurface samples can be taken with a subsurface sampling chamber, called a sampling bomb, or with a formation pressure testing tool (RFT, MDT, RCI), all of which are devices run on wireline to the reservoir depth. These tools are further described in Section 6.3.6.

The sampling bomb requires the well to be flowing, and the flowing bottom hole pressure (P_{wf}) should preferably be above the bubble point pressure of the fluid to avoid phase segregation. If this condition can be achieved, a sample of oil containing the correct amount of gas (R_{si} scf/stb) will be collected. If the reservoir pressure is close to the bubble point, this means sampling at low rates to maximise the sampling pressure. The valves on the sampling bomb are open to allow the fluid to flow through the tool and are then hydraulically or electrically closed to trap a volume (typically 600 cm^3) of fluid. This small sample volume is one of the drawbacks of subsurface sampling (Figure 6.24).

Sampling saturated reservoirs with this technique requires special care to attempt to obtain a representative sample, and in any case when the flowing bottom hole pressure is lower than the bubble point, the validity of the sample remains doubtful. Multiple subsurface samples are usually taken by running sample bombs in tandem or performing repeat runs. The samples are checked for consistency by measuring their bubble point pressure at surface temperature. Samples whose bubble point lie within 2% of each other may be sent to the laboratory for PVT analysis.

Samples taken from *formation pressure testers* (FPTs) can be taken immediately after drilling and do not require the well to be flowing. A small probe is sealed

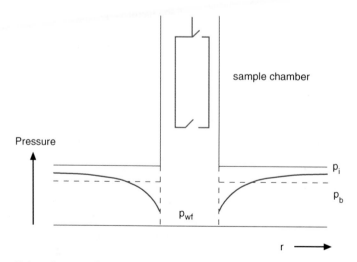

Figure 6.24 Subsurface sampling apparatus (after Dake, 1978).

against the borehole wall through which reservoir fluids flow into a series of chambers which are then sealed to maintain reservoir pressure conditions (see Section 6.3.6). A variety of downhole analysis techniques are used to identify the fluids and estimate flowing potential. This type of sampling allows reservoir fluids to be characterized without resorting to more expensive well-testing.

6.2.6.2. Surface samples

Surface sampling involves taking samples of the two phases (gas and liquid) flowing through the surface separators, and recombining the two fluids in an appropriate ratio such that the *recombined sample* is representative of the reservoir fluid.

The oil and gas samples are taken from the appropriate flowlines of the same separator, whose pressure, temperature and flowrate must be carefully recorded to allow the recombination ratios to be calculated. In addition, the pressure and temperature of the stock tank must be recorded to be able to later calculate the shrinkage of oil between the point at which it is sampled and the stock tank. The oil and gas samples are sent separately to the laboratory where they are recombined before PVT analysis is performed. A quality check on the sampling technique is that the bubble point of the liquid sample at the temperature of the separator from which the samples were taken should be equal to the separator pressure.

The advantages of surface sampling and recombination are that large samples may be taken, that stabilised conditions can be established over a number of hours prior to sampling and that costly wireline entry into the well is avoided. The subsurface sampling requirements also apply to surface sampling; if P_{wf} is below P_b, then it is probable that an unrepresentative volume of gas will enter the wellbore, and even good surface sampling practice will not obtain a true reservoir fluid sample.

6.2.6.3. PVT analysis

Typical analysis in the laboratory consists of sample validation, a compositional analysis of the individual and recombined samples, measurement of oil and gas density and viscosity over a range of temperatures and determination of the basic PVT parameters B_o, R_s and B_g.

For the details of PVT analysis refer to *Fundamentals of Reservoir Engineering* by L.P. Dake.

It is of particular interest to note the different data requirements of the disciplines when the laboratory tests are performed. During the compositional analysis, petroleum engineers are satisfied with a compositional analysis of the hydrocarbons which extend up to around the C_6 components, with C_{7+} components being lumped together and characterised by a pseudo-component. Process engineers require a more detailed compositional analysis, typically extending up to C_{30}. This is because the heavy ends play a more important role in the phase behaviour at the lower temperatures and pressures experienced during surface processing. For example, the long-chain hydrocarbons will form solids (such as wax) at surface conditions, but will remain in solution at reservoir conditions.

Part of the PVT analysis will include passing the reservoir fluid sample through a series of expansions to simulate the separator conditions. At the design stage, process

Table 6.2 PVT table for input to reservoir simulation

Pressure (psia)	B_o (rb/stb)	B_g (rb/Mscf)	R_s (scf/stb)	μ_o (cP)	μ_g (cP)
6500	1.142	0.580	213	1.41	0.0333
6000	1.144	0.609	213	1.32	0.0317
5000	1.150	0.670	213	1.18	0.0282
4000	1.158	0.768	213	1.08	0.0248
3000	1.169	0.987	213	0.99	0.0215
2000	1.177	1.302	213	0.93	0.0180
1200	1.189	2.610	213	0.85	0.0144
980[a]	1.191	3.205	213	0.83	0.0138
500	1.147	6.607	130	1.03	0.0125
100	1.015	33.893	44	1.07	0.0120

[a]Saturation pressure or bubble point.

engineers will design a combination of surface separator conditions which will meet the predicted temperatures and pressures at the wellhead, whilst trying to maximise the oil yield (i.e. minimise the shrinkage of oil). In general, the larger the number of separators which are operated in series, the less the shrinkage of oil occurs, as more of the light ends of the mixture remain in the liquid phase. There is clearly a cost–benefit relationship between the incremental cost of separation facilities and the benefit of the lighter oil attained.

Table 6.2 is a typical oil PVT table which is the result of PVT analysis, and which would be used by the reservoir engineer in calculation of reservoir fluid properties with pressure. The initial reservoir pressure is 6000 psia, and the bubble point pressure of the oil is 980 psia.

6.2.7. Properties of formation water

In Section 6.2.8, we shall look at pressure–depth relationships, and will see that the relationship is a linear function of the density of the fluid. Since water is the one fluid which is always associated with a petroleum reservoir, an understanding of what controls formation water density is required. Additionally, reservoir engineers need to know the fluid properties of the formation water to predict its expansion and movement, which can contribute significantly to the drive mechanism in a reservoir, especially if the volume of water surrounding the hydrocarbon accumulation is large.

Data gathering in the water column should not be overlooked at the appraisal stage of the field life. Assessing the size and flow properties of the aquifer is essential in predicting the pressure support which may be provided. Sampling of the formation water is necessary to assess the salinity of the water for use in the determination of hydrocarbon saturations.

6.2.7.1. Water density and formation volume factor (B_w)

Formation water density is a function of its salinity (which ranges from 0 to 300,000 ppm), amount of dissolved gas and the reservoir temperature and pressure.

As pressure increases, so does *water density*, though the compressibility is small (typically $2–4 \times 10^{-6}\,\mathrm{psi}^{-1}$). Small amounts of gas (typically CO_2) are dissolved in water. As temperature increases, the density reduces due to expansion, and the opposing effects of temperature and pressure tend to offset each other. Correlations are available in the chartbooks available from logging companies.

The formation volume factor for water (B_w, reservoir volume per stock tank volume) is close to unity (typically between 1.00 and 1.07 rb/stb) depending on amount of dissolved gas and reservoir conditions, and is greater than unity due to the thermal contraction and evolution of gas from reservoir to stock tank conditions.

6.2.7.2. Formation water viscosity

This parameter is important in the prediction of aquifer response to pressure drops in the reservoir. As for liquids in general, water viscosity reduces with increasing temperature. Water viscosity is in the order of 0.5–1.0 cP, and is usually lower than that of oil.

The fluid properties of formation water may be looked up on correlation charts, as may most of the properties of oil and gas so far discussed. Many of these correlations are also available as computer programmes. It is always worth checking the range of applicability of the correlations, which are often based on empirical measurements and are grouped into fluid types (e.g. California light gases).

6.2.8. Pressure–depth relationships

The relationship between reservoir fluid pressure and depth may be used to define the interface between fluids (e.g. gas–oil or oil–water interface) or to confirm the observations made directly by wireline logs. This is helpful in determining the volumes of fluids in place, and in distinguishing between areas of a field which are in different pressure regimes or contain different fluid contacts. If different pressure regimes are encountered within a field, this is indicative of areas which are isolated from each other either by sealing faults or by lack of reservoir continuity. In either case, the development of the field will have to reflect this lack of communication, often calling for dedicated wells in each separate fault block. This is important to understand during development planning, as later realisation is likely to lead to a sub-optimal development (either loss of recovery or increase in cost).

Normal pressure regimes follow a hydrostatic fluid gradient from surface, and are approximately linear. Abnormal pressure regimes include overpressured and underpressured fluid pressures, and represent a discontinuity in the normal pressure gradient. Drilling through abnormal pressure regimes requires special care, as discussed in Section 4.7, Chapter 4.

6.2.8.1. Fluid pressure

Assuming a normal pressure regime, at a given depth below ground level, a certain pressure must exist which just balances the overburden pressure (OBP) due to the weight of rock (which forms a matrix) and fluid (which fills the matrix) overlying this point. The OBP is in fact balanced by a combination of the fluid pressure in the pore space (FP) and the stress between the rock grains of the matrix (σ_g).

$$OBP = FP + \sigma_g$$

At a given depth, the OBP remains constant (at a gradient of approximately 1 psi/ft), so that with production of the reservoir fluid, the fluid pressure decreases, creating an increase in the grain-to-grain stress. This may result in the grains of rock crushing closer together, providing a small amount of drive energy (compaction drive) to the production. In extreme cases of pressure depletion in poorly compacted rocks this can give rise to a reduction in the thickness of the reservoir, leading ultimately to surface subsidence. This has been experienced in the Groningen gas field in the Netherlands (approximately 1 m of subsidence), and more dramatically in the Ekofisk Field in the Norwegian sector of the North Sea (around 6 m subsidence), as mentioned in Section 6.1.3.

In a normal pressure regime, the pressure in a hydrocarbon accumulation is determined by the pressure gradient of the overlying water $(dP/dD)_w$, which ranges from 0.435 psi/ft (10 kPa/m) for freshwater to around 0.5 psi/ft (11.5 kPa/m) for salt-saturated brine. At any depth (D), the water pressure (P_w) can be determined from the following equation, assuming that the pressure at the surface datum is 14.7 psia (1 bara):

$$P_w = \left\{ \frac{dP}{dD} \right\}_w D \quad \text{(psia) or (bara)}$$

The water pressure gradient is related to the water density (ρ_w, kg/m^3) by the following equation:

$$\left\{ \frac{dP}{dD} \right\}_w \rho_w g \quad \text{(Pa/m)}$$

where g is acceleration due to gravity (9.81 m/s^2).

Hence it can be seen that from the density of a fluid, the pressure gradient may be calculated. Furthermore, the densities of water, oil and gas are so significantly different that they will show quite different gradients on a pressure–depth plot.

This property is useful in helping to define the interface between fluids. The intercept between the gas and oil gradients indicates the GOC, whilst the intercept between the oil and water gradients indicates the free water level (FWL) which is related to the oil–water contact (OWC) via the transition zone, as described in Section 6.2.9.

The gradients may be calculated from surface fluid densities, or may be directly measured by downhole pressure measurements using a formation pressure testing tool (discussed in Section 6.3.6). The interfaces predicted can be used to confirm wireline measurements of fluid contact, or to predict interfaces when no logs have directly found the contacts.

For example, in the following situation, two wells have penetrated the same reservoir sand. The updip well finds the sand gas bearing, with *gas down to* (GDT) the base of the sands, whilst the downdip well finds the same sand to be fully oil bearing, with an *oil up to* (OUT) at the top of the sand. Pressures taken at intervals in

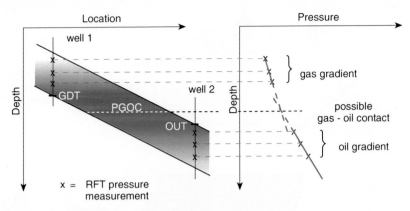

Figure 6.25 The gradient intercept technique.

each well may be used to predict where the *possible gas–oil contact* (PGOC) lies. This method is known as the *gradient intercept technique* (Figure 6.25).

6.2.8.2. Normal and abnormal pressure regimes

In a normally pressured reservoir, the pressure is transmitted through a continuous column of water from the surface down to the reservoir. At the datum level at surface the pressure is 1 atm. The datum level for an offshore location is the mean sea level (MSL), and for an onshore location, the groundwater level.

In abnormally pressured reservoirs, the continuous pressure–depth relationship is interrupted by a sealing layer, below which the pressure changes. If the pressure below the seal is higher than the normal (or hydrostatic) pressure, the reservoir is termed overpressured. Extrapolation of the fluid gradient in the overpressured reservoir back to the surface datum would show a pressure greater than 1 atm. The actual value by which the extrapolated pressure exceeds 1 atm defines the level of overpressure in the reservoir. Similarly, an underpressured reservoir shows a pressure less than 1 atm when extrapolated back to the surface datum.

In order to maintain underpressure or overpressure, a pressure seal must be present. In hydrocarbon reservoirs, there is by definition a seal at the crest of the accumulation, and the potential for abnormal pressure regimes therefore exists (Figure 6.26).

The most common causes of abnormally pressured reservoirs are

- uplift/burial of rock, whereby permeable rock, encapsulated by thick layers of shale or salt, is either uplifted (causing overpressure) or downthrown (causing underpressure). The OBP is altered, but the fluid in the pores cannot escape, and therefore absorb the change in overburden stress
- thermal effects, causing the expansion or contraction of water which is unable to escape from an encapsulated system
- rapid burial of sediments consisting of layers of clay and sand, the speed of which does not allow the fluids to escape from the pore space as the rock compacts – this leads to overpressures. Most deltaic sequences show this to some degree

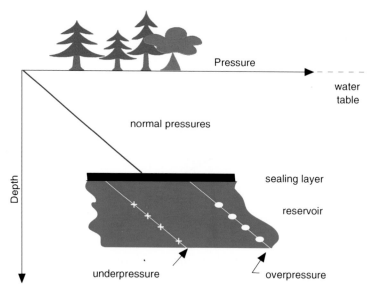

Figure 6.26   Normal and abnormal pressure regimes.

- depletion of a sealed or low-permeability reservoir due to production within the reservoir
- depletion due to production in an adjacent field whose pressure drops, with pressure connection via a common aquifer
- phase changes, for example anhydrite into gypsum or alteration of clay mineralogy
- overpressures as a result of hydrocarbon columns
- inflation of pressure as a result of seal failure, for example a fault between blocks. This can result in uncontrolled cross-flow between reservoirs.

6.2.8.3. Drilling through abnormal pressures

When drilling through normally pressured formations, the mud weight in the well is usually controlled to maintain a pressure greater than the formation pressure to prevent the influx of formation fluid. A typical overbalance would be in the order of 200 psi. A larger overbalance would encourage excessive loss of mud into the formation, which is both costly, and may damage the reservoir properties. If an influx of formation fluid into the borehole did occur due to insufficient overbalance, the lighter formation fluid would reduce the pressure of the mud column, thus encouraging further influx, and an unstable situation would occur, possibly leading to a blowout. Hence, it is important to avoid the influx of formation fluid by using the correct mud weight in the borehole.

When drilling through a shale into an overpressured formation, the mud weight must be increased to prevent influx. If this increased mud weight would cause large losses in shallower, normally pressured formations, it is necessary to isolate the normally pressured formation behind casing before drilling into the overpressured formation. The prediction of overpressures is therefore important in well design.

Similarly, when drilling into an underpressured formation, the mud weight must be reduced to avoid excessive losses into the formation. If the rate of loss is greater than the rate at which mud can be made up, then the level of fluid in the wellbore will drop and there is a risk of influx from the normally pressured overlying formations. Again, it may be necessary to set a casing before drilling into underpressures.

6.2.9. Capillary pressure and saturation–height relationships

In a reservoir at initial conditions, an equilibrium exists between buoyancy forces and capillary forces. These forces determine the initial distribution of fluids, and hence the volumes of fluid in place. An understanding of the relationship between these forces is useful in calculating volumetrics, and in explaining the difference between *FWL* and OWC introduced in Section 6.2.8.1.

A well-known example of capillary–buoyancy equilibrium is the experiment in which a number of glass tubes of varying diameter are placed into a tray of water. The water level rises up the tubes, reaching its highest point in the narrowest of the tubes. The same observation would be made if the fluids in the system were oil and water rather than air and water (Figure 6.27). The capillary tubes of differing diameters can be likened to different sizes of pore throats in a connecting porous system.

The capillary effect is apparent whenever two non–miscible fluids are in contact, and is a result of the interaction of attractive forces between molecules in the two liquids (surface tension effects), and between the fluids and the solid surface (*wettability* effects).

Surface tension arises at a fluid-to-fluid interface as a result of the unequal attraction between molecules of the same fluid and the adjacent fluid. For example, the molecules of water in a water droplet surrounded by air have a larger attraction to each other than to the adjacent air molecules. The imbalance of forces creates an inward pull which causes the droplet to become spherical, as the droplet minimises its surface area. A surface tension exists at the interface of the water and air, and a

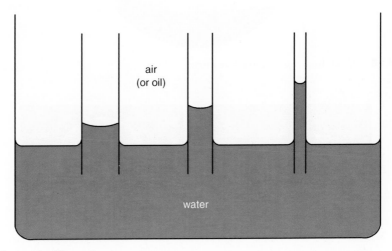

Figure 6.27   Capillary tubes in a tray.

pressure differential exists between the water phase and the air. The pressure on the water side is greater due to the net inward forces (Figure 6.28).

The relationship between the pressure drop across the interface ΔP, the *interfacial tension* σ and the radius of the droplet r is

$$\Delta P = \frac{2\sigma}{r}$$

Wettability describes the relationship between the contact of two fluids and a solid. The type of contact is characterised by the contact angle (θ) between the fluids and the solid, and is measured by convention through the denser fluid. If the contact angle measured through a liquid is less than $90°$, the surface is said to be wetting to that fluid. Figure 6.29 shows the difference in contact angles for water-wet and oil-wet reservoir rock surfaces. The measurement of wettability at reservoir conditions is very difficult, since the property is affected by the drilling and recovery of the samples. It is believed that the majority of clastic reservoir rocks are water wet, but the subject of wettability is a contentious one.

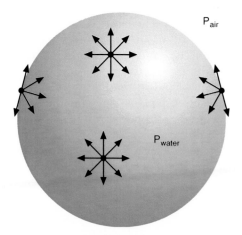

Figure 6.28   Water droplet with attractive forces.

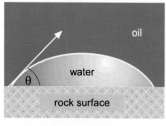 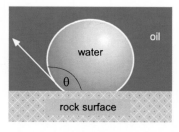

Water wet   Oil wet

Figure 6.29   Rock wettability types.

6.2.9.1. Capillary pressure

Returning to the experiment with the oil, water and the glass capillaries (Figure 6.27), the *interfacial tension* and wettability lead to a pressure differential across the liquid interface and a contact angle with the glass. The pressure in the water phase is greater than the pressure in the oil phase, and the glass is water wet, as determined by the contact angle. The pressure difference between the water phase and the oil phase is called the *capillary pressure* (P_c), and is related to the interfacial tension (σ), the radius of the capillary tube (r_t) and the contact angle (θ) by

$$P_c = \frac{2\sigma \cos \theta}{r_t}$$

Notice that the capillary pressure is greater for smaller capillaries (or throat sizes), and that when the capillary has an infinite radius, as on the outside of the capillaries in the tray of water, P_c is zero.

6.2.9.2. Capillary–buoyancy equilibrium

Consider the pressure profile in just one of the capillaries in the experiment (see Figure 6.30). Inside the capillary tube, the capillary pressure (P_c) is the pressure difference between the oil phase pressure (P_o) and the water phase pressure (P_w) at the interface between the oil and the water.

$$P_c = P_o - P_w$$

The capillary pressure can be related to the height of the interface above the level at which the capillary pressure is zero (called the free water level) by using the hydrostatic pressure equation. Assuming the pressure at FWL is P_i:

$$P_w = P_i - \rho_w gh, \quad \text{where } \rho_w \text{ is water density}$$
$$P_o = P_i - \rho_o gh, \quad \text{where } \rho_o \text{ is oil density}$$

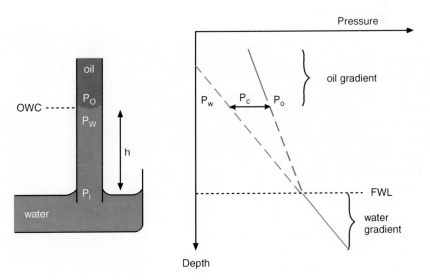

Figure 6.30   Pressure–depth plot for one capillary.

by subtraction

$$P_o - P_w = (\rho_w - \rho_o)gh = P_c$$

and remember that

$$P_c = \frac{2\sigma \cos \theta}{r_t}$$

This is consistent with the observation that the largest difference between the oil–water interface and FWL occurs in the narrowest capillaries, where the capillary pressure is greatest. In the tighter reservoir rocks, which contain the narrower capillaries, the difference between the oil–water interface and the FWL is larger.

If a pressure-measuring device were run inside the capillary, an oil gradient would be measured in the oil column. A pressure discontinuity would be apparent across the interface (the difference being the capillary pressure), and a water gradient would be measured below the interface. If the device also measured resistivity, a contact would be determined at this interface, and would be described as the OWC. Note that if oil and water pressure measurements alone were used to construct a pressure–depth plot (Figure 6.30), and the gradient intercept technique was used to determine an interface, it is the *FWL* which would be determined, not the OWC.

The difference between the OWC and the FWL is greater in tight reservoirs, and may be up to 100 m difference. A difference between GOC and free oil level exists for the same reasons, but is much smaller, and is often neglected.

For the purpose of calculating oil in place in the reservoir, it is the OWC, not the FWL, which should be used to define to what depth oil has accumulated. Using the FWL would overestimate the oil in place, and could lead to a significant error in tight reservoirs.

6.2.9.3. Saturation–height relationships

Saturation is the proportion of one fluid phase in a pore system to the total amount of fluid. Initially, the pores in the structure are filled with water. As oil migrates into the structure, it displaces water downwards, and starts with the larger pore throats where lower pressures are required to curve the oil–water interface sufficiently for oil to enter the pore throats. As the process of accumulation continues, the pressure difference between the oil and water phases increases above the FWL because of the density difference between the two fluids. As this happens the narrower pore throats now begin to fill with oil, and the smallest pore throats are the last to be filled. When no more water is able to be removed, the reservoir is at *irreducible water saturation*.

The reservoir is composed of pores of many different sizes, and can be compared to a system of capillary tubes of widely differing diameters, as shown in Figure 6.31.

The narrowest capillaries determine the level above which only the irreducible (or connate) water remains. Typical irreducible water saturations are in the range 10–40%. The largest capillaries determine the level below which the water saturation is 100%, that is the OWC. Between the two points there is a gradual change in the water saturation, and the interval is called the *transition zone*. The height of the transition zone depends on the distribution of pore sizes, but can be

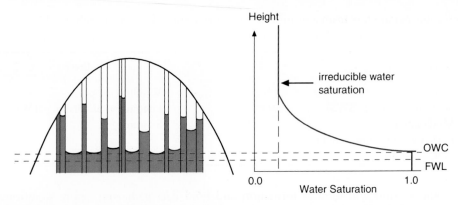

Figure 6.31 Saturation–height with capillaries.

many tens of metres. When taking pressure samples with a FPT to construct a pressure–depth plot, it is advisable to obtain pressures outside the transition zone, where the gradients are truly representative of the single fluid, rather than of a mixture of the two fluids (oil and water in this example).

The change of saturation with height above a FWL can have a significant effect on the volume of hydrocarbons. It is therefore important to accurately model this change mathematically in order to estimate the correct volume of hydrocarbons in the reservoir. A saturation–height function is the mathematical term that is derived to model this change and is calibrated using log data and *special core analysis* (SCAL) tests which are discussed further in Section 6.3.2.

6.3. Data Gathering

Introduction and Commercial Application: Data gathering is an activity which provides the geologist and engineer with the information required to estimate the volume of the reservoir, its fluid content, productivity and potential for development. Data gathering is not only carried out at the appraisal and development planning stage of the field life cycle, but continues throughout the field life. This section will focus on the data gathered for field development planning; data gathering for managing the field during the production period is discussed in Chapter 16.

The timely acquisition of static and dynamic reservoir data is critical for the optimisation of development options and production operations. Reservoir data enable the description and quantification of fluid and rock properties. The amount and accuracy of the data available will determine the range of uncertainty associated with estimates made by the subsurface engineer.

6.3.1. Classification of methods

The basic data gathering methods are direct methods, which allow visual inspection or at least direct measurement of properties, and indirect methods whereby we infer reservoir parameters from a number of measurements taken in a borehole. The main techniques available within these categories are summarised in the following table:

Direct	Indirect
Coring	Wireline logging
Sidewall sampling (SWS)	Logging while drilling (LWD)
Mudlogging	Seismic
Formation pressure sampling	
Fluid sampling	

This section will look at formation and fluid data gathering before significant amounts of fluid have been produced, hence describing how the static reservoir is sampled. Data gathered prior to production provide vital information, used to predict reservoir behaviour under dynamic conditions. Without these baseline data no meaningful reservoir simulation can be carried out. The other major benefit of data gathered at initial reservoir conditions is that pressure and fluid distribution are in equilibrium; this is usually not the case once production commences. Data gathered at initial conditions are therefore not complicated by any pressure disturbance or fluid redistribution, and offer a unique opportunity to describe the condition prior to production.

6.3.2. Coring and core analysis

To gain an understanding of the composition of the reservoir rock, inter-reservoir seals and the reservoir pore system, it is desirable to obtain an undistorted and continuous reservoir core sample. Cores are also used to establish physical rock properties by direct measurements in a laboratory. They allow description of the depositional environment, sedimentary features and the diagenetic history of the sequence.

In the pre-development stage, core samples can be used to test the compatibility of injection fluids with the formation, to predict borehole stability under various drilling conditions and to establish the probability of formation failure and sand production.

Coring is performed in between drilling operations. Once the formation for which a core is required has been identified on the mud log, the drilling assembly is pulled out of hole. For coring operations, a special assembly is run on drill pipe comprising a core bit and a core barrel (Figure 6.32).

Unlike a normal drill bit which breaks down the formation into small cuttings, a core bit can be visualised as a hollow cylinder with an arrangement of cutters on the outside. These cut a circular groove into the formation. Inside the groove remains an intact cylinder of rock which moves into the inner core barrel as the coring process progresses. Eventually, the core is cut free (broken) and prevented from falling out of the barrel whilst being brought to surface by an arrangement of steel fingers or 'catchers'. Core diameters vary typically from 3 to 7 in. and are usually

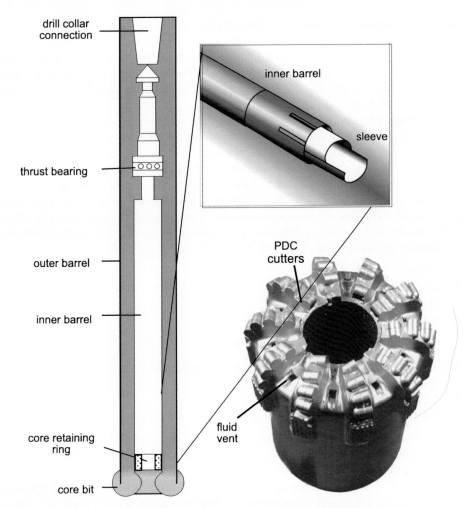

drill collar connection

inner barrel

sleeve

thrust bearing

PDC cutters

outer barrel

inner barrel

core retaining ring

fluid vent

core bit

Figure 6.32 Coring assembly and core bit (courtesy of Security DBS and Corepro).

about 90 ft long. However, in favourable hole/formation conditions longer sections may be achievable.

If a conventional core has been cut, it will be retrieved from the barrel on the rig floor, cut into sections and crated. A lithologic description can be done at this stage. To avoid drying out of core samples and the escape of light hydrocarbons some sections will be immediately sealed in a coating of hot wax and foil.

Commonly, a fibre glass or aluminium sleeve is inserted into the steel inner core barrel and the core is retrieved within the sleeve. At the surface the gap (annulus) between the inner sleeve and core is injected with an inert stabilizing material which 'sets' to hold the core in place. The core is cut into sections (typically 1 m) and shipped to the laboratory. Core handling is a delicate procedure and it is important to minimise any alteration or damage to the cored sample or the contained fluids. Any changes in original core properties through alteration of formation clay mineralogy, precipitation of minerals or evaporation of pore fluids will cause inaccuracies in petrophysical measurements. Additionally, mechanical damage can render the whole core useless for testing purposes. Significant technological and procedural advances have been made in the past 10 years to reduce friction, minimize filtrate invasion, retain fluids and maintain rock integrity.

The borehole section which has been cored will subsequently be logged using wireline tools (see Section 6.3.4). Upon arrival in the laboratory a GR measurement will be taken from the core itself, thus allowing calibration of wireline logs with core data. Additionally, if the core is still in a sleeve a *CT scan* (X-ray scan) may be made of the whole core to identify optimum locations to sample and slice the core.

In addition to a geological evaluation on a macroscopic and microscopic scale, plugs (small cylinders of 3 cm diameter and 5 cm length) are cut from the whole core, usually at about 30 cm intervals. Core analysis is carried out on these samples.

Routine core analysis of plugs will include determination of

- porosity
- horizontal air permeability
- fluid saturation
- grain density.

SCAL will include measurements of

- electrical tests (cementation and saturation exponents)
- relative permeability
- capillary pressure
- strength tests.

Finally, the core will be sectioned (one third:two thirds) along its entire length (slabbed) and photographed under normal and ultraviolet light (UV light will reveal hydrocarbons not visible under normal light, as shown in Figure 6.33).

The main cost factor of coring is usually the rig time spent on the total operation rather than the follow-up investigations in the laboratory. Core analysis is complex and may involve different laboratories. It may therefore take months before final results are available. As a result of the relatively high costs and a long lead time of some core evaluations, the technique is only used in selected intervals in a number of wells drilled. It is therefore vital that efforts are made to maximize core recovery and integrity so that all the core material cut can be used.

Mudlogging is another important direct data gathering technique. The returns to surface (drill cuttings and gas levels) and ROP are continuously recorded and analysed to establish the nature of the formation and fluid fill.

6.3.3. Sidewall sampling

The sidewall sampling tool (SWS) can be used to obtain small plugs (2 cm diameter, 5 cm length, often less) directly from the borehole wall. The tool is run on wireline after the hole has been drilled and logged. Some 20–30 individual bullets are fired from each gun (Figure 6.34) at different depths. The hollow bullet will penetrate the

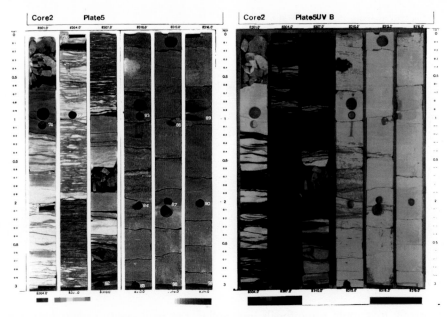

Figure 6.33   Photograph of core (left = normal light, right = UV).

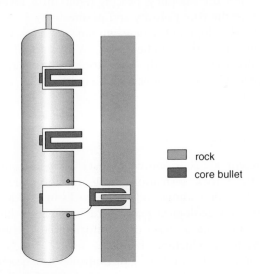

Figure 6.34   Sidewall sampling gun.

formation and a rock sample will be trapped inside the steel cylinder. By pulling the tool upwards, wires connected to the gun pull the bullet and sample from the borehole wall.

Sidewall samples are useful to obtain direct indications of hydrocarbons (under UV light) and to differentiate between oil and gas. The technique is applied

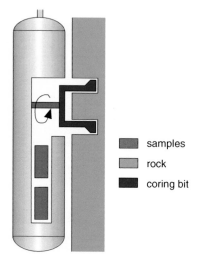

Figure 6.35   Sidewall coring tool.

extensively to sample microfossils and pollen for *stratigraphic analysis* (age dating, correlation, depositional environment). Qualitative inspection of porosity is possible, but very often the sampling process results in a severe crushing of the sample, thus obscuring the true porosity and permeability.

In a more recent development a new wireline tool has been developed that actually drills a plug out of the borehole wall. With *sidewall coring* (Figure 6.35), some of the main disadvantages of the SWS tool are mitigated, in particular the crushing of the sample. Up to 20 samples can be individually cut and are stored in a container inside the tool.

6.3.4. Wireline logging

Wireline logs represent a major source of data for geoscientists and engineers investigating subsurface rock formations. Logging tools are used to look for reservoir quality rock, hydrocarbons and source rocks in exploration wells, support volumetric estimates and geological/geophysical modelling during field appraisal and development, and provide a means of monitoring the distribution of remaining hydrocarbons during the production lifetime.

A large investment is made by oil and gas companies in acquiring *openhole log data*. Logging activities can represent between 5 and 15% of total well cost. It is important therefore to ensure that the cost of acquisition can be justified by the value of information generated and that thereafter the information is effectively managed.

Wells can be broadly divided into two groups in terms of how logging operations should be prioritised: information wells and development wells. Exploration and appraisal wells are drilled for information and failure to acquire log data will compromise well objectives. Development wells are primarily drilled as production and injection conduits and whilst information gathering is an important

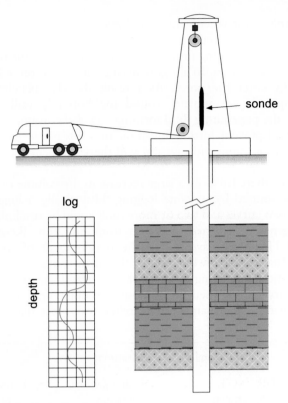

Figure 6.36   Principle of wireline logging.

secondary objective it should normally remain subordinate to well integrity considerations. In practical terms this means that logging operations will be curtailed in development wells if hole conditions deteriorate. This need not rule out further data acquisition, as *logging through casing* options still exist.

Figure 6.36 depicts the basic setup of a wireline logging operation. A *sonde* is lowered downhole after the drill string has been removed. The sonde is connected via an insulated and reinforced electrical cable to a winch unit at the surface. At a speed of about 600 m/h the cable is spooled upward and the sonde continuously records formation properties like natural GR radiation, formation resistivity or formation density. The measured data are electrically transmitted through the cable and are recorded and processed in a sophisticated logging unit at the surface. Offshore, this unit will be located in a cabin, whilst on land it is truck mounted. Today, the important acquisition and processing software is carried to the well site by the logging engineer in a laptop. The data acquisition can be observed in real time by the client using a secure Internet site. In some cases, specialist tools can be run and processed 'remotely' by offsite experts at logging company headquarters.

A vast variety of logging tools are in existence and Section 6.4 will cover only those which enable the evaluation of essential reservoir parameters, specifically net reservoir thickness, lithology, porosity and hydrocarbon saturation.

A complicating factor when acquiring downhole data is the contamination of the measured formation by mud filtrate, which is discussed in detail at the end of Section 6.4. During the drilling process, *mud filtrate* will enter the newly penetrated formation to varying degrees. In a highly permeable formation, a large quantity of fluid will initially enter the pores. As a result the clay platelets suspended in the mud will quickly accumulate around the borehole wall. The formation effectively filters the penetrating fluid forming a *mudcake* around the borehole wall which in turn will prevent further invasion. In a less permeable formation, this process will take more time and invasion will therefore penetrate deeper into the formation.

In recent years, there has been a large increase in the volume of data that can be acquired and transmitted by wireline logging. Historically, a logging tool would generate one or two curves and two or more tools could be run linked together in a 'toolstring' to generate a series of curves from one run. Recent advances in technology and IT power have led to the development of sophisticated tools which record a data array at any one point rather than a single value. Downhole and surface mathematical processing is conducted to transform the array data into a product set of curves to be used by all subsurface disciplines. The following table provides a summary of the mainstream wireline tool types that are routinely run today.

Generic Device	Tool Examples	Measurement Type	Application
Gamma	GR, NGT, Spectralog	Natural gamma radiation	Lithology, correlation
SP	SP	Spontaneous potential	Lithology, permeability (indicator)
Density	LDL, ZDL, SDL	Bulk density	Porosity, lithology
Neutron	CN, CNL, DSN	Hydrogen index	Lithology, porosity, gas indicator
Acoustic	BHC, XMAC, DSI	Travel time, acoustic waveform	Porosity, seismic calibration
Resistivity	DLL, HRLA, HDLL	Electrical resistance of formation	Saturation, permeability indicator
Induction	ILD, AIT, HILT, HDIL, HRAI	Induced electrical current	Saturation (OBMs)
Image	FMI, STAR, CBIL, EI, OBMI, CAST	Resistivity or acoustic pixellated image	Sedimentology, fracture/fault analysis
NMR	MRIL, MREX, CMR	Nuclear magnetic resonance	Porosity, permeability, saturation
Formation tester	RFT, MDT, RCI pressure	Pore pressure	Fluid types, pressures and contacts

Several disadvantages are related to wireline logging. We already mentioned mud invasion. Some logging jobs may last several days and as the 'openhole time' increases the quality of acquired data and the stability of the borehole will deteriorate. Wireline logging is also expensive both in terms of service charges by the logging company and in terms of rig time. It may therefore be desirable to measure formation properties whilst drilling is in progress. Not only would this eliminate the drawbacks of wireline operations but the availability of real time data allows operational decisions, for example selection of completion intervals, or sidetracking to be taken at a much earlier stage.

6.3.5. Logging/measurement while drilling (LWD/MWD)

Basic MWD technology was first introduced in the 1980s by drilling companies, and was initially restricted to retrievable inserts for directional measurements and then natural GR logs. These developments were quickly followed by logging tools integrated into drill collars (DCs) (LWD). Recently, LWD development has progressed to the stage where most of the conventional wireline logging tools can be effectively replaced by a LWD equivalent. Early LWD technology was often considered to be inferior to wireline. However, recent mergers between wireline and drilling companies has resulted in technology-transfer in R&D which has led to a significant improvement in LWD log quality. A lazy use of terminology within the industry means that LWD and MWD can be considered as synonymous. A more appropriate term for today's sophisticated devices is formation evaluation while drilling (FEWD).

Perhaps the greatest stimulus for the development of such tools has been the proliferation of high-angle wells in which deviation surveys are difficult and wireline logging services are impossible (without some sort of pipe conveyance system), and where LWD logging can minimise formation damage by reducing openhole exposure times.

Whilst providing deviation and logging options in high-angle wells is a considerable benefit, the greatest advantage offered by LWD technology, in either conventional or high-angle wells, is the acquisition of *real time data* at surface. Most of the LWD applications which are now considered standard, exploit this feature in some way, and include

- real time correlation for picking coring and casing points
- real time overpressure detection in exploration wells
- real time logging to minimise 'out of target' sections (geosteering)
- real time formation evaluation to facilitate 'stop drilling' decisions.

Although there are a wide range of LWD services available, not all are required in every situation and the full LWD logging suites which include directional and formation logging sensors are run much less frequently than gamma/resistivity/directional combinations. An example of an LWD tool configuration is given in Figure 6.37.

All LWD tools have both a power supply and data transmission system, often combined in one purpose-built collar and usually located above the measurement sensors as shown in Figure 6.38 (a Baker–Hughes multicombination tool).

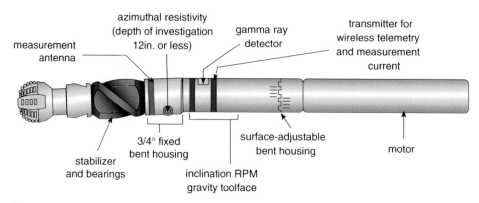

Figure 6.37   Schlumberger geosteering tool with LWD.

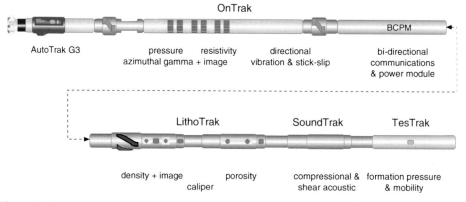

Figure 6.38   BakerHughes Inteq 'Pentacombo' tool.

Data transmission may be within the downhole assembly from the sensors to a memory device or from the sensors to surface. The latter is usually achieved by *mud pulse telemetry,* a method by which data are transmitted from the tool in real time, that is as data are being acquired. Positive or negative pressure pulses created in the mudstream downhole travel through the mud (inside the drill pipe) to surface and are detected by a pressure transducer in the flowline. Positive pressure pulses are created by extending a plunger into a choke orifice, momentarily restricting flow (as shown in the top of Figure 6.38), an operation which is repeated to create a binary data string. Negative pulses are created by opening a bypass valve and venting mud to the annulus, momentarily reducing the drill pipe pressure.

Data transmission rate per foot is a function of both pulse frequency and ROP. Sensors acquire and transmit data samples at fixed time intervals and therefore the sampling per foot is a function of ROP. Current tools allow a real time sampling and transmission rate similar to wireline tools as long as the penetration rate does not exceed about 100 ft/h. If drilling progresses faster or if there are significant variations in penetration rate, resampling by depth as opposed to time intervals may be required. The quantity of data that can be transmitted in real time is limited and focused towards responses used for geosteering and other drilling decisions. A full data set is stored in the downhole memory which is retrieved when the tool is brought to the surface.

Electrical power is supplied to LWD tools either from batteries run in the downhole assembly or from an alternator coupled to a turbine set in the mudstream.

In terms of log data quality and tool response modern LWD/FEWD tools can be as good as their wireline counterparts. However, the biggest issue when comparing the two technologies is *depth control*. Wireline depth is accurately measured at the surface by a spooling wheel in front of the cable drum which records the length of cable that has been reeled in. LWD depth is measured as 'drillers depth' where the driller records the length of drill pipe that has been run in the hole. Individual lengths of pipe can differ and the 'pipe tally' (record of pipe length) is not always accurate. Additionally, the actual length of pipe in a long borehole may change depending upon the amount of compression or tension within the string. On occasions when LWD and wireline logs over the same section are compared, the depth differences can be up to several metres. Currently, LWD companies are designing depth-adjustment software to overcome this problem.

6.3.6. Pressure measurements and fluid sampling

A common objective of a data gathering programme is the acquisition of fluid samples. The detailed composition of oil, gas and water is to some degree required by almost every discipline involved in field development and production.

One method of sampling reservoir fluids and taking formation pressures under reservoir conditions in openhole is by using a wireline FPT. A number of wireline logging companies provide such a tool under the names such as RFT (repeat formation tester) and FMT (formation multitester), so called because they can take a series of pressure samples in the same logging run. Newer versions of the tool are called a *modular dynamic tester* or MDT (Schlumberger tool), shown in Figure 6.39 and *reservoir characterisation instrument* or RCI. The latest LWD/FEWD technology includes FPTs.

The tool is positioned across the objective formation and set against the side of the borehole by either two packers or by up to three probes (the configuration used will depend on the test requirements). The probes are pushed through the mudcake and against the formation. A pressure drawdown can now be created at one probe and the drawdown be observed in the two observation probes. This will enable an estimate of vertical and horizontal permeability and hence indicate reservoir heterogeneities as well as recording a pore pressure. Alternatively fluids can be sampled. In this case, a built-in resistivity tool will determine when uninvaded formation fluid (hydrocarbons or formation water) is entering the sample module. The flow can be diverted back into the wellbore until only the desired fluid is flowing, thus providing fluid samples uncontaminated with mud. The pressure drawdown can be controlled from surface, enhancing the chance of creating a monophasic flow by keeping the drawdown above bubble point.

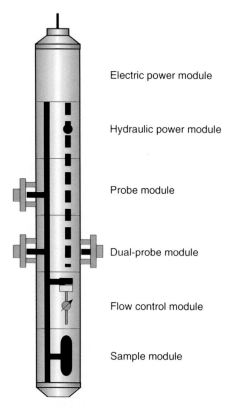

Electric power module

Hydraulic power module

Probe module

Dual-probe module

Flow control module

Sample module

Figure 6.39 MDT tool configuration for permeability measurement.

The pressure measurement and fluid sampling procedures can be repeated at multiple depths in the reservoir.

In some cases, when drilling fluids invade a very low permeability zone, pressure equalisation in the formation can take a considerable time. The pressure recorded by the tool will then be close to the pressure of the mud and much higher than the true formation pressure. This is known as supercharging. Supercharging pressures indicate tight formation, but are not useful in establishing the true fluid pressure gradient.

The use of these pressure data at various depths to determine fluid gradients and contacts is discussed in Section 6.2.8.

6.4. Data Interpretation

Introduction and Commercial Application: This section introduces the main methods used to convert raw well data into useful information – information with which to characterise the reservoir. A huge volume of data is generated by drilling and logging a typical well. Collecting and storing data require substantial investment but unless it is processed and presented appropriately much of the potential value is not realised. Describing a reservoir can be a simple task if it has been laid down as a thick blanket of sand, but becomes increasingly complex where hydrocarbons are found in, for example, ancient estuarine or reef deposits. In all cases, however, there are two main issues which need to be resolved: firstly, how much oil does the reservoir contain (the HCIIP), and secondly, how much can be recovered (the UR). There are a number of ways to determine these volumes (which will be explained in Chapter 7) but the basic physical parameters for describing the reservoir remain the same:

- net reservoir thickness
- porosity
- hydrocarbon saturation
- permeability.

At each stage of a field life cycle raw data have to be converted into information, but for the information to have value it must influence decision making and profitability.

6.4.1. Well correlation

Well correlation is used to establish and visualise the lateral extent and the variations of reservoir parameters. In carrying out a correlation we subdivide the objective sequence into *lithologic units* and follow those units or their generic equivalent laterally through the area of interest. As we have seen earlier, the reservoir parameters such as N/G, porosity, permeability, etc. are to a large extent controlled by the reservoir geology, in particular the depositional environment. Thus, by correlation we can establish lateral and vertical trends of those parameters throughout the structure. This will enable us to calculate hydrocarbon volumes in different parts of a field, predict production rates and optimise the location for appraisal and development wells.

Usually well logs are only one type of data used to establish a correlation. Any meaningful interpretation will need to be supported by *palaeontological data* (micro fossils) and *palynological data* (pollen of plants). The logs most frequently for correlation are *GR, density logs, sonic log, dipmetre* and *formation imaging tools*. On a detailed scale, these curves should always be calibrated with core data as described below.

On a larger scale, for example in a regional context, seismic stratigraphy will help to establish a reliable correlation. It is employed in combination with the concept of *sequence stratigraphy*. This technique, initially introduced by Exxon Research, and since then considerably refined, postulates that global (eustatic) sea level changes create unconformities which can be used to subdivide the stratigraphic record. These unconformities are modified and affected by more local (relative) changes in sea level as a result of local tectonic movements, climate and the resulting impact on sediment supply. The most significant stratigraphic discontinuities used in a sequence stratigraphic approach are

- regressive surfaces of erosion, caused by a lowering of sea level
- transgressive surfaces of erosion, caused by an increase in sea level
- maximum flooding surfaces at times of 'highest' sea level.

Relative sea level changes affect many shallow marine and coastal depositional environments.

Sequence stratigraphy integrates information gleaned from seismic, cores, well logs and often outcrops. In many cases, it has increased the understanding of reservoir geometry and heterogeneity and improved the correlation of individual drainage units. Sequence stratigraphy has also proved a powerful tool to predict presence and regional distribution of reservoirs. For instance, shallow marine regressive surfaces may indicate the presence of turbidites in a nearby, deeper marine area.

In preparation for a field wide 'quick look' correlation, all well logs need to be corrected for borehole inclination. This is done routinely with software which uses the *measured depth* (MD) below the derrick floor ('alonghole depth' below derrick floor [AHBDF] or MD) and the acquired directional surveys to calculate the *true vertical depth* subsea (TVSS). This is the vertical distance of a point below a common reference level, for instance chart datum (CD) or MSL. Figure 6.40 shows the relationship between the different depth measurements.

To start the correlation process, we take the set of logs and select a *datum plane*. This is a marker which can be traced through all data points (three wells in the example of Figure 6.41). A good datum plane would be a continuous shale because we can assume that it represents a 'flooding surface' present over a wide area. Since shales are low-energy deposits, we may also assume that they have been deposited mostly horizontally, blanketing the underlying sediments thus 'creating' a true datum plane.

Next, we align all logs at the datum plane which now becomes a straight horizontal line. Note that by doing so we ignore all structural movements to which the sequence has been exposed.

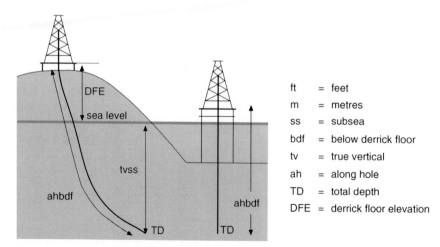

Figure 6.40 Depth measurements used.

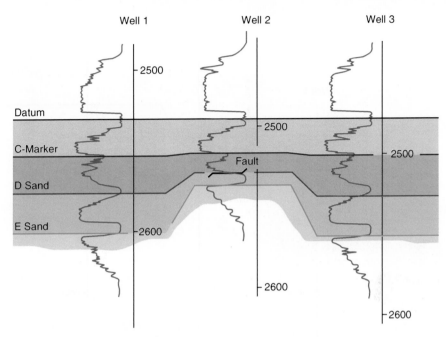

Figure 6.41   Datum plane correlation.

We can now correlate all 'events' below or above the datum plane by comparing the log response. In many instances, correlations are ambiguous. Where two or more correlation options seem possible, the problem may be resolved by checking whether an interpretation is consistent with the geological model and by further validating it with other data. This could be, for instance, pressure data which will indicate whether or not sands in different wells communicate. In cases where correlation is difficult to establish, a detailed palaeontological zonation may be useful.

If correlation is 'lost', that is if no similarity exists any more between the log shapes of two wells (such as in well 2 in our example), this could be for a number of reasons:

- faulting: the well has intersected a fault and part of the sequence is missing. Faulting can also cause a duplication of sequences!
- unconformity: parts of the sequence have been eroded.

These events will need to be marked on the correlation panel. In case of faults, the thickness of the missing section or 'cut out' should be quantified.

Correlations on paper panels are made easier if a *type log* has been created of a typical and complete sequence of the area. If this log is available as a transparency, it can be easily compared against the underlying paper copy. Type logs are also handy if the reservoir development has to be documented in reports or presentations.

To make the correlation results applicable for the field development process, it may be desirable to display the correlated units in their true structural position. For instance, if water injection is planned for the field, water should enter the structure

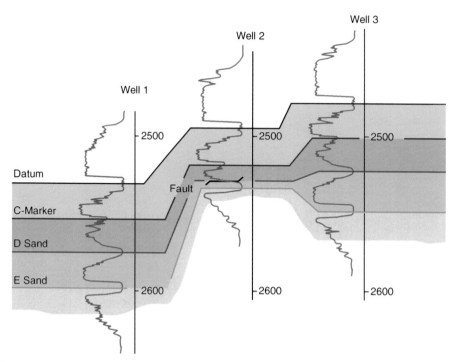

Figure 6.42   Structural correlation.

at or below the OWC and move upwards. Hence the correlation panel should visually show the sand development in the same direction. For this, all markers on the panel are displayed and connected at their TVSS position (Figure 6.42). This is called a *structural correlation*.

If appropriate, correlation panels may contain additional information such as depositional environments, porosities and permeabilities, saturations, lithological descriptions and indications of which intervals have been cored.

6.4.2. Maps and sections

Having gathered and evaluated relevant reservoir data, it is desirable to present these data in a way that allows easy visualisation of the subsurface situation. With a workstation it is easy to create a 3D picture of the reservoir, displaying the distribution of a variety of parameters, for example reservoir thickness or saturations. All realisations need to be in line with the geological model.

We have all used maps to orientate ourselves in an area on land. Likewise, a reservoir map will allow us to find our way through an oil or gas field if, for example we need to plan a well trajectory or if we want to see where the best reservoir sands are located. However, maps will only describe the surface of an area. To get the third dimension we need a section which cuts through the surface. This is the function of a *cross-section*. Figure 6.43 shows a reservoir map and the corresponding cross–section.

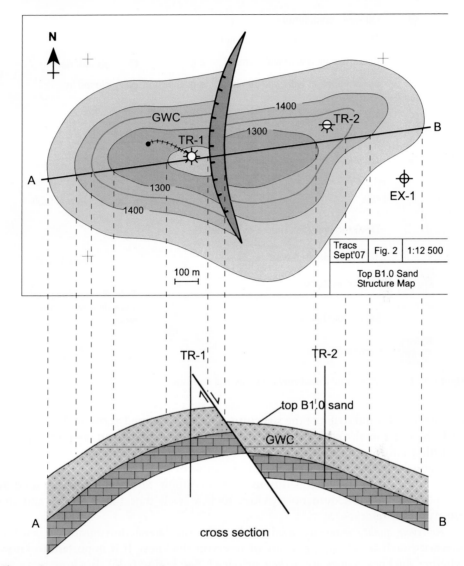

Figure 6.43 Structural map and section.

The maps most frequently consulted in field development are structural maps and reservoir quality maps. Commonly a set of maps will be constructed for each drainage unit.

To construct a section as shown in Figure 6.43, a set of maps (one per horizon) is needed.

Structural maps display the top (and sometimes the base) of the reservoir surface below the datum level. The depth values are always true vertical subsea. One could say that the contours of structure maps provide a picture of the subsurface topography. They display the shape and extent of a hydrocarbon accumulation and

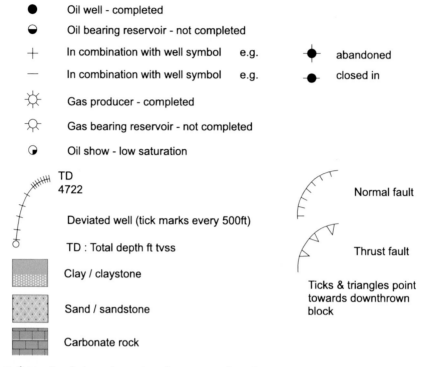

Figure 6.44 Symbols used on subsurface maps and sections.

indicate the dip and strike of the structure. The *dip* is defined as the angle of a plane with the horizontal, and is perpendicular to the *strike*, which runs along the plane.

Other information that can be obtained from such maps is the location of faults, the status and location of wells and the location of the fluid contacts. Figure 6.44 shows some of the most frequently used map symbols. Structural maps are used in the planning of development activities such as well trajectories/targets and the estimation of reserves.

Reservoir quality maps are used to illustrate the lateral distribution of *reservoir parameters* such as net sand, porosity or reservoir thickness. It is important to know whether thickness values are *isochore* or *isopach* (see Figure 6.45). Isochore maps are useful if properties related to a fluid column are contoured, for example net oil sand (NOS). Isopach maps are used for sedimentological studies, for example to show the lateral thinning out of a sand body. In cases of low structural dip ($< 12°$), isochore and isopach thickness are virtually the same.

By adding or subtracting parameter maps (see Figure 7.3) additional information can be obtained. They show *trends* in the parameters and are used to optimise reserves development and management.

Because of the nature of subsurface data, maps and sections are only models or approximations of reality, and always contain a degree of uncertainty. Reduction of these uncertainties is one of the tasks of the geoscientists, and will be further discussed in Section 7.2, Chapter 7.

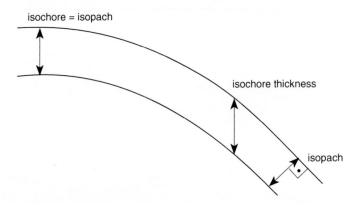

Figure 6.45   Isochore and isopach thickness.

Maps can be created by hand or by computer mapping packages. The latter has become standard. Nevertheless, care should be taken that the mapping process reflects the geological model. Highly complex areas may require considerable manual input to the maps which can subsequently be digitised.

6.4.3. Net to gross ratio

In nearly all oil or gas reservoirs there are layers which do not contain, or will not produce reservoir fluids. These layers may have no porosity or limited permeability and are generally defined as '*non-reservoir*' *intervals*. The thickness of productive (net) reservoir rock within the total (gross) reservoir thickness is termed the *net to gross* or *N/G ratio*.

The most common method of determining the N/G ratio is by using wireline GR logs. Non-productive layers such as shales can be differentiated from clean (non-shaly) formation by measuring and comparing natural radioactivity levels along the borehole. Shales contain small amounts of radioactivity elements such as thorium, potassium and uranium which are not normally present in clean reservoir rock, therefore high levels of natural radioactivity indicate the presence of shale, and by inference non-productive formation layers (Figure 6.46).

If a 'sand line' (0% shale) and a 'shale line' (100% shale) are defined on the GR log, a cut-off limit of 50% shale can be used to differentiate the reservoir from non-reservoir intervals. This type of cut-off is often used in preliminary log evaluations and is based on the assumption that reservoir permeability is destroyed once a rock contains more than 50% shale (Figures 6.47 and 6.48).

Other logs employed to determine N/G ratio include the *spontaneous potential* (SP) *log* and the *microlog*, which differentiate permeable from non-permeable intervals. In geologically complex rocks, the N/G ratio can be established using a combination of density and neutron log responses. The N/G ratio can also be measured directly on cores if there is visible contrast between the reservoir and non-reservoir sections, or from permeability measurements on core samples, provided sample coverage is sufficient. Direct core measurement is important in

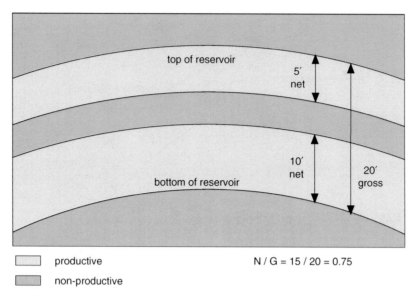

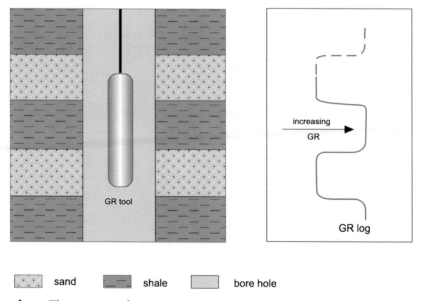

Figure 6.46   Net to gross ratio.

Figure 6.47   The gamma ray log.

laminated rocks where shale and sand beds are very thin as it is difficult to calculate N/G from logs.

The N/G ratio is usually not constant across a reservoir and may change over quite short distances from 1.0 (100% reservoir) to 0.0 (no reservoir) in some depositional

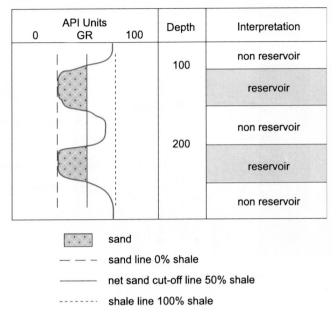

API Units 0 GR 100	Depth	Interpretation
	100	non reservoir
		reservoir
	200	non reservoir
		reservoir
		non reservoir

⌗⌗⌗⌗ sand

– – – sand line 0% shale

———— net sand cut-off line 50% shale

······· shale line 100% shale

Figure 6.48   GR log interpretation.

environments. Reservoirs with a low or unpredictable N/G ratios often require large numbers of wells to access reserves and are therefore more expensive to develop.

6.4.4. Porosity

Reservoir porosity can be measured directly from core samples or indirectly using logs. However, as core coverage is rarely complete, logging is the most common method employed, and the results are compared against measured core porosities where core material is available.

The *formation density log* is the main tool for measuring porosity. It measures the bulk density of a small volume of formation in front of the logging tool, which is a mixture of minerals and fluids. Provided the rock matrix and fluid densities are known, the relative proportion of rock and fluid (and hence porosity) can be determined (Figure 6.49).

The density tool is constructed so that medium-energy GRs are directed from a radioactive source into the formation. These GRs interact with the formation by a process known as Compton scattering, in which GRs lose energy each time they collide with an electron. The number of GRs reaching detectors in the tool is inversely proportional to the number of electrons (or the electron density) in the formation, which is related to the formation bulk density. A low GR count implies a high electron (and bulk) density and therefore a low porosity.

The bulk density measured by the logging tool is the weighted average of the rock matrix and fluid densities, so that

$$\rho_b = \rho_f \phi + \rho_{ma}(1 - \phi)$$

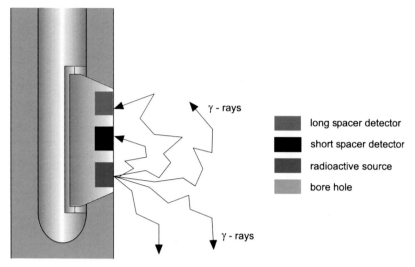

Figure 6.49   Formation density measurement.

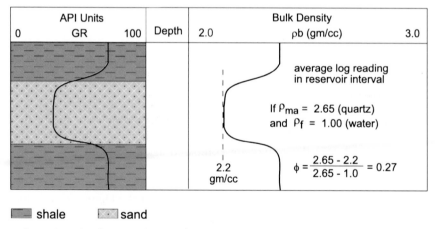

shale   sand

Figure 6.50   Porosity from the density log.

The formation bulk density (ρ_b) can be read directly from the density log (see Figure 6.50) and the matrix density (ρ_{ma}) and fluid density (ρ_f) found in tables, assuming we have already identified lithology and fluid content from other measurements. The equation can be rearranged for porosity (ϕ) as follows:

$$\phi = \frac{\rho_{ma} - \rho_b}{\rho_{ma} - \rho_f}$$

Other logging tools which can be used to determine porosity include the *neutron*, *sonic* and *NMR tools*. The *neutron tool* has a design similar to the density tool except that it employs neutrons instead of GRs. The neutrons are slowed down as they travel through the formation and some are captured. Of the common reservoir elements, hydrogen has the greatest stopping power. A low count rate at the detector indicates large number of hydrogen atoms in the formation and, as hydrogen is present in water and oil in similar amounts, implies high porosity.

Because the neutron tool responds to hydrogen it can be used to differentiate between gas and liquids (oil or water) in the formation. A specific volume of gas will contain a lot fewer hydrogen atoms than the same volume of oil or water (at the same pressure), and therefore in a gas-bearing reservoir the neutron porosity (which assumes the tool is investigating fluid-filled formation) will register an artificially low porosity. A large apparent decrease in porosity in the upper section of a homogenous reservoir interval is often indicative of entering gas-bearing formation.

The *sonic tool* measures the time taken for a sound wave to pass through the formation. Sound waves travel in high-density (i.e. low porosity) formation faster than in low-density (high porosity) formation. The porosity can be determined by measuring the *transit time* for the sound wave to travel between a transmitter and receiver, provided the rock matrix and fluid are known.

The *NMR tool* magnetically aligns hydrogen protons and then measures the time taken for this alignment to decay. In a reservoir the hydrogen atoms occur chiefly in the fluid as either water or hydrocarbon within the pore space. The speed of decay is proportional to the size of the pore. Hence the NMR tool can not only determine porosity but also indicate the pore size distribution.

6.4.5. Hydrocarbon saturation

Nearly all reservoirs are water bearing prior to hydrocarbon charge. As hydrocarbons migrate into a trap they displace the water from the reservoir, but not completely. Water remains trapped in small pore throats and pore spaces. In 1942, *Archie* developed an equation describing the relationship between the electrical conductivity of reservoir rock and the properties of its pore system and pore fluids (Figure 6.51).

The relationship was based on a number of observations, firstly, that the conductivity (C_o) of a water-bearing formation sample is dependent primarily upon pore water conductivity (C_w) and porosity (ϕ) distribution (as the rock matrix does not conduct electricity) such that

$$C_o = \phi^m C_w$$

The pore system is described by the volume fraction of pore space (the fractional porosity) and the shape of the pore space which is represented by m, known as the cementation exponent. The cementation exponent describes the complexity of the pore system, that is how difficult it is for an electric current to find a path through the reservoir.

Secondly, it can be observed that as water is displaced by (non-conductive) oil in the pore system, the conductivity (C_t) of an oil-bearing reservoir sample decreases. As the water saturation (S_w) reduces so does the electrical conductivity of the

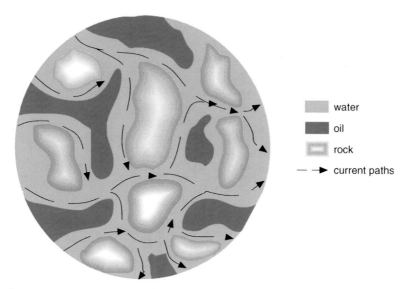

Figure 6.51   Passage of electric current through the reservoir.

sample, such that

$$C_t = S_w^n \phi^m C_w$$

The volume fraction of water (S_w) and the saturation exponent n can be considered as expressing the increased difficulty experienced by an electrical current passing through a partially oil-filled sample. (Note: C_o is only a special case of C_t; when a reservoir sample is fully water bearing $C_o = C_t$.)

In practice, the logging tools are often used to measure the resistivity of the formation rather than the conductivity and therefore the equation above is more commonly inverted and expressed as

$$R_t = S_w^{-n} \phi^{-m} R_w$$

where R_t is the formation resistivity (ohm m), S_w the water saturation (fraction), ϕ the porosity (fraction), R_w the water resistivity (ohm m), m the cementation exponent, and n the saturation exponent.

Formation resistivity is measured using a logging tool, porosity is determined from logs or cores and *water resistivity* can be determined from logs in water-bearing sections or measured on produced samples. In a large range of reservoirs, the saturation and cementation exponents can be taken as $m = n = 2$. The remaining unknown is the water saturation and the equation can be rearranged so that

$$S_w = \sqrt[n]{\frac{R_w}{\phi^m R_t}}$$ and hydrocarbon saturation (fraction) $S_h = 1 - S_w$

The most common method for measuring formation resistivity and hence determining hydrocarbon saturation is by logging with a resistivity tool such as the *laterolog*. The tool is designed to force electrical current through the formation

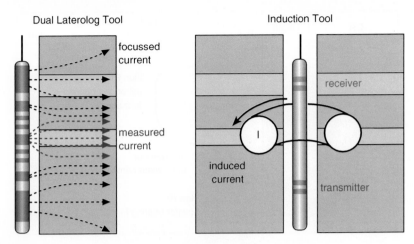

Figure 6.52 Resistivity measurements using the laterolog.

adjacent to the borehole and measure the potential difference across the volume investigated. With this information the formation resistivity can be calculated and output every foot as a resistivity log (Figure 6.52).

The laterolog tool needs a conductive environment to operate. Therefore, in oil based mud (OBM) other types of tools are used. The most common is the *induction log tool*, based upon the principles of mine detection. A transmitting coil induces currents in the formation which in turn induce a current in the receiver coil.

The majority of FEWD resistivity tools use the electro-magnetic (EM) wave resistivity, as the signal is not affected by steel DCs. The EM wave response is a function of conductivity and distance. The tool has two receivers of known spacing, therefore conductivity (the inverse of resistivity) can be deduced.

More recent resistivity tools are array devices which measure resistivities at different distances into the formation. Additionally, there are also tools which measure resistivity in 3D which are used in thinly bedded rocks.

The resistivity log can also be used to define oil–water or gas–water contacts. Figure 6.53 shows that the fluid contact can be defined as the point at which the resistivity begins to increase in the reservoir interval, inferring the presence of hydrocarbons above that point.

6.4.6. Permeability

All the parameters discussed above are needed to calculate the volume of hydrocarbons in the reservoir. The *formation permeability* is a measure of the ease with which fluids can pass through the reservoir, and hence is needed for estimating well productivity, reservoir performance and hydrocarbon recovery.

Formation permeability around the wellbore can be measured directly on core samples from the reservoir or from well testing (see Section 10.4, Chapter 10), or indirectly (estimated) from logs.

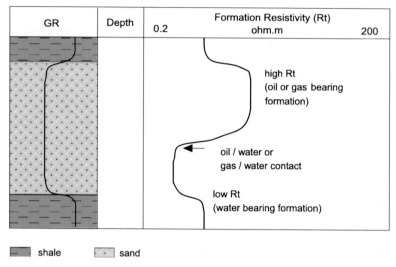

Figure 6.53   The formation resistivity log.

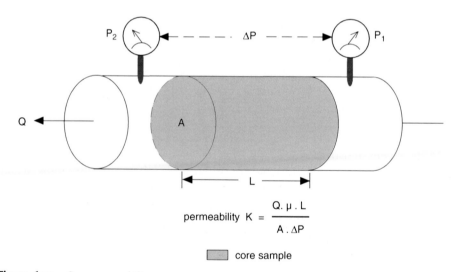

$$permeability \ K = \frac{Q.\mu.L}{A.\Delta P}$$

core sample

Figure 6.54   Core permeability measurement.

For direct measurement from core samples, the samples are mounted in a holder and gas is flowed through the core. The pressure drop across the core and the flowrate are measured. Provided the gas viscosity (μ) and sample dimensions are known, the permeability can be calculated using the *Darcy* equation shown in Figure 6.54.

Permeabilities measured on small core samples, whilst accurate, are not necessarily representative of the reservoir. Averaging a number of samples can allow comparisons with well test permeabilities to be made.

Permeable intervals can be identified from a number of logging tool measurements, the most basic of which is the caliper tool. The caliper tool is used to measure the borehole diameter which, in a gauge hole, is a function of the bit size and the mudcake thickness. Mudcake will only build-up across permeable sections of the borehole where mud filtrate has invaded the formation and mud solids (which are too big to enter the formation pore system) plate out on the borehole wall. Therefore, the presence of mudcake implies permeability.

Mud filtrate invasion is normally restricted to within a few inches into the formation, after which the build up of mudcake prevents further filtrate loss. If resistivity tools with different depths of investigation (in the invaded and non-invaded zones) are used to measure formation resistivity over the same vertical interval, then separation of the log curves can indicate invasion and hence permeability (Figures 6.55 and 6.56).

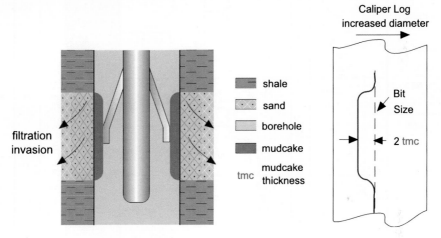

Figure 6.55   Measurement of mudcake.

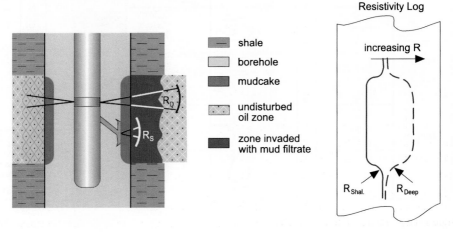

Figure 6.56   Permeability indications from resistivity logs.

The NMR tool (discussed in Section 6.4.4) provides the best estimate of permeability from any logging tool. The relaxation of the hydrogen proton magnetic alignment is dependant upon the pore size. The NMR signal will indicate a pore size distribution which is manipulated using empirical equations into a

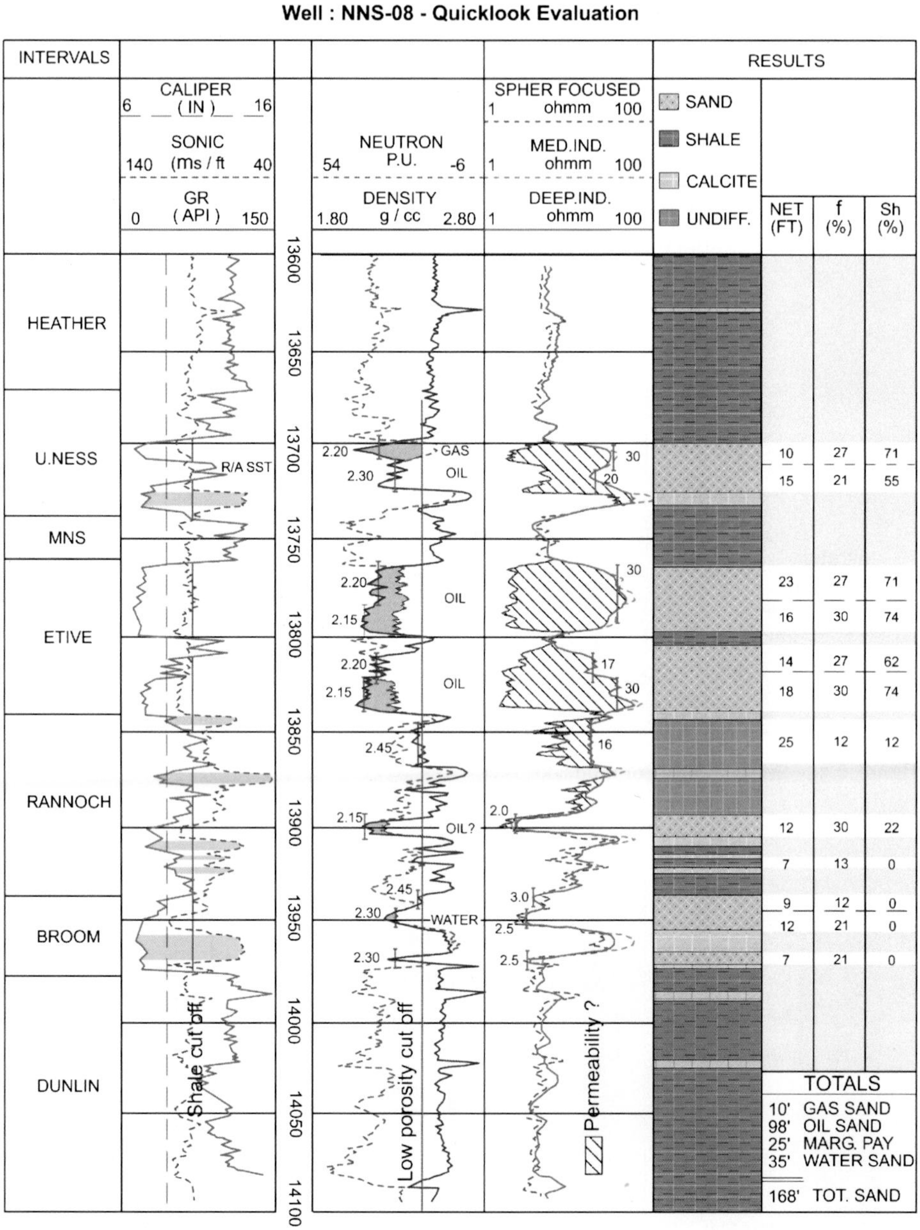

Figure 6.57   Quick look interpretation.

permeability 'log' based upon the assumption that reservoirs with smaller pores have a lower permeability. This type of measurement is usually calibrated with core measurements.

The methods discussed above only give an indication of permeability near the wellbore. Reservoir permeability is usually estimated from production tests and is described in Section 10.4, Chapter 10.

6.4.7. Quick look evaluation

Using the logs and processes described in Sections 6.4.3–6.4.5, a quick look evaluation of the reservoir can be performed on an individual well fairly quickly.

An example of a 'quick look interpretation' employing the logs discussed so far is shown in Figure 6.57.

1. The lithologies (sand, shale, etc.) are picked using the GR/sonic logs. Where GR is greater than the 'shale line' indicates a shale. Otherwise, the lithology is a sandstone. The high sonic layers (dotted shading) indicate the sandstone is calcite cemented, and are not net sand. The 'net' thickness is recorded in the results column.
2. The density log is subdivided into smaller intervals within the sandstone and an average value assigned. These values of RHOB (bulk density) are used to calculate the average porosity over the interval and recorded in the results column.
3. The average values of the deep induction log over the sand intervals are assigned and are taken to represent R_t. These values along with the calculated porosity are input into the Archie equation (Section 6.4.5) to calculate saturation.
4. Water is interpreted in the sand at 13,950 ft and oil is still present at 13,900 ft, albeit at a low saturation. The OWC is somewhere between the two, but cannot be observed in this well due to poor reservoir quality.
5. A small gas cap is interpreted at the very top of the reservoir where the neutron log has a very low value.
6. In the oil leg the separation between the deep resistivity and shallow indicates that the shallow resistivity is measuring conductive mud filtrate which has invaded the formation. This suggests the reservoir has a good permeability. In the waterleg there is no separation, which only indicates the deep resistivity is now reading conductive water, but does not indicate low permeability.

6.4.8. Integration of core and logs

After the quick look evaluation has been undertaken further, data such as core analysis and image logs will become available which can be used to build a geological interpretation and refine the evaluation.

The *routine core analysis* data are plug porosity and permeability measurements plus a GR log. The GR and plug porosities are compared to the log results to improve the accuracy of the interpretation. A relationship between the logs and plug permeabilities is established so that permeability can be inferred from the logs in sections of reservoir that have not been cored.

The core itself is examined and *logged* by specialist sedimentologists and structural geologists to identify key geological features. Many of these important features can be observed from *image logs* which have been run through the whole borehole and not just over the cored interval.

Image logs are pixellated, high-resolution resistivity or acoustic images of the borehole wall. Resistivity images are generated from arrays up to 192 electrode buttons located on either four or six articulated arms pressed against the borehole wall (Figure 6.58). Image resolution is 0.1 in. (compared to 6 or 3 in. for standard logging tools). Acoustic images are gathered by a rotating ultrasonic transducer which measures the amplitude and travel time of the signal, again at very high resolution. Both resistivity and acoustic image tools are run with an inclinometry device which enables the log to be oriented.

Geological features are observed in the image log (Figure 6.59) and can be matched to observations from the core. Planar features such as faults, fractures and

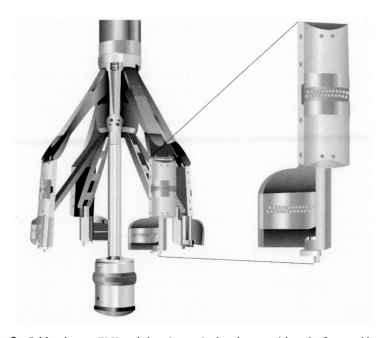

Figure 6.58 Schlumberger FMI tool showing articulated arms with pads, flaps and buttons.

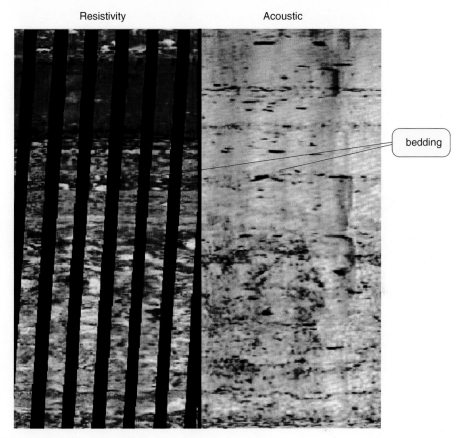

Figure 6.59   Example of resistivity and acoustic images.

bedding can be measured to determine their orientation and help build a geological model. Additionally, the high resolution of the images allows detailed interpretation to be made of difficult, thinly laminated rocks, which cannot be detected using standard logs. Considerable quantities of additional reserves can be identified using this *thin-bed evaluation* technique from image logs.

CHAPTER 7

VOLUMETRIC ESTIMATION

Introduction and Commercial Application: Volumetric estimation is concerned with quantifying how much oil and gas exists in an accumulation. The estimate will vary throughout the field lifetime as more information becomes available and as the technology for gathering and interpreting the data improves. A volumetric estimate is therefore a current estimate, and should be expected to change over time. Two main methods of estimating volumetrics are used: deterministic and probabilistic. Deterministic methods average the data gathered at various points in the reservoir, from well logs, cores and seismic to estimate the field-wide properties. Probabilistic methods use predictive tools, statistics, analogue field data and input regarding the geological model to predict trends in reservoir properties away from the sample points. This section will concentrate on the deterministic methods and the techniques used for expressing uncertainty in these volumetric estimates.

The volumetrics of a field, along with the anticipated recovery factors (RFs), control the reserves in the field – those hydrocarbons which will be produced in the future. The value of an oil or gas company lies predominantly in its hydrocarbon reserves which are used by shareholders and investors as one indication of the strength of the company, both at present and in the future. A reliable estimate of the reserves of a company is therefore important to the current value as well as the longer term prospects of an oil or gas company.

7.1. DETERMINISTIC METHODS

Volumetric estimates are required at all stages of the field life cycle. In many instances, a first estimate of 'how big' an accumulation could be is requested. If only a 'back of the envelope' estimate is needed or if the data available is very sparse, a quick look estimation can be made using field-wide averages.

The formulae to calculate volumes of oil or gas are

$$STOIIP = GRV \frac{N}{G} \phi S_o \frac{1}{B_o} \quad (stb)$$

$$GIIP = GRV \frac{N}{G} \phi S_g \frac{1}{B_g} \quad (scf)$$

$$UR = HCIIP \times Recovery\ factor \quad (stb)\ or\ (scf)$$

$$Reserves = UR - Cumulative\ production \quad (stb)\ or\ (scf)$$

STOIIP is a term which normalises volumes of oil contained under high pressure and temperature in the subsurface to surface conditions (e.g. 1 bar, 15°C). In the early days of the industry this surface volume was referred to as *stock tank oil* and since measured prior to any production having taken place it was the volume *initially in place*.

GIIP is the equivalent expression for gas initially in place.

HCIIP is the hydrocarbons initially in place – a general term covering STOIIP and GIIP.

Ultimate recovery (UR) and reserves are linked to the volumes initially in place by the RF, or fraction of the in-place volume which will be produced. Before production starts, reserves and UR are the same.

GRV is the *gross rock volume* of the hydrocarbon-bearing interval and is the product of the area (*A*) containing hydrocarbons and the interval thickness (*H*), hence

$$GRV = AH \quad (ft^3) \text{ or (acre ft) or } (m^3)$$

The area can be measured from a map. Figure 7.1 clarifies some of the reservoir definitions used in reserves estimation:

H is the isochore thickness of the total interval (gross thickness), regardless of lithology

Net sand is the height of the lithologic column with reservoir quality, that is the column that can potentially store hydrocarbons

Net oil sand (NOS) is the length of the net sand column that is oil bearing

The other parameters used in the calculation of STOIIP and GIIP have been discussed in Section 6.4, Chapter 6. The formation volume factors (B_o and B_g) were introduced in Section 6.2, Chapter 6. We can therefore proceed to the 'quick and

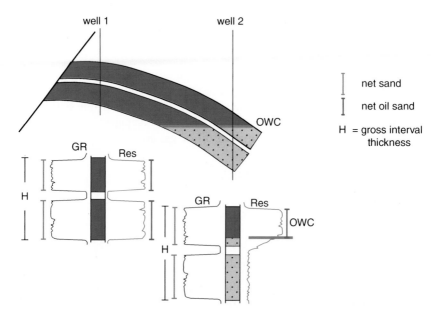

Figure 7.1   Definitions used for volumetric estimates.

easy' deterministic method most frequently used to obtain a volumetric estimate. It can be done on paper or by using available software. The latter is only reliable if the software is constrained by the geological reservoir model.

7.1.1. The area–depth method

From a top reservoir map (Figure 7.2) the area within a selected depth interval is measured. This is done using a planimeter, a hand-operated device that measures areas.

The stylus of the planimeter is guided around the depth to be measured and the respective area contained within this contour can then be read off. The area is now plotted for each depth as shown in Figure 7.2 and entered onto the *area–depth graph*. Since the structure is basically cut into slices of increasing depth, the area measured for each depth will also increase. We are essentially integrating area with thickness.

Connecting the measured points will result in a curve describing the area–depth relationship of the top of the reservoir. If we know the *gross thickness* (H) from logs, we can establish a second curve representing the area–depth plot for the base of the reservoir. The area between the two lines will equal the volume of rock between the two markers. The area above the OWC is the oil-bearing *GRV*. The other parameters to calculate STOIIP can be taken as averages from our petrophysical evaluation (see Section 6.4, Chapter 6). Note that this method assumes that the

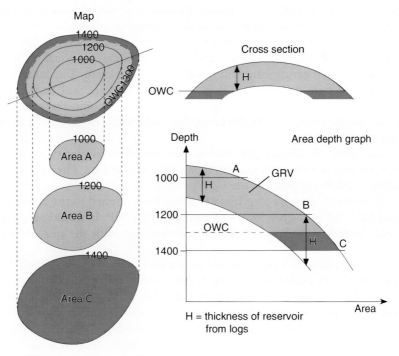

Figure 7.2 The area–depth method.

reservoir thickness is constant across the whole field. If this is not a reasonable approximation, then the method is not applicable, and an alternative such as the area–thickness method must be used (see Section 7.1.2).

This procedure can be easily carried out for a set of reservoirs or separate reservoir blocks. It is especially practical if stacked reservoirs with common contacts are to be evaluated. In cases where parameters vary across the field, we could divide the area into sub-blocks of equal values, which we measure and calculate separately.

7.1.2. The area–thickness method

In some depositional environments, for example fluviatile channels, marked differences in reservoir thickness will be encountered. Hence the assumption of a constant thickness, or a linear trend in thickness across the field will no longer apply. In those cases, a set of additional maps will be required. Usually, a *NOS map* will be prepared by the production geologist and then used to evaluate the hydrocarbon volume in place.

In the following example (Figure 7.3), well 1 had found an oil-bearing interval in a structure (1). An OWC was established from logs and had been extrapolated across the structure assuming continuous sand development. However, the core (in reality cores from a number of wells) and 3D seismic have identified a channel depositional environment. The channel had been mapped using specific field data and possibly analogue data from similar fields resulting in a net sand map (2). In this case, the hydrocarbon volume is *constrained* by the structural feature of the field *and* the distribution of reservoir rock, that is the channel geometry.

Hence we need to *combine* the two maps to arrive at a NOS map (3). The 'odd shape' is a result of that combination and actually it is easy to visualise: at the fault the thickness of oil-bearing sand will rapidly decrease to zero. The same is the case at the OWC. Where the net sand map indicates 0 m there will be 0 m of NOS. Where the channel is best developed showing maximum thickness we will encounter the maximum NOS thickness, but only until the channel cuts through the fault or the OWC.

We can now planimeter the *thickness* of the different NOS contours, plot thickness vs. area and then integrate both with the planimeter. The resulting value is the *volume of* NOS (4) and not the GRV!

It is clear that if the area–depth method had been applied to the above example, it would have led to a gross over-estimation of STOIIP. It would also have been impossible to target the best developed reservoir area with the next development well.

It should be noted that our example used a very simple reservoir model to show the principle. NOS mapping is usually a fairly complex undertaking.

As will be shown in the next section, the methods discussed so far do not take account of the uncertainties and lateral variations in reservoir parameters. Hence the accuracy of the results is not adequate for decision making. The next section introduces a more comprehensive approach to volumetric estimation.

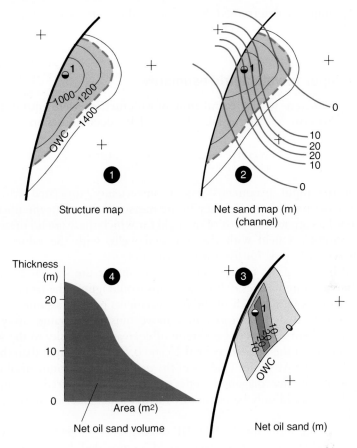

Figure 7.3   Net oil sand mapping and area–thickness method.

7.2. EXPRESSING UNCERTAINTY

As shown in Section 7.1, the calculation of volumetrics for a field involves the combination of a number of input parameters. It should be realised that each of these has a range of uncertainty in its estimation. The extent of this range of uncertainty will depend upon the amount of data available, and the accuracy of that data. The value in combining ranges of uncertainty in the input parameters to give a range of estimates for STOIIP, GIIP and UR is that both upside potential and downside risks can be quantified. Using a single figure to represent, say STOIIP, may lead to missed opportunities, or unrecognised risk.

The range of uncertainty in the UR may be too large to commit to a particular development plan, and field appraisal may be required to reduce the uncertainty and allow a more suitable development plan to be formed. Unless the range of uncertainty is quantified using statistical techniques and representations, the need for appraisal cannot be determined. Statistical methods are used to express ranges of values of STOIIP, GIIP, UR and reserves.

7.2.1. The input to volumetric estimates

The input parameters to the calculation of volumetrics were introduced at the beginning of Section 7.1. Let us take the STOIIP calculation as an example.

$$\text{STOIIP} = \text{GRV}\frac{N}{G}\phi S_o \frac{1}{B_o} \quad \text{(stb)}$$

Each of the input parameters has an uncertainty associated with it. This uncertainty arises from the inaccuracy in the measured data, plus the uncertainty as to what the values are for the parts of the field for which there are no measurements. Take, for example, a field with five appraisal wells, with the values of average porosity shown in Figure 7.4 for a particular sand.

It would be unrealistic to represent the porosity of the sand as the arithmetic average of the measured values (0.20), since this would ignore the range of measured values, the volumes which each of the measurements may be assumed to represent and the possibility that the porosity may move outside the range away from the control points. There appears to be a trend of decreasing porosity to the south-east, and the end points of the range may be 0.25 and 0.15, that is larger than the range of measurements made. An understanding of the geological environment of deposition and knowledge of any diagenetic effects would be required to support this hypothesis, but it could only be proven by further data gathering in the extremities of the field.

When providing input for the STOIIP calculation, a range of values of porosity (and all of the other input parameters) should be provided, based on the measured data and estimates of how the parameters may vary away from the control points. The uncertainty associated with *each* parameter may be expressed in terms of a probability density function (PDF), and these may be combined to create a PDF for STOIIP.

It is common practice within oil companies to use *expectation curves* to express ranges of uncertainty. The relationship between PDFs and expectation curves is a simple one.

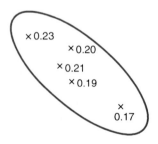

Figure 7.4   Porosity distribution in a field.

7.2.2. Probability density functions and expectation curves

A well-recognised form of expressing uncertainty is the *probability density* function. For example, if one measured the heights of a class of students and plotted them on a histogram of height ranges against the number of people within that height range, one might expect a relative frequency distribution plot, also known as a probability density function (PDF) with discrete values, such as that in the upper diagram in Figure 7.5. Each person measured is represented by one square, and the squares are placed in the appropriate height category. The number of squares or area under the curve represents the total population.

If the value on the x-axis is continuous rather than split into discrete ranges, the discrete PDF would become a continuous function. This is useful in predicting what fraction of the population has property X (height in our example) greater than a chosen value (X_1).

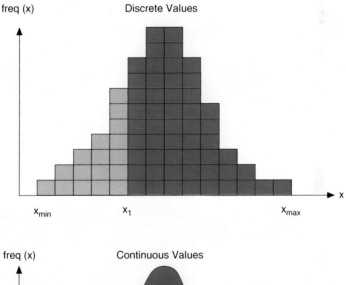

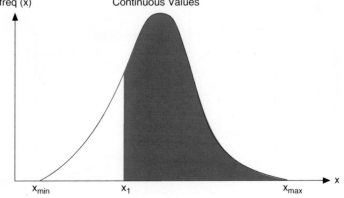

Figure 7.5   A probability density function.

From the continuous PDF one would estimate that approximately 70% of the population sampled were of height greater than or equal to X_1. In other words, if one were to randomly pick a person from the sample population, there is a 70% probability that the height of that person is greater than or equal to X_1. There is a 100% probability that the height of the person is greater than or equal to X_{min}, and a 0% chance that the height of the person is greater than X_{max}. The expectation curve is simply a representation of the cumulative PDF (Figure 7.6).

For oil field use, the x-axis on expectation curves is typically the STOIIP, GIIP, UR or reserves of a field.

Expectation curves are alternatively known as 'probability of excedence curves' or 'reverse cumulative probability curves'. This text will use the term 'expectation curve' for conciseness.

The slope of the expectation curve indicates the range of uncertainty in the parameter presented: a broad expectation curve represents a large range of uncertainty and a steep expectation curve represents a field with little uncertainty (typical of fields which have much appraisal data or production history).

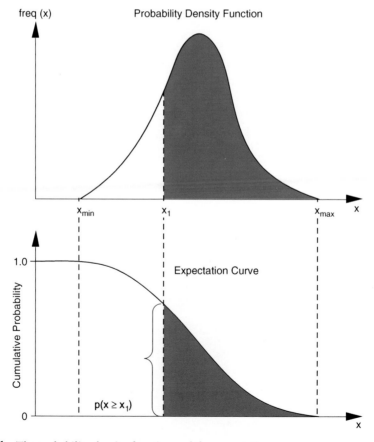

Figure 7.6   The probability density function and the expectation curve.

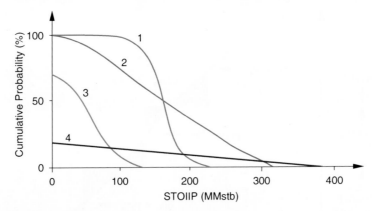

Figure 7.7 Types of expectation curve.

In Figure 7.7, expectation curves 1 and 2 represent discoveries, since they both have a 100% probability of containing a finite amount of oil (greater than zero). Case 1 is a well-defined discovery since the range of uncertainty in STOIIP values is small (at least 100 MMstb, but less than 220 MMstb). By contrast, case 2 represents a poorly defined discovery, with a much broader range of STOIIP, and would probably require appraisal activity to reduce this range of uncertainty before committing to a development plan.

Cases 3 and 4 are both exploration prospects, since the volumes of potential oil present are multiplied by a chance factor, the probability of success (POS), which represents the probability of there being oil there at all. For example, case 3 has an estimated probability of oil present of 65%, that is low risk of failure to find oil (35%). However, even if there is oil present, the volume is small, no greater than 130 MMstb. This would be a low risk, low reward prospect.

Case 4 has a high risk of failure (85%) to find any oil, but if there is oil there then the volume in place might be quite large (up to 400 MMstb). This would class as a high risk, high reward prospect.

7.2.2.1. Expectation curves for a discovery
For a discovery, a typical expectation curve for UR is shown in Figure 7.8.

For convenience, the probability axis may be split into three equal sectors in order to be able to represent the curve by just three points. Each point represents the average value of reserves within the sector. Again for convenience, the three values correspond to chosen cumulative probabilities (85, 50 and 15%), and are denoted by the values:

Low estimate   = 85% cumulative probability
                   (i.e. 85% probability of at least these reserves)
Medium estimate = 50% cumulative probability
High estimate   = 15% cumulative probability

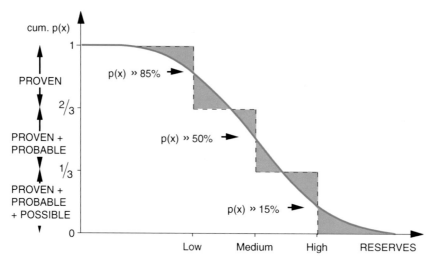

Figure 7.8 Expectation curve for a discovery.

The percentages chosen are often denoted as the p85, p50 and p15 values. Because they each approximately represent one-third of the distribution, their discrete probabilities may each be assigned as 1/3. This approximation is true for a normal (or symmetrical) PDF.

If the whole range is to be represented by just one value (which of course gives no indication of the range of uncertainty), then the 'expectation value' is used:

$$\text{Expectation value} = \frac{\text{High} + \text{Medium} + \text{Low}}{3}$$

An alternative and commonly used representation of the range of reserves is the proven, proven plus probable and proven plus probable plus possible definition. The exact cumulative probability which these definitions correspond to on the expectation curve for UR varies from country to country, and sometimes from company to company. However, it is always true that the values lie within the following ranges:

- proven: between 100 and 66%
- proven + probable: between 66 and 33%
- proven + probable + possible: between 33 and 0%.

The *annual reporting requirements* to the US Securities and Exchange Commission (SEC) legally oblige listed oil companies to state their proven reserves.

Many companies choose to represent a continuous distribution with discrete values using the p90, p50 and p10 values. The discrete probabilities which are then attached to these values are then approximately 30, 40 and 30%, respectively, for a normal distribution.

7.2.2.2. Expectation curves for an exploration prospect

When an explorationist constructs an expectation curve, the above approach for the volumetrics of an accumulation is taken, but one important additional parameter must be taken into account: the probability of there being hydrocarbons present at all. This probability is termed the *probability of success* (POS), and is estimated by multiplying together the probability of there being

- a source rock where hydrocarbons were generated
- a reservoir in a structure which may trap hydrocarbons
- a seal on top of the structure to stop the hydrocarbons migrating further
- a migration path for the hydrocarbons from source rock to trap
- the correct sequence of events in time (trap present as hydrocarbons migrated).

The estimated probabilities of each of these events occurring are multiplied together to estimate the POS, since they must *all* occur simultaneously if a hydrocarbon accumulation is to be formed. If the POS is estimated at say 30%, then the probability of failure must be 70%, and the expectation curve for an exploration prospect may look as shown in Figure 7.9.

As for the expectation curve for discoveries, the 'success' part of the probability axis can be divided into three equal sections, and the average reserves for each section calculated to provide a low, medium and high estimate of reserves, if there are hydrocarbons present.

More detail of this approach is given in Chapter 15.

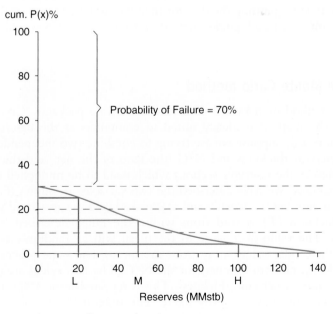

Figure 7.9 Expectation curve for an exploration prospect.

7.2.3. Generating expectation curves

Returning to the input parameters for an UR calculation, we have established that

$$UR = GRV \frac{N}{G} \phi S_o \frac{1}{B_o} RF \quad (stb)$$

Each of the input parameters requires an estimate of the range of values, which can itself be represented by a distribution, or expectation curve. Ideally, the expectation curves for the input parameters are combined together statistically.

Some variables often have dependencies, such as reservoir porosity and permeability (a positive correlation) or the capital cost of a specific equipment item and its lifetime maintenance cost (a negative correlation). We can test the linear dependency of two variables (say x and y) by calculating the covariance between the two variables (σ_{xy}) and the correlation coefficient (r):

$$\sigma_{xy} = \frac{1}{n} \sum_{i=1}^{n} (x_i - \mu_x)(y_i - \mu_y) \text{ and } r = \frac{\sigma_{xy}}{\sigma_x \sigma_y}$$

where μ is the mean value of the variable.

The value of r varies between $+1$ and -1, the positive values indicating a positive correlation (as x increases, so does y) and the negative values indicating a negative correlation (as x increases, y decreases). The closer the absolute value of r is to 1.0, the stronger the correlation. A value of $r = 0$ indicates that the variables are unrelated. Once we are satisfied that a true dependency exists between variables, we can generate equations which link the two using methods such as the least squares fit technique. If a correlation coefficient of 1.0 was found, it would make more sense to represent the relationship in a single line entry in the economic model. There is always value in cross-plotting the data for the two variables to inspect the credibility of a correlation. As a rough guide, correlation factors above 0.8 would suggest good correlation.

7.2.4. The Monte Carlo method

This is the method used by the commercial software packages 'Crystal Ball' and '@RISK'. The method is ideally suited to computers as the description of the method will reveal. Suppose we are trying to combine two independent variables, say gross reservoir thickness and N/G (the ratio of the net sand thickness to the gross thickness of the reservoir section) which need to be multiplied to produce a net sand thickness. We have described the two variables as follows (Figure 7.10).

A random number (between 0 and 1) is picked, and the associated value of gross reservoir thickness (T) is read from within the range described by the above distribution. The value of T close to the mean will be randomly sampled more frequently than those values away from the mean. The same process is repeated (using a different random number) for the N/G. The two values are multiplied to obtain one value of net sand thickness. This is repeated some 1000–10,000 times, with each outcome being equally likely. The outcomes are used to generate a distribution of values of net sand thickness. This can be performed simultaneously

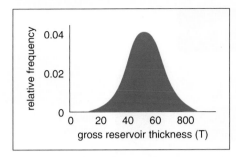 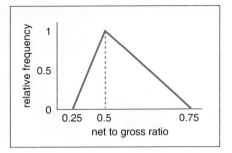

Figure 7.10 Probability distributions for two variables, input for Monte Carlo.

for more than two variables. For example, in estimating the UR for an oil reservoir, one would need to use the following variables:

$$UR = Area \times Thickness \times \frac{N}{G} \phi S_o \frac{1}{B_o} RF$$

The undefined variables so far in the text are

ϕ porosity
S_o the oil saturation in the pore space
B_o the formation volume factor of the oil (rb/stb), linked to the shrinkage of oil as it
         is brought from the subsurface to the surface
RF   recovery factor: the recoverable fraction of oil initially in place

Shown schematically in Figure 7.11, the Monte Carlo simulation is generating a limited number of possible combinations of the variables which approximates a distribution of all possible combinations. The more sets of combinations are made, the closer the Monte Carlo result will be to the theoretical result of using every possible combination. Using 'Crystal Ball' or '@Risk', one can watch the distribution being constructed as the simulation progresses. When the shape ceases to change significantly, the simulation can be halted. Of course, one must remember that the result is only a combination of the ranges of input variables defined by the user; the actual outcome could still lie outside the simulation result if the input variable ranges are constrained.

If two variables are dependent, the value chosen in the simulation for the dependent variable can be linked to the randomly selected value of the first variable using the defined correlation.

A Monte Carlo simulation is fast to perform on a computer, and the presentation of the results is attractive. However, one cannot guarantee that the outcome of a Monte Carlo simulation run twice with the same input variables will yield exactly the same output, making the result less auditable. The more simulation runs performed, the less of a problem this becomes. The simulation as described does not indicate which of the input variables the result is most sensitive to, but one of the routines in 'Crystal Ball' and '@Risk' does allow a sensitivity analysis to be performed as the simulation is run. This is done by calculating the coefficient of variation of each input variable with the outcome (for example, between area and UR). The higher the coefficient, the stronger the dependence between the input variable and the outcome.

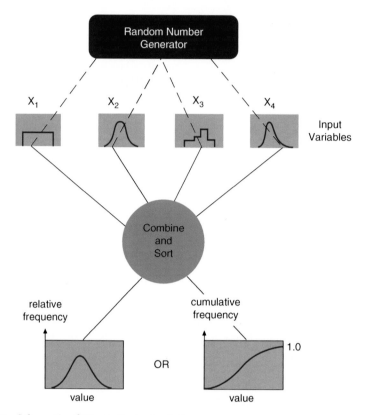

Figure 7.11   Schematic of Monte Carlo simulation.

7.2.5. The parametric method

The parametric method is an established statistical technique used for combining variables containing uncertainties, and has been advocated for use within the oil and gas industry as an alternative to Monte Carlo simulation. The main advantages of the method are its simplicity and its ability to identify the sensitivity of the result to the input variables. This allows a ranking of the variables in terms of their impact on the uncertainty of the result, and hence indicates where effort should be directed to better understand or manage the key variables in order to intervene to mitigate downside and/or take advantage of upside in the outcome.

The method allows variables to be added or multiplied using basic statistical rules, and can be applied to dependent as well as independent variables. If input distributions can be represented by a mean and standard deviation, then the following rules are applicable *for independent variables*:

Sums (say $c_i = a_i + b_i$, where a_i and b_i are distributions)

1. the sum of the distributions tends towards a normal distribution
2. the mean of the sum of distributions is the sum of the means:

$$\mu_c = \mu_a + \mu_b$$

3. the variance of the sum of distributions is the sum of the variances:

$$\sigma_c^2 = \sigma_a^2 + \sigma_b^2$$

Products (say $c_i = a_i b_i$, where a_i and b_i are distributions)

4. the product of the distributions tends towards a log-normal distribution
5. the mean of the product of distributions is the product of the means:

$$\mu_c = \mu_a \mu_b$$

For the final rule, another parameter, K, the coefficient of variation, is introduced,

$$K = \frac{\sigma}{\mu}$$

6. the value of $(1+K^2)$ for the product is the product of the individual $(1+K^2)$ values:

$$(1 + K_c^2) = (1 + K_a^2)(1 + K_b^2)$$

Having defined some of the statistical rules, we can refer back to our example of estimating UR for an oil field development. Recall that

$$UR = Area \times Thickness \times \frac{N}{G} \phi S_o \frac{1}{B_o} RF$$

From the probability distributions for each of the variables on the right hand side, the values of K, μ and σ can be calculated. Assuming that the variables are independent, they can now be combined using the above rules to calculate K, μ and σ for UR. Assuming the distribution for UR is log-normal, the value of UR for any confidence level can be calculated. This whole process can be performed on paper, or quickly written on a spreadsheet. The results are often within 10% of those generated by Monte Carlo simulation.

One significant feature of the parametric method is that it indicates, through the $(1 + K_i^2)$ value, the relative contribution of each variable to the uncertainty in the result. Subscript i refers to any individual variable. $(1 + K_i^2)$ will be greater than 1.0; the higher the value, the more the variable contributes to the uncertainty in the result. In the following example, we can rank the variables in terms of their impact on the uncertainty in UR. We could also calculate the relative contribution to uncertainty (Figure 7.12).

The purpose of this exercise is to identify what parameters need to be further investigated if the current range of uncertainty in reserves is too great to commit to a development. In this example, the engineer may recommend more appraisal wells or better definition seismic to reduce the uncertainty in the reservoir area and the N/G, plus a more detailed study of the development mechanism to refine the understanding of the RF. A fluid properties study to reduce uncertainty in B_o (linked to the shrinkage of oil) would have little impact on reducing the uncertainty

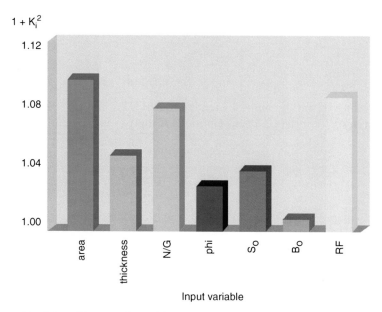

Figure 7.12   Ranking of impact of variables on uncertainty in reserves.

in reserves. This approach can thus be used for

- planning data gathering activities
- planning how to mitigate the effects of downside in key variables
- planning how to take advantage of upside in key variables.

7.2.6. Three-point estimates: a short-cut method

If there is insufficient data to describe a continuous probability distribution for a variable (as with the area of a field in an earlier example), we may be able to make a subjective estimate of high, medium and low values. If those are chosen using the p85, p50 and p15 cumulative probabilities described in Section 7.2.2, then the implication is that the three values are equally likely, and therefore each has a probability of occurrence of 1/3. Note that the low and high values are not the minimum and maximum values.

To estimate the product of the two variables shown in Figure 7.13, a short-cut method is to multiply the low, medium and high values in a matrix (in which numbers have been selected).

Note that the low value of the combination is not the absolute minimum (which would be 4, and is still a possible outcome), just as the high value is not the maximum. The three values (which are calculated by taking the mean of the three lowest values in the matrix, etc.) represent equally likely outcomes of the product $A*B$, each with a probability of occurrence of 1/3.

This short-cut method could be repeated to include another variable, and could therefore be an alternative to the previous two methods introduced. This method

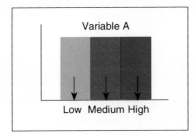 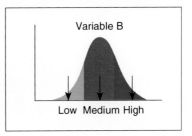

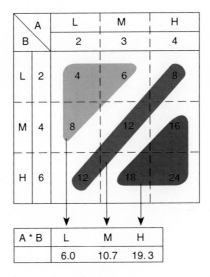

Figure 7.13 Combining three-point estimates.

can always be used as a last resort, but beware that the range of uncertainty narrows each time the process is repeated because the tails of the input variables are always neglected. This can lead to a false impression of the range of uncertainty in the final result.

Further detail on managing uncertainty is discussed in Chapter 15.

CHAPTER 8

FIELD APPRAISAL

Introduction and Commercial Application: The objective of performing appraisal activities on discovered accumulations is to reduce the uncertainty in the description of the hydrocarbon reservoir, and to provide information with which to make a decision on the next action. The next action may be, for example, to undertake more appraisal, to commence development, to stop activities or to sell the discovery. In any case, the appraisal activity should lead to a decision which yields a greater value than the outcome of a decision made in the absence of the information from the appraisal. The improvement in the value of the action, given the appraisal information, should be greater than the cost of the appraisal activities, otherwise the appraisal effort is not worthwhile.

Appraisal activity should be prioritised in terms of the amount of reduction of uncertainty it provides, and its impact on the value derived from the subsequent action.

The objective of appraisal activity is not necessarily to prove more hydrocarbons. For example, appraisal activity which determines that a discovery is non-commercial should be considered as worthwhile, since it saves a financial loss which would have been incurred if development had taken place without appraisal.

This section will consider the role of appraisal in the field life cycle, the main sources of uncertainty in the description of the reservoir and the appraisal techniques used to reduce this uncertainty. The value of the appraisal activity will be compared with its cost to determine whether such activity is justified.

8.1. THE ROLE OF APPRAISAL IN THE FIELD LIFE CYCLE

Appraisal activity, if performed, is the step in the field life cycle between the discovery of a hydrocarbon accumulation and its development. The role of appraisal is to provide *cost-effective information* with which the subsequent decision can be made. Cost effective means that the value of the decision with the appraisal information is greater than the value of the decision without the information. If the appraisal activity does not add more value than its cost, then it is not worth doing. This can be represented by a simple flow diagram (Figure 8.1), in which the cost of appraisal is $A, the profit (net present value, NPV) of the development with the appraisal information is $(D2−A), and the profit of the development without the appraisal information is $D1.

The appraisal activity is only worthwhile if the value of the outcome with the appraisal information is greater than the value of the outcome without the

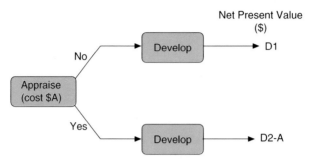

Figure 8.1 Net present value with and without appraisal.

information

$$\text{i.e.} \quad D2 - A > D1$$

$$\text{or} \quad A \quad < D2 - D1$$

In other words, the cost of the appraisal must be less than the improvement in the value of the development which it provides. It is often necessary to assume outcomes of the appraisal in order to estimate the value of the development with these outcomes.

8.2. IDENTIFYING AND QUANTIFYING SOURCES OF UNCERTAINTY

Field appraisal is most commonly targeted at reducing the range of uncertainty in the volumes of hydrocarbons in place, where the hydrocarbons are and the prediction of the performance of the reservoir during production.

The parameters which are included in the estimation of STOIIP, GIIP and UR, and the controlling factors are shown in the following table.

Input Parameter	Controlling Factors
Gross rock volume	Shape of structure; dip of flanks; position of bounding faults; position of internal faults; depth of fluid contacts (e.g. OWC)
Net:gross ratio	Depositional environment; diagenesis
Porosity	Depositional environment; diagenesis
Hydrocarbon saturation	Reservoir quality; capillary pressures
Formation volume factor	Fluid type; reservoir pressure and temperature
Recovery factor (initial conditions only)	Physical properties of the fluids; formation dip angle; aquifer volume; gas cap volume

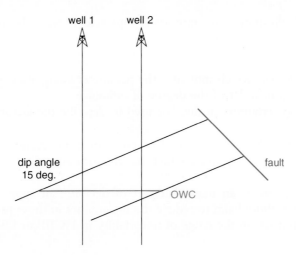

Figure 8.2 Partially appraised structure.

It should be noted that the RF for a reservoir is highly dependent on the development plan, and that initial conditions alone cannot be used to determine this parameter.

In determining an estimate of reserves for an accumulation, all of the above parameters will be used. When constructing an expectation curve for STOIIP, GIIP or UR, a range of values for each input parameter should be used, as discussed in Section 7.2, Chapter 7. In determining an appraisal plan, it is necessary to determine which of the parameters contributes most to the uncertainty in STOIIP, GIIP or UR.

Take an example of estimating GRV, based on seismic data and the results of two wells in a structure (Figure 8.2). The cross-section has been generated, and a base case GRV has been calculated.

The general list of factors influencing the uncertainty in the GRV included the shape of structure, dip of flanks, position of bounding faults, position of internal faults and depth of fluid contacts (in this case the OWC). In the above example, the OWC is penetrated by two wells, and the dip of the structure can be determined from the measurements made in the wells, which in turn will allow calibration of the 3D seismic. The most significant sources of uncertainty in GRV are probably the position and dip of the bounding fault, and the extent of the field in the plane perpendicular to this section. By looking at the quality of the seismic data, an estimate may be made of the uncertainty in the position of the fault, and any indications of internal faulting which may affect the volumetrics. The determination of geological uncertainties requires knowledge of the environment of deposition, diagenesis and the structural pattern of the field. The quantification often starts with a subjective estimate based on regional knowledge of the geology. In cases where little data is available, 'guesstimates' may need to be supplemented with data or reservoir trends observed in neighbouring fields.

The example illustrates some important steps in identifying the uncertainties and then beginning to quantify them

- consider the factors which influence the parameter being assessed
- rank the factors in order of the degree of influence
- consider the uncertainties in the data used to describe the factor.

The same procedure may be used to rank the parameters themselves (GRV, N/G, ϕ, S_h, B_o, RF), to indicate which has the greatest influence on the HCIIP or UR.

The *ranking process* is an important part of deciding an appraisal programme, since the activities should aim to reduce the uncertainty in those parameters which have the most impact on the range of uncertainty in HCIIP or UR.

8.3. APPRAISAL TOOLS

The main tools used for appraisal are those which have already been discussed for exploration, namely *drilling wells* and shooting 2D or 3D *seismic surveys*. Appraisal activity may also include *re-processing* an existing old seismic survey (again, 2D or 3D) using new processing techniques to improve the definition. It is not necessary to re-process the whole survey dataset; a sample may be re-processed to determine whether the improvement in definition is worthwhile. In the majority of cases where only 2D seismic is available, time and money will be better spent on shooting a new 3D seismic survey.

Seismic surveys are traditionally an exploration and appraisal (E&A) tool. However, 3D seismic is now being used more widely as a development tool, that is applied for assisting in selecting well locations, and even in identifying remaining oil in a mature field. This was discussed in Chapter 3. Seismic data acquired at the appraisal stage of the field life is therefore likely to find further use during the development period.

Appraisal activity should be based on the information required. The first step is therefore to determine what uncertainties appraisal is trying to reduce, and then what information is required to tie down those uncertainties. For example, if fluid contacts are a major source of uncertainty, drilling wells to penetrate the contacts is an appropriate tool; seismic data or well testing may not be. Other examples of appraisal tools are

- an *interference test* between two wells to determine pressure communication across a fault
- a well drilled in the *flank* of a field to improve the control of the dips seen on seismic
- a well drilled with a long enough *horizontal section* to emerge from the flanks of the reservoir, and determine the extent of the reservoir in the flanks (horizontal

wells may provide significantly more appraisal information about reservoir continuity than vertical wells)

- a *production test* on a well to determine the productivity from future development wells
- *coring* and production testing of the water leg in a field to predict *aquifer behaviour* during production, or to test for injectivity in the water leg
- *deepening* a well to investigate possible underlying reservoirs
- coring a well to determine diagenetic effects.

It is worth noting that if field development using horizontal wells is under consideration, then *horizontal appraisal wells* will help to gather representative data and determine the benefits of this technique, which is further discussed in Section 10.3, Chapter 10.

8.4. EXPRESSING REDUCTION OF UNCERTAINTY

The most informative method of expressing uncertainty in HCIIP or UR is by use of the expectation curve, as introduced in Section 7.2, Chapter 7. The high (H), medium (M) and low (L) values can be read from the expectation curve. A mathematical representation of the *uncertainty* in a parameter (e.g. STOIIP) can be defined as

$$\%\text{Uncertainty} = \frac{H - L}{2M} \times 100\%$$

The stated objective of appraisal activity is to reduce uncertainty. The impact of appraisal on uncertainty can be shown on an expectation curve, if an outcome is assumed from the appraisal. The following illustrates this process.

Suppose that four wells have been drilled in a field, and the geologist has identified three possible top sands maps based on the data available. These maps, along with the ranges of data for the other input parameters (N/G, S_o, ϕ, B_o) have been used to generate an expectation curve for STOIIP (Figure 8.3).

If well A is oil bearing, then the low case must increase, though the high case may not be affected. If well A is water bearing (dry), then the medium and high cases must reduce, though the low case may remain the same. For both outcomes, the post-appraisal expectation curve becomes steeper, and the range of uncertainty is reduced.

Note that it is not the objective of the appraisal well to find more oil, but to reduce the range of uncertainty in the estimate of STOIIP. Well A being dry does not imply that it is an unsuccessful appraisal well.

The choice of the location for well A should be made on the basis of the position which most effectively reduces the range of uncertainty. It may be for example, that a location to the north of the existing wells would actually be more

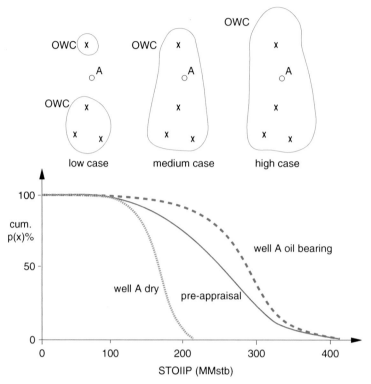

Figure 8.3   Impact of appraisal well A on expectation curve.

effective in reducing uncertainty. Testing the appraisal well proposal using this method will help to identify where the major source of uncertainty lies.

 ## 8.5. COST-BENEFIT CALCULATIONS FOR APPRAISAL

As discussed at the beginning of this section, the value of information from appraisal is the difference between the outcome of the decision with the information and the outcome of the decision without the information.

The determination of the value of the information is assisted by the use of *decision trees*. Consider the following decision tree as a method of justifying how much should be spent on appraisal. Suppose the range of uncertainty in STOIIP prior to appraisal is (20, 48, 100 MMstb; L, M, H values). One can perform appraisal that will determine which of the three cases is actually true, and then tailor a development plan to the STOIIP, or one can go ahead with a development in the absence of the appraisal information, only finding out which of the three STOIIPs truly exist, after committing to the development.

There are two types of nodes in the decision tree: *decision nodes* (rectangular) and *chance nodes* (circular). Decision nodes branch into a set of possible actions, whilst chance nodes branch into all possible results or situations.

The decision tree can be considered as a road map which indicates the chronological order in which a series of actions will be performed, and shows several possible courses, only one of which will actually be followed.

The tree is drawn by starting with the first decision to be taken, asking which actions are possible, and then considering all possible results from these actions, followed by considering future actions to be taken when these results are known, and so on. The tree is constructed in chronological order, from left to right.

Then the values of the 'leaves' are placed on the diagram, starting in the far-most future; the right hand side. The values on the leaves represent the NPVs of the cashflows which correspond to the individual outcomes.

The *probabilities* of each branch from chance nodes are then estimated and noted on the diagram.

Finally, the evaluation can be performed by 'rolling back' the tree, starting at the leaves and working backwards towards the root of the tree.

For chance nodes it is not possible to foretell which will be the actual outcome, so each result is considered with its corresponding probability. The value of a chance node is the statistical (weighted) average of all its results.

For decision nodes, it is assumed that good management will lead us to decide on the action which will result in the highest NPV. Hence the value of the decision node is the optimum of the values of its actions (Figure 8.4).

In the example, the first decision is whether or not to appraise. If one appraises, then there are three possible outcomes represented by the chance node: the high,

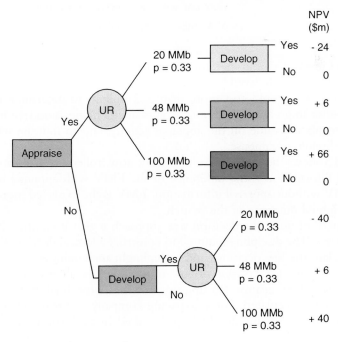

Figure 8.4 Decision tree for appraisal.

medium or low STOIIP. On the branches from the chance node, the estimated probability of these outcomes is noted (0.33 in each case). The sum of the probabilities on the branches from a chance node must be 1.0, since the branches should describe all possible outcomes. The next decision is whether to develop or not. The development plan in each case will be tailored to the STOIIP, and will have different costs and production profiles. It can be seen that for the low case STOIIP, development would result in a negative NPV.

If no appraisal was performed, and the development was started based, say, on the medium case STOIIP of 48 MMstb, then the actual STOIIP would not be found until the facilities were built and the early development wells were drilled. If it turned out that the STOIIP was only 20 MMstb, then the project would lose $40 million, because the facilities were oversized. If the STOIIP is actually 48 MMstb, then the NPV is assumed to be the same as for the medium case after appraisal. If the STOIIP was actually 100 MMstb, then the NPV of +$40 million is lower than for the case after appraisal (+$66 million) since the facilities are too small to handle the extra production potential.

In the example, development without appraisal leads to an NPV which is the weighted average of the outcomes: $million $(-40+6+40)/3 = +\$2$ million. Development after appraisal allows the decision not to develop in the case of the low STOIIP, and the weighted average of the outcomes is $million $(0+6+66)/3 = +\$24$ million.

Value of appraisal information = Value of outcome with appraisal information − value of

outcome without appraisal information

$$= \$24 \text{ million} - \$2 \text{ million} = \$22 \text{ million}$$

In this example it would therefore be justifiable to spend up to $22 million on appraisal activity which would distinguish between the high, medium and low STOIIP cases. If it would cost more than $22 million to determine this, then it would be better to go ahead without the appraisal. The decision tree has therefore been used to place a value on the appraisal activity, and to indicate when it is no longer worthwhile to appraise.

Figure 8.5 shows the same decision tree but now 'rolled back' to show the value of appraisal information – the difference in the EMV with appraisal information, and the EMV without appraisal information. EMV is the expected monetary value; the risk weighted outcome of the branch.

The benefit of using the decision tree approach is that it clarifies the decision-making process. The discipline required to construct a logical decision tree may also serve to explain the key decisions and to highlight uncertainties.

The *fiscal regime* (or tax system) in some countries allows the cost of E&A activity to be offset against existing income as a fiscal allowance before the taxable income is calculated. For a taxpaying company, the real cost of appraisal is therefore reduced, and this should be recognised in performing the cost-benefit calculations.

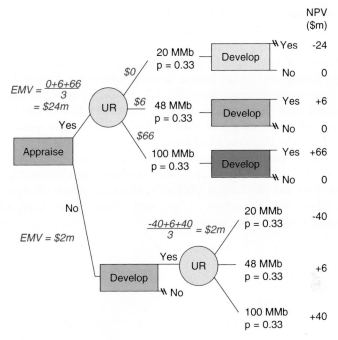

Figure 8.5 Rolled-back decision tree.

8.6. PRACTICAL ASPECTS OF APPRAISAL

In addition to the cost-benefit aspects of appraisal activities, there are frequently other practical considerations which affect appraisal planning, such as

- time pressure to start development (e.g. resulting from PSCs which limit the E&A period)
- the views of the partners in the block
- availability of funds of operator and partners
- increased incentives to appraise due to tax relief available on appraisal
- rig availability.

Appraisal wells are often abandoned after the required data has been collected, by placing cement and mechanical plugs in the well and capping the well with a sealing device. If development of the field appears promising, consideration should be given to *suspending* the appraisal wells. This entails securing the well in an approved manner using safety devices which can later be removed, allowing the well to be used for production or injection during the field development, such a well is often referred to as a 'keeper'. Approval must normally be given by the host government authority to temporarily suspend the well. Such action may save some of the cost of drilling a new development well, though in offshore situations the cost of re–using an appraisal well by later installing a subsea wellhead, a tie–back flowline and a riser may be comparable with that of drilling a new well.

In locations where the addition of facilities for production is relatively cheap, *phased development* of a field may be an option. Instead of reducing the uncertainty to optimise the development plan before development starts, appraisal and development may be performed simultaneously. The results of appraisal during the early development are used to determine the next part of the development plan. This has the advantage of combining the data gathering with early production, which considerably helps the cashflow of a project. Phased development with simultaneous appraisal is more appropriate in onshore and shallow water developments, where facilities costs are lower. In deep water offshore developments, using single integrated drilling and production platforms, there is a much stronger incentive to get the facilities design correct at an early stage, since later additions and modifications are much more expensive.

CHAPTER 9

RESERVOIR DYNAMIC BEHAVIOUR

Introduction and Commercial Application: The reservoir and well behaviour under dynamic conditions are key parameters in determining what fraction of the HCIIP will be produced to surface over the lifetime of the field, at what rates they will be produced and which unwanted fluids such as water are also produced. This behaviour will therefore dictate the revenue stream which the development will generate through sales of the hydrocarbons. The reservoir and well performance are linked to the surface development plan, and cannot be considered in isolation; different subsurface development plans will demand different surface facilities. The prediction of reservoir and well behaviour are therefore crucial components of field development planning, as well as playing a major role in reservoir management during production.

This section will consider the behaviour of the reservoir fluids in the bulk of the reservoir, away from the wells, to describe what controls the displacement of fluids towards the wells. Understanding this behaviour is important when estimating the RF for hydrocarbons, and the production forecast for both hydrocarbons and water. In, Chapter 10, the behaviour of fluid flow at the wellbore will be considered; this will influence the number of wells required for development, and the positioning of the wells.

9.1. THE DRIVING FORCE FOR PRODUCTION

Reservoir fluids (oil, water, gas) and the rock matrix are contained under high temperatures and pressures; they are compressed relative to their densities at standard temperature and pressure. Any reduction in pressure on the fluids or rock will result in an increase in the volume, according to the definition of *compressibility*. As discussed in Section 6.2, Chapter 6, isothermal conditions are assumed in the reservoir. Isothermal compressibility (c) is defined as

$$c = -\frac{1}{V}\frac{dV}{dP}$$

Applying this directly to the reservoir, when a volume of fluid (dV) is removed from the system through production, the resulting drop in pressure (dP) will be determined by the compressibility and volume (V) of the components of the reservoir system (fluids plus rock matrix). Assuming that the compressibility of the rock matrix is negligible (which is true for all but under-compacted, loosely

201

consolidated reservoir rocks and very low porosity systems)

$$dV = [c_o V_o + c_g V_o + c_w V_w]dP$$

where the subscripts refer to oil, gas and water. The term dV represents the underground withdrawal of fluids from the reservoir, which may be a combination of oil, water and gas. The exact compressibilities of the fluids depend on the temperature and pressure of the reservoir, but the following ranges indicate the relative compressibilities

$$c_o = 10 \times 10^{-6} \text{ to } 20 \times 10^{-6} \text{ psi}^{-1}$$
$$c_g = 500 \times 10^{-6} \text{ to } 1500 \times 10^{-6} \text{ psi}^{-1}$$
$$c_w = 3 \times 10^{-6} \text{ to } 5 \times 10^{-6} \text{ psi}^{-1}$$

Gas has a much higher compressibility than oil or water, and therefore expands by a relatively large amount for a given pressure drop. As underground fluids are withdrawn (i.e. production occurs), any free gas present expands readily to replace the voidage, with only a small drop in reservoir pressure. If only oil and water were present in the reservoir system, a much greater reduction in reservoir pressure would be experienced for the same amount of production.

The expansion of the reservoir fluids, which is a function of their volume and compressibility, act as a source of drive energy which can act to support *primary production* from the reservoir. Primary production means using the natural energy stored in the reservoir as a drive mechanism for production. *Secondary recovery* would imply adding some energy to the reservoir by injecting fluids such as water or gas, to help to support the reservoir pressure as production takes place.

Figure 9.1 shows how the expansion of fluids occurs in the reservoir to replace the volume of fluids produced to the surface during production.

The relationship between the underground volumes (measured in reservoir barrels) and the volumes at surface conditions is discussed in Section 6.2, Chapter 6. The relationships were denoted by

		Typical range
Oil formation volume factor	B_o (rb/stb)	1.1–2.0
Gas formation volume factor	B_g (rb/scf)	0.002–0.0005
Water formation volume factor	B_w (rb/stb)	1.0–1.1

One additional contribution to drive energy is by *pore compaction*, introduced in Section 6.2, Chapter 6. As the pore fluid pressure reduces due to production the grain to grain stress increases, which leads to the rock grains crushing closer together, thereby reducing the remaining pore volume, and effectively adding to the drive energy. The effect is usually small (less than 3% of the energy contributed by fluid expansion), but can lead to reservoir compaction and surface subsidence in cases where the pore fluid pressure is dropped considerably and the rock grains are loosely consolidated.

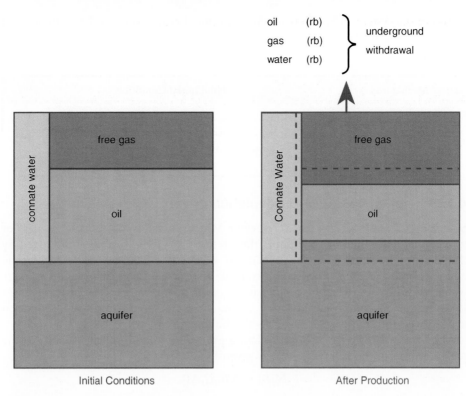

Figure 9.1   Expansion of fluids to replace produced volumes.

Reservoir engineers describe the relationship between the volume of fluids produced, the compressibility of the fluids and the reservoir pressure using *material balance* techniques. This approach treats the reservoir system like a tank, filled with oil, water, gas and reservoir rock in the appropriate volumes, but without regard to the distribution of the fluids (i.e. the detailed distribution or movement of fluids inside the system). Material balance uses the PVT properties of the fluids described in Section 6.2.6, Chapter 6, and accounts for the variations of fluid properties with pressure. The technique is firstly useful in predicting how reservoir pressure will respond to production. Secondly, material balance can be used to reduce uncertainty in volumetrics by measuring reservoir pressure and cumulative production during the producing phase of the field life. An example of the simplest material balance equation for an oil reservoir above the bubble point will be shown in the next section.

9.2. Reservoir Drive Mechanisms

The previous section showed that the fluids present in the reservoir, their compressibilities, and the reservoir pressure all determine the amount of energy stored in the system. Three sets of the fluid initial conditions can be distinguished for an oil, and reservoir and production behaviour may be characterised in each case

Drive Mechanism	Fluid Initial Condition
Solution gas drive (or depletion drive)	Undersaturated oil (no gas cap)
Gas cap drive	Saturated oil with a gas cap
Water drive via injection or with a large underlying aquifer	Saturated or undersaturated oil

9.2.1. Solution gas drive (or depletion drive)

Solution gas drive occurs in a reservoir which contains no initial gas cap or underlying active aquifer to support the pressure and therefore oil is produced by the driving force due to the expansion of oil and connate water, plus any compaction drive. The contribution to drive energy from compaction and connate water is small, so the oil compressibility initially dominates the drive energy. Because the oil compressibility itself is low, pressure drops rapidly as production takes place, until the pressure reaches the bubble point.

The material balance equation relating produced volume of oil (N_p stb) to the pressure drop in the reservoir (ΔP) is given by

$$N_p B_o = N B_{oi} C_e \Delta P$$

where B_o is the oil formation volume factor at the reduced reservoir pressure (rb/stb); B_{oi} the oil formation volume factor at the original reservoir pressure (rb/stb); C_e the volume averaged compressibility of oil, connate water and rock (psi^{-1}); N the STOIIP (stb).

Once the bubble point is reached, solution gas starts to become liberated from the oil, and since the liberated gas has a high compressibility, the rate of decline of pressure per unit of production slows down.

Once the liberated gas has overcome a critical gas saturation in the pores, below which it is immobile in the reservoir, it can either migrate to the crest of the reservoir under the influence of buoyancy forces, or move toward the producing wells under the influence of the hydrodynamic forces caused by the low pressure created at the producing well. In order to make use of the high compressibility of the gas, it is preferable that the gas forms a *secondary gas cap* and contributes to the drive energy. This can be encouraged by reducing the pressure sink at the producing wells (which means less production per well) and by locating the producing wells away from the crest of the field. In a steeply dipping field, wells would be located down-dip. However, in a field with low dip, the wells must be perforated as low as possible to keep away from a secondary gas cap (Figure 9.2). The problem of water coning, discussed in Section 10.2, Chapter 10 is a constraint on just how low down the perforation can be placed without producing excessive amounts of water.

The characteristic production profile for a reservoir developed by solution gas drive is shown in Figure 9.3.

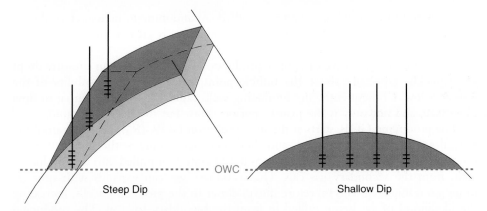

Figure 9.2   Location of wells for solution gas drive.

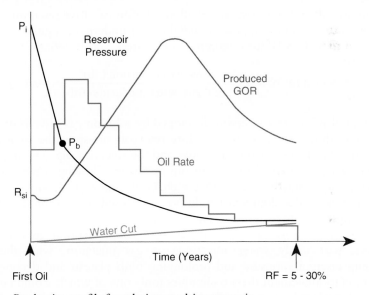

Figure 9.3   Production profile for solution gas drive reservoir.

As for all production profiles, there are three distinct phases, defined by looking at the oil production rate (for an oil field). After the *first production date*, there is a *build-up* period, during which the development wells are being drilled and brought on stream, and its shape is dependent on the drilling schedule. Once the *plateau* is reached, the facilities are filled and any extra production potential from the wells is choked back. The facilities are usually designed for a plateau rate which provides an optimum offtake from the field, where the optimum is a balance between producing oil as early as possible and avoiding unfavourable displacement in the reservoir, caused by producing too fast, and thereby losing UR. Typical production rates during the plateau period vary between 2 and 5% of the STOIIP per year. Once the well potential can no longer sustain the plateau oil rate, the *decline* period begins and continues until the *abandonment* rate is reached. Abandonment, better referred to as decommissioning, occurs when the cost of production is greater than the revenues from the production.

In the solution gas drive case, once production starts the reservoir pressure drops very quickly, especially above the bubble point, since the compressibility of the system is low. Consequently, the producing wells rapidly lose the potential to flow to surface, and not only is the plateau period short, but the decline is rapid.

The producing GOR starts at the initial solution GOR (R_{si}), decreases until the critical gas saturation is reached, and then increases rapidly as the liberated gas is produced into the wells, either directly as it is liberated, or pulled into the producing wells from the secondary gas cap. The secondary gas cap expands with time, as more gas is liberated, and therefore moves closer to the producing wells, increasing the likelihood of gas being pulled in from the secondary gas cap. The producing GOR may decline in later years as the remaining volume of gas in the reservoir diminishes.

Commonly the *water cut* remains small in solution gas drive reservoirs, assuming that there is little pressure support provided by the underlying aquifer. Water cut is also referred to as *BS&W* (base sediment and water), and is defined as

$$\text{Water cut (or BS\&W)} = \frac{\text{Water production (stb)}}{\text{Oil plus water production (stb)}} \times 100(\%)$$

The typical RF from a reservoir developed by solution gas drive is in the range 5–30%, depending largely on the absolute reservoir pressure, the solution GOR of the crude, the abandonment conditions and the reservoir dip. The upper end of this range may be achieved by a high dip reservoir (allowing segregation of the secondary gas cap and the oil), with a high GOR, light crude and a high initial reservoir pressure. Abandonment conditions are caused by high producing GORs and lack of reservoir pressure to sustain production.

This rather low RF may be boosted by implementing *secondary recovery* techniques, particularly water injection, or gas injection, with the aim of maintaining reservoir pressure and prolonging both plateau and decline periods. The decision to implement these techniques (only one of which would be selected) is both technical and economic. Technical considerations would be the external supply of gas, and the feasibility of injecting the fluids into the reservoir. Figure 9.4 indicates how these techniques may be applied. Note again, that it is unlikely that both gas and water injection would be simultaneously adopted – one or the other secondary recovery technique would normally be chosen.

9.2.2. Gas cap drive

The initial condition required for gas cap drive is an initial gas cap. The high compressibility of the gas provides drive energy for production, and the larger the gas cap, the more energy is available. The well positioning follows the same reasoning as for solution gas drive; the objective being to locate the producing wells

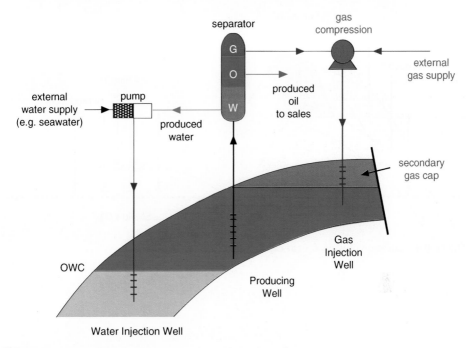

Figure 9.4 Secondary recovery: gas or water injection schemes.

and their perforations as far away from the gas cap (which will expand with time) as possible (Figure 9.5), but not so close to the OWC to allow significant water production via coning (see Section 10.2, Chapter 10).

Compared to the solution gas drive case, the typical production profile for gas cap drive shows a much slower decline in reservoir pressure, due to the energy provided by the highly compressible gas cap, resulting in a more prolonged plateau and a slower decline (Figure 9.6). The producing GOR increases as the expanding gas cap approaches the producing wells, and gas is coned or cusped into the producers. Again, it is assumed that there is negligible aquifer movement, and water cut remains low (in the order of 10% at the end of field life). Typical RFs for gas cap drive are in the range 20–60%, influenced by the field dip and the gas cap size. A small gas cap would be 10% of the oil volume (at reservoir conditions), whilst a large gas cap would be upwards of 50% of the oil volume. Abandonment conditions are caused by very high producing GORs, or lack of reservoir pressure to maintain production, and can be postponed by reducing the production from high GOR wells, or by *recompleting* these wells to produce further away from the gas cap. Recompletion of wells is further discussed in Section 10.7, Chapter 10.

Natural gas cap drive may be supplemented by *reinjection* of produced gas, with the possible addition of make-up gas from an external source. The gas injection well would be located in the crest of the structure, injecting into the existing gas cap.

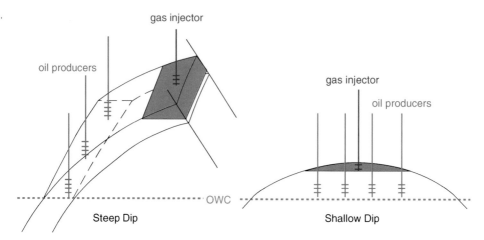

Figure 9.5   Location of wells for gas cap drive.

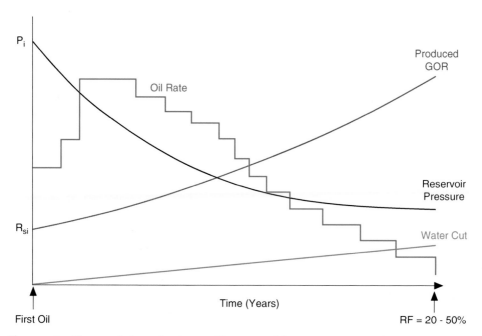

Figure 9.6   Characteristic production profile: gas cap drive.

9.2.3. Water drive

Natural water drive occurs when the underlying aquifer is both large (typically greater than ten times the oil volume) and the water is able to flow into the oil column, that is it has a *communication path* and *sufficient permeability*. If these conditions are satisfied, then once production from the oil column creates a pressure drop the aquifer responds by expanding, and water moves into the oil column to replace the voidage created by production. Since the water compressibility is low, the volume of water must be large to make this process effective, hence the need for the large connected aquifer. In this context, 'large' would be 10 to 100 × the volume of oil in place.

The prediction of the size and permeability of the aquifer is usually difficult, since there is typically little data collected in the water column; exploration and appraisal wells are usually targeted at locating oil. Hence the prediction of aquifer response often remains a major uncertainty during reservoir development planning. In order to see the reaction of an aquifer, it is necessary to produce from the oil column, and measure the response in terms of reservoir pressure and fluid contact movement; use is made of the material balance technique to determine the contribution to pressure support made by the aquifer. Typically 5% of the STOIIP must be produced to measure the response; this may take a number of years.

Water drive may be imposed by *water injection* into the reservoir, preferably by injecting into the water column to avoid by-passing down-dip oil (Figure 9.7). If the permeability in the water leg is significantly reduced due to compaction or diagenesis, it may be necessary to inject into the oil column. Once water injection is adopted, the potential effect of any natural aquifer is usually negated. Clearly if it were possible to predict the natural aquifer response at the development planning stage, the decision to install water injection facilities would be made easier. A common solution is to initially produce the reservoir using natural depletion, and to install water injection facilities in the event of little aquifer support.

The aquifer response (or impact of the water injection wells) may maintain the reservoir pressure close to the initial pressure, providing a long plateau period and slow decline of oil production (Figure 9.8). The producing GOR may remain approximately at the solution GOR if the reservoir pressure is maintained above the bubble point. The outstanding feature of the production profile is the large *increase in water cut* over the life of the field, which is usually the main reason for

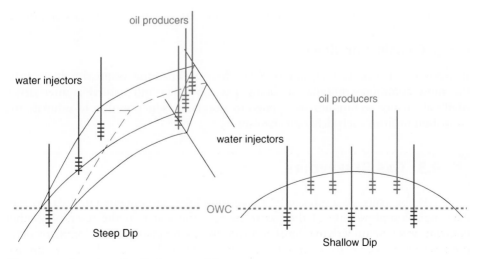

Figure 9.7 Location of wells for water drive.

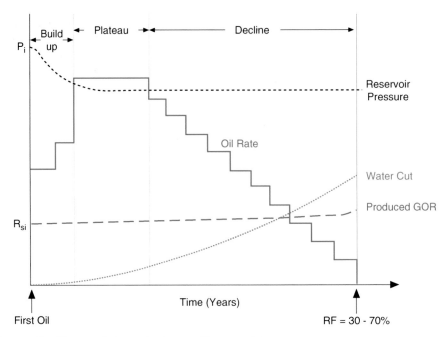

Figure 9.8 Characteristic production profile: water drive.

abandonment. Water cut may exceed 90% in the final part of the field life. This is important to the process engineers, who need to design a facility capable of handling large water throughputs in later field life. As water cut increases, so oil production typically declines; a constant gross liquids (oil plus water) production may be maintained.

The RF is in the range 30–70%, depending on the strength of the natural aquifer, or the efficiency with which the injected water sweeps the oil. The high RF is an incentive for water injection into reservoirs which lack natural water drive.

9.2.4. Combination drive

It is possible that more than one of these drive mechanisms occur simultaneously; the most common combination being gas cap drive and natural aquifer drive. Material balance techniques are applied to historic production data to estimate the contribution from each drive mechanism.

9.3. GAS RESERVOIRS

Gas reservoirs are produced by expansion of the gas contained in the reservoir. The high compressibility of the gas relative to the water in the reservoir (either connate water or underlying aquifer) make the gas expansion the dominant drive mechanism. Relative to oil reservoirs, the material balance calculation for gas reservoirs is rather simple. A major challenge in gas field development is to ensure a long sustainable plateau (typically 10 years) to attain a good sales price for the gas; the customer usually requires a reliable supply of gas at an agreed rate over many years. The RF for gas reservoirs depends on how low the abandonment pressure can be reduced, which is why compression facilities are often provided on surface. Typical RFs are in the range 50–80%.

9.3.1. Major differences between oil and gas field development

The main differences between oil and gas field development are associated with

- the economics of transporting gas
- the market for gas
- product specifications
- the efficiency of turning gas into energy.

Per unit of energy generated, the transportation of gas is significantly more expensive than transporting oil, due to the volumes required to yield the same energy. In other words, the *energy density of gas* is low compared to oil. On a calorific basis approximately 6000 scf of gas is equivalent to one barrel (5.6 scf) of oil. The compression costs of transporting gas at sufficient pressure to make transportation more economic are also high. This means that unless there are sufficiently large quantities of gas in the reservoir to take advantage of economies of scale, development may be uneconomic.

For an offshore field requiring significant infrastructure for development, recoverable volumes of less than 0.5 trillion scf (Tcf) are typically uneconomic to develop. This would equate to an oil field with recoverable reserves of approximately 80 MMstb. If close to existing offshore infrastructure, this threshold is closer to 50 Bcf.

For the above reasons, gas is typically economic to develop only if it can be used locally, that is if a local demand exists. The exception to this is where a sufficient quantity of gas exists to provide the economy of scale to make transportation of gas or liquefied gas attractive. As a guide, approximately 5 Tcf of recoverable gas would be required to justify building a liquefied natural gas (LNG) plant. Globally there are around 30 such plants, but an example would be the LNG plant in Malaysia which liquefies gas and transports it by refrigerated tanker to Japan. The investment capital required for an LNG plant is very large; typically in the order of $5 billion.

Whereas a 'spot market' has always existed for oil, gas sales traditionally require a contract to be agreed between the producer and a customer. This forms an important part of gas field development planning, since the price agreed between producer and customer will vary, and will depend on the quantity supplied, the plateau length and the flexibility of supply. Whereas oil price is approximately the same across the globe, gas prices can vary very significantly (by a factor of two or more) from region to region.

When a customer agrees to purchase gas, product quality is specified in terms of the calorific value of the gas, measured by the Wobbe Index (WI) (MJ/m^3 or Btu/scf), the hydrocarbon dew point and the water dew point, and the fraction of other gases such as N_2, CO_2, H_2S. The WI specification ensures that the gas the customer receives has a predictable calorific value and hence predictable burning characteristics. If the gas becomes lean, less energy is released, and if the gas becomes too rich there is a risk that the gas burners 'flame out'. Water and hydrocarbon dew points (the pressure and temperature at which liquids start to drop out of the gas) are specified to ensure that over the range of temperature and pressure at which the gas is handled by the customer, no liquids will drop out (these could cause possible slugging, corrosion and/or hydrate formation).

H_2S is undesirable because of its toxicity and corrosive properties. CO_2 can cause corrosion in the presence of water, and N_2 simply reduces the calorific value of the gas as it is inert.

9.3.2. Gas sales profiles; influence of contracts

If the gas purchaser is a company which distributes gas to domestic and industrial end users, he typically wants the producer to provide

- a guaranteed minimum quantity of gas for as long a duration as possible (for ease of planning and the comfort of being able to guarantee supply to the end user) *and*
- peaks in production when required (e.g. when the weather unexpectedly turns cold).

The better the producer can meet these two requirements, the higher the price paid by the purchaser is likely to be.

In contrast to an oil production profile, which typically has a plateau period of 2–5 years, a gas field production profile will typically have a much longer plateau period, producing around 2/3 of the reserves on plateau production in order to satisfy the needs of the distribution company to forecast their supplies. Figure 9.9 compares typical oil and gas field production profiles.

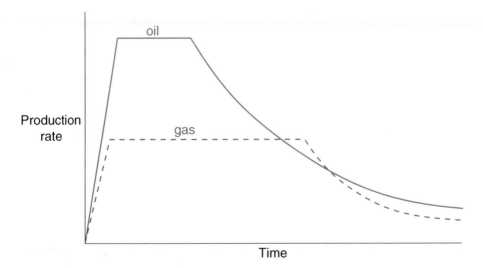

Figure 9.9   Comparison of typical oil and gas field production profiles.

If the distribution of gas in a country is run by a nationalised or state owned company, there is effectively a monopoly on this service, and prices for gas distributed through a grid system will have to be negotiated with the distribution company. If the market for distribution is not regulated then opportunities arise to sell gas to other customers and directly to consumers, perhaps including a tariff payment for transport through a national grid.

This situation has arisen in the UK where competition for gas sales has been encouraged. Gas producers can enter into direct agreements with consumers (ranging from power stations to domestic users), using the national distribution grid if necessary. Such a de-regulated market increases competition between the distribution companies and thus regulates prices.

When a contract is agreed with a consumer, some delivery quantities will usually be specified such as

Daily contract quantity (DCQ)	The daily production which will be supplied; usually averaged over a period such as a quarter year.
Swing factor	The amount by which the supply must exceed the DCQ if the customer so requests (e.g. $1.4 \times DCQ$).
Take or pay agreement	If the buyer chooses not to accept a specified quantity, he will pay the supplier anyway.
Penalty clause	The penalty which the supplier will pay if he fails to deliver the quantity specified within the DCQ and swing factor agreements.

Figure 9.10 shows the relationship between DCQ and the swing factor. If, for example a swing factor of 1.4 is agreed, then on any one day the customer may

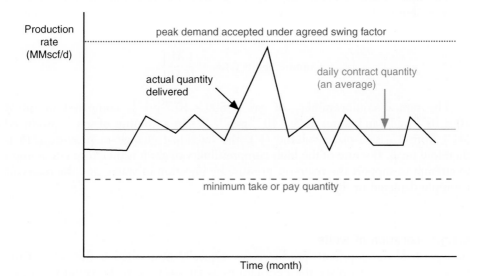

Figure 9.10 Typical delivery quantities specified in a gas sales contract.

request the producer to provide 1.4 times the DCQ. This means that the producer has to be confident that there is sufficient well potential and transport capacity to meet this demand, otherwise a penalty will be incurred. For most of the time this means that the producer is providing a production potential (sometimes called deliverability) which is not being realised. As compensation to the producer for investing in additional capital to provide this level of redundancy, a higher gas price would be expected.

9.3.3. Subsurface development of gas reservoirs

One of the major differences in fluid flow behaviour for gas fields compared to oil fields is the mobility difference between gas and oil or water. Recall that *mobility* is an indicator of how fast fluid will flow through the reservoir, and is defined as

$$\text{Mobility} = \frac{k}{\mu}$$

Permeability (k) is a rock property, whilst viscosity (μ) is a fluid property. A typical oil viscosity is 0.5 cP, whilst a typical gas viscosity is 0.01 cP, water being around 0.3 cP. For a given reservoir, gas is therefore around two orders of magnitude more mobile than oil or water. In a gas reservoir underlain by an aquifer, the gas is highly mobile compared to the water and flows readily to the producers, provided that the permeability in the reservoir is continuous. For this reason, production of gas with zero water cut is common, at least in the early stages of development when the perforations are distant from the gas–water contact.

The other main physical property of gas which distinguishes it from oil is its compressibility; the fractional change in volume (V) per unit of change in pressure (P) at constant temperature (T). Recall that

$$\text{Compressibility } (c) = -\frac{1}{V}\frac{\delta V}{\delta P}\bigg|_{T}$$

The typical compressibility of gas is $500 \times 10^{-6}\,\text{psi}^{-1}$, compared to oil at $10 \times 10^{-6}\,\text{psi}^{-1}$ and water at $3 \times 10^{-6}\,\text{psi}^{-1}$. When a volume of gas is produced (δV) from a gas-in-place volume (V), the fractional change in pressure (δP) is therefore small. Because of the high compressibility of gas it is therefore uncommon to attempt to support the reservoir pressure by injection of water, and the reservoir is simply depleted or 'blown down'.

9.3.3.1. Location of wells

In a gas field development, producers are typically positioned at the crest of the reservoir, in order to place the perforations as far away from the rising gas–water contact as possible.

9.3.3.2. Movement of gas–water contact during production

As the gas is produced, the pressure in the reservoir drops, and the aquifer responds to this by expanding and moving into the gas column. As the gas–water contact moves up, the risk of coning water into the well increases, hence the need to initially place the perforations as high as possible in the reservoir.

The above descriptions may suggest that rather few wells, placed in the crest of the field are required to develop a gas field. There are various reasons why gas field development requires additional wells

- the need to provide excess deliverability to meet swing requirements as agreed in the sales contract
- the reservoir will not be homogeneous and certain areas will require closer well spacing to drain tighter parts of the reservoir in the same time frame as the more permeable areas are drained
- the reservoir may not be continuous and dedicated producers will be required to drain isolated fault blocks
- the reservoir may have a flat structure and therefore it may be impossible to place perforations at sufficient height above the water contact to avoid water coning. In this case, a lower production rate is necessary, implying more wells to meet the required production rate.

9.3.3.3. Pressure response to production

The primary drive mechanism for gas field production is the expansion of the gas contained in the reservoir. Relative to oil reservoirs, the material balance calculation for gas reservoirs is rather simple; the RF is linked to the drop in reservoir pressure in an almost linear manner. The non-linearity is due to the changing z-factor (introduced in Section 6.2.4, Chapter 6) as the pressure drops. A plot of (P/z) against the RF is linear if aquifer influx and pore compaction are negligible. The material balance may therefore be represented by the following plot (often called the 'P over z' plot) (Figure 9.11).

The subscript 'i' refers to the initial pressure, and the subscript 'ab' refers to the abandonment pressure; the pressure at which the reservoir can no longer produce gas to the surface. If the abandonment conditions can be predicted, then an estimate of the RF can be made from the plot. G_p is the cumulative gas produced, and G the GIIP. This is an example of the use of PVT properties and reservoir pressure data being used in a material balance calculation as a predictive tool.

From the above plot, it can be seen that the RF for gas reservoirs depends on how low an abandonment pressure can be achieved. To produce at a specified delivery pressure, the reservoir pressure has to overcome a series of pressure drops; the drawdown pressure (refer to Figure 10.2), and the pressure drops in the tubing, processing facility and export pipeline (refer to Figure 10.13). To improve recovery of gas, compression facilities are often provided on surface to boost the pressure to overcome the pressure drops in the export line and meet the delivery pressure specified.

Typical RFs for gas field development are in the range 50–80%, depending on the continuity and quality of the reservoir, and the amount of compression installed (i.e. how low an abandonment pressure can be achieved).

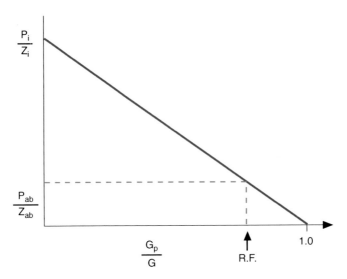

Figure 9.11   The 'P over z' plot for gas reservoirs.

9.3.4. Surface development for gas fields

The amount of processing required in the field depends on the composition of the gas and the temperature and pressure to which the gas will be exposed during transportation. The process engineer is trying to avoid liquid drop-out during transportation, since this may cause *slugging*, *corrosion* and possibly *hydrate formation* (refer to Section 11.1.3, Chapter 11). For dry gases (refer to Section 6.2.2, Chapter 6) the produced fluids are often exported with very little processing. Wet gases may be dried of the heavier hydrocarbons by dropping the temperature and pressure through a Joule–Thompson expansion valve. Gas containing water vapour may be dried by passing the gas through a molecular sieve, or through a glycol-contacting tower. Hydrate inhibition may be achieved by glycol injection.

One of the main surface equipment items typically required for gas fields is compression, which is installed to allow a low reservoir pressure to be attained. Gas compression takes up a large space and is expensive. If gas compression is not initially required on a platform, then its installation is usually delayed until it becomes necessary. This reduces the initial capital investment and capital exposure. Figure 9.12 indicates when gas compression is typically installed.

A comfortable margin is maintained between the flowing tubing head pressure (FTHP, downstream of compression) and the minimum pressure required for export, since the penalties for not meeting contract quantities can be severe. The decision not to install a fourth stage of compression in the above example is dictated by economics. During the final part of the pressure decline above, the field production is of course also declining.

Another method of maintaining production potential from the field is to drill more wells, and it is common for wells to be drilled in batches, just as the compression is added in stages, to reduce early expenditure.

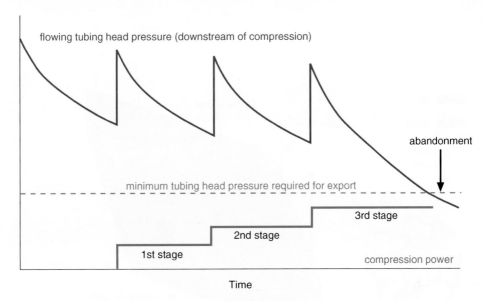

Figure 9.12   Installing compression in stages on a gas field.

9.3.5. Alternative uses for gas

A gas discovery may be a useful source of energy for supporting pressure in a neighbouring oil field, or for a miscible gas drive. Selling the gas is not the only method of exploiting a gas field. Gas reservoirs may also be used for storage of gas. For example, a neighbouring oil field may be commercial to develop for its oil reserves, but the produced associated gas may not justify a dedicated export pipeline. This associated gas can be injected into a gas reservoir, which can act as a storage facility, and possibly back-produced at a later date if sufficient additional gas is discovered to justify building a gas export system. As gas supplies in Western Europe become more scarce, such *gas storage* projects have become more common.

9.4. FLUID DISPLACEMENT IN THE RESERVOIR

The RFs for oil reservoirs mentioned in the previous section varied from 5 to 70%, depending on the drive mechanism. The explanation as to why the other 95 to 30% remains in the reservoir is not only due to the abandonment necessitated by lack of reservoir pressure or high water cuts, but also to the displacement of oil in the reservoir.

Figure 9.13 indicates a number of situations in which oil is left in the reservoir, using a water drive reservoir as an example.

On a macroscopic scale, oil is left behind due to *by-passing*; the oil is displaced by water in the more permeable parts of the reservoir, leaving oil in the less permeable areas at the initial oil saturation. This by-passing can occur in three dimensions. In the areal plane oil in lenses of tighter sands remains unswept. In the vertical plane, oil in the tighter layers is displaced less quickly than the oil in the more permeable

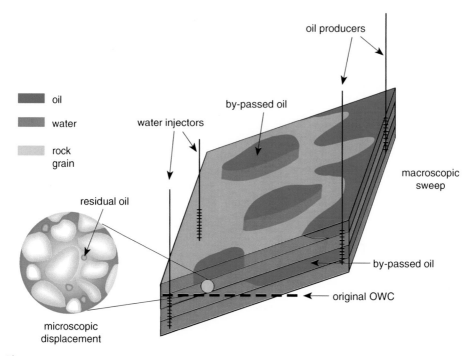

Figure 9.13   Oil remaining in the reservoir at abandonment.

layers, and if the wells are abandoned due to high water cut arising from water breakthrough in the permeable layers, then oil will remain in the yet unswept parts of the less permeable layers.

The *macroscopic sweep efficiency* is the fraction of the total reservoir which is swept by water (or by gas in the case of gas cap drive). This will depend on the *reservoir quality* and *continuity*, and the *rate* at which the displacement takes place. At higher rates, displacement will take place even more preferentially in the high permeability layers, and the macroscopic displacement efficiency will be reduced.

This is why an *offtake limit* on the plateau production rate is often imposed, to limit the amount of by-passed oil, and increase the macroscopic sweep efficiency.

On a microscopic scale (the inset represents about 1–$2\,\text{mm}^2$), even in parts of the reservoir which have been swept by water, some oil remains as *residual oil*. The surface tension at the oil–water interface is so high that as the water attempts to displace the oil out of the pore space through the small capillaries, the continuous phase of oil breaks up, leaving small droplets of oil (snapped off, or capillary trapped oil) in the pore space. Typical *residual oil saturation* (S_{or}) is in the range 10–40% of the pore space, and is higher in tighter sands, where the capillaries are smaller.

The *microscopic displacement efficiency* is the fraction of the oil which is recovered in the swept part of the reservoir. If the initial oil saturation is S_{oi}, then

$$\text{Microscopic displacement efficiency} = \frac{S_{oi} - S_{or}}{S_{oi}} \times 100(\%)$$

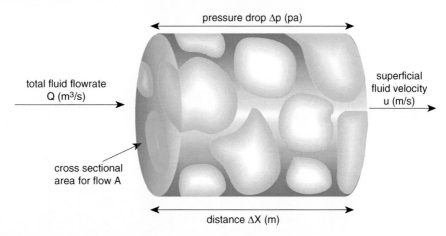

Figure 9.14 Single fluid flowing through a section of reservoir rock.

This must be combined with the macroscopic sweep efficiency to determine the RF for oil (in this example)

RF = Macroscopic displacement efficiency × microscopic sweep efficiency

On a microscopic scale, the most important equation governing fluid flow in the reservoir is *Darcy's law*, which was derived from the following situation (Figure 9.14).

For a single fluid flowing through a section of reservoir rock, Darcy showed that the superficial velocity of the fluid (u) is proportional to the pressure drop applied (the hydrodynamic pressure gradient), and inversely proportional to the viscosity of the fluid. The constant of proportionality is called the *absolute permeability* (k_{abs}) which is a rock property, and is dependent on the pore size distribution. The superficial velocity is the average flowrate per unit area.

$$\mu = \frac{Q}{A} = \frac{k_{abs}}{\mu}\frac{\Delta P}{\Delta X} \quad \text{(m/sec) units of } k_{abs} \text{ (Darcy) or (m}^2)$$

The field unit for permeability is the Darcy (D) or millidarcy (mD). For clastic oil reservoirs, a good permeability would be greater than 0.1 D (100 mD), whilst a poor permeability would be less than 0.01 D (10 mD). For practical purposes, the millidarcy is commonly used (1 mD = 10^{-15} m^2). For gas reservoirs 1 mD would be a reasonable permeability; because the viscosity of gas is much lower than that of oil, this permeability would yield an acceptable flowrate for the same pressure gradient. Typical fluid velocities in the reservoir are less than 1 m per day.

The above experiment was conducted for a single fluid only. In hydrocarbon reservoirs there is always connate water present, and commonly two fluids are competing for the same pore space (e.g. water and oil in water drive). The permeability of one of the fluids is then described by its *relative permeability* (k_r), which is a function of the saturation of the fluid. Relative permeabilities are measured in the laboratory on reservoir rock samples using reservoir fluids.

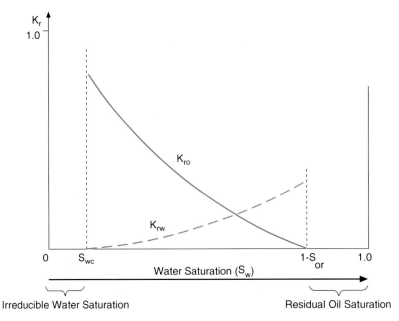

Figure 9.15   Relative permeability curve for oil and water.

The following diagram shows an example of a relative permeability curve for oil and water. For example, at a given water saturation (S_w), the permeability to water (k_w) can be determined from the absolute permeability (k) and the relative permeability (k_{rw}) as follows (Figure 9.15)

$$k_w = kk_{rw}$$

The mobility of a fluid is defined as the ratio of its permeability to viscosity

$$\text{Mobility} = \frac{kk_r}{\mu}$$

When water is displacing oil in the reservoir, the mobility ratio determines which of the fluids moves preferentially through the pore space. The *mobility ratio* for water displacing oil is defined as

$$\text{Mobility ratio } (M) = \frac{k_{rw}/\mu_w}{k_{ro}/\mu_o}$$

If the mobility ratio is greater than 1.0, then there will be a tendency for the water to move preferentially through the reservoir, and give rise to an *unfavourable displacement* front which is described as viscous fingering. If the mobility ratio is less than unity, then one would expect *stable displacement*, as shown in Figure 9.16. The mobility ratio may be influenced by altering the fluid viscosities, and this is further discussed in Section 9.8, when enhanced oil recovery (EOR) is introduced.

Unstable displacement is clearly less preferable, since a mixture of oil and water is produced much earlier than in the stable displacement situation, and some oil may

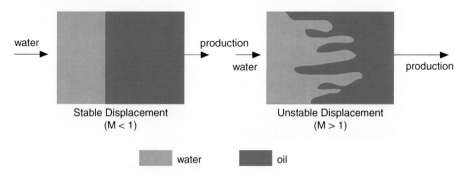

Figure 9.16 Stable and unstable displacement in the horizontal plane.

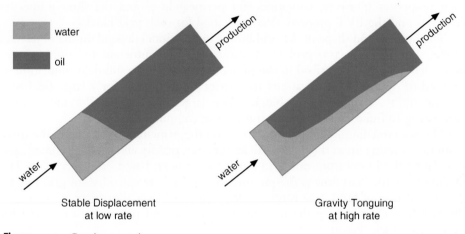

Figure 9.17 Gravity tonguing.

be left unrecovered at the abandonment condition which may be dictated by a maximum water cut.

So far we have looked only at the *viscous forces* (which are a measure of the resistance to flow) acting on reservoir fluids. Another important force which determines flow behaviour is the *gravity force*. The effect of the gravity force is to separate fluids according to their density. During displacement in the reservoir, both gravity forces and viscous forces play a major role in determining the shape of the displacement front. Consider the following example of water displacing oil in a dipping reservoir. Assuming a mobility ratio less than 1.0, the viscous forces will encourage water to flow through the reservoir faster than oil, whilst the gravity forces will encourage water to remain at the lowest point in the reservoir (Figure 9.17).

At low injection rates the displacement is stable; the gravity forces are dominating the viscous forces. At higher rates of injection, the viscous forces are dominating, and the water underruns the oil, forming a so-called 'gravity tongue'. This is a less favourable situation, since the produced fluid will be a mixture of oil and water long before all of the oil is produced. If high water cut is an abandonment constraint this could lead to a reduction in recovery. The steeper the dip of the reservoir, the more influence the gravity force will have, meaning that high dip reservoirs are more likely to yield stable displacement. The above is an example of a *rate dependent process*, in which the displacement rate affects the shape of the displacement front, and possibly the UR. Physical effects such as this are the reason for limiting the offtake rate from producing fields.

9.5. Reservoir Simulation

Reservoir simulation is a technique in which a computer-based mathematical representation of the reservoir is constructed and then used to predict its dynamic behaviour. The reservoir is gridded up into a number of grid blocks. The reservoir rock properties (porosity, saturation and permeability), and the fluid properties (viscosity and the PVT properties) are specified for each grid block.

The number and shape of the grid blocks in the model depend on the objectives of the simulation. A 100 grid block model may be sufficient to confirm rate-dependent processes described in the previous section, but a full field simulation to be used to optimise well locations and perforation intervals for a large field may contain up to a million grid blocks. The larger the model, the more time-consuming to build, and slower to run on the computer.

The reservoir simulation operates based on the principles of balancing the three main forces acting upon the fluid particles (viscous, gravity and capillary forces), and calculating fluid flow from one grid block to the next, based on Darcy's law. The driving force for fluid flow is the pressure difference between adjacent grid blocks. The calculation of fluid flow is repeatedly performed over short time steps, and at the end of each time step the new fluid saturation and pressure is calculated for every grid block (Figure 9.18).

The amount of detail input, and the type of simulation model depend on the issues to be investigated, and the amount of data available. At the exploration and appraisal stage it would be unusual to create a simulation model, since the lack of data make simpler methods cheaper and as reliable. Simulation models are typically constructed at the field development planning stage of a field life, and are continually updated and increased in detail as more information becomes available.

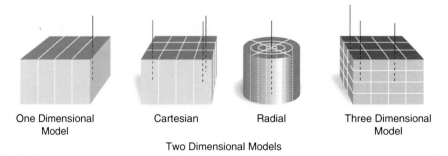

| One Dimensional | Cartesian | Radial | Three Dimensional |
| Model | | | Model |

Two Dimensional Models

Figure 9.18   Typical grid block configurations for reservoir simulation.

At the field development planning stage, reservoir simulation may be used to look at questions such as

- most suitable drive mechanism (gas injection, water injection)
- number and location of producers and injectors
- rate dependency of displacement and RF
- estimating RF and predicting production forecast for a particular development proposal
- reservoir management policy (offtake rates, perforation intervals).

Once production commences, data such as reservoir pressure, cumulative production, GOR, water cut and fluid contact movement are collected, and may be used to 'history match' the simulation model. This entails adjusting the reservoir model to fit the observed data. The updated model may then be used for a more accurate prediction of future performance. This procedure is cyclic, and a full field reservoir simulation model will be updated whenever a significant amount of new data becomes available (say, every 2–5 years).

9.6. Estimating the Recovery Factor

Recall that the RF defines the relationship between the HCIIP and the UR for the field

$$\text{Ultimate recovery} = \text{HCIIP} \times \text{recovery factor} \quad \text{(stb) or (scf)}$$

$$\text{Reserves} = \text{UR} - \text{cumulative production} \quad \text{(stb) or (scf)}$$

Section 9.2 indicated the ranges of RFs which can be anticipated for different drive mechanisms, but these were too broad to use when trying to establish a range of RFs for a specific field. The main techniques for estimating the RF are

- field analogues
- analytical models (displacement calculations, material balance)
- reservoir simulation.

These are listed in order of increasing complexity, reliability, data input requirements and effort required.

Field analogues should be based on reservoir rock type (e.g. tight sandstone, fractured carbonate), fluid type and environment of deposition. This technique should not be overlooked, especially where little information is available, such as at the exploration stage. Summary charts such as the one shown in Figure 9.19 may be used in conjunction with estimates of macroscopic sweep efficiency (which will depend on well density and positioning, reservoir homogeneity, offtake rate and fluid type) and microscopic displacement efficiency (which may be estimated if core measurements of residual oil saturation are available).

Analytical models using classical reservoir engineering techniques such as material balance, aquifer modelling and displacement calculations can be used in combination with field and laboratory data to estimate RFs for specific situations. These methods are most applicable when there is limited data, time and resources,

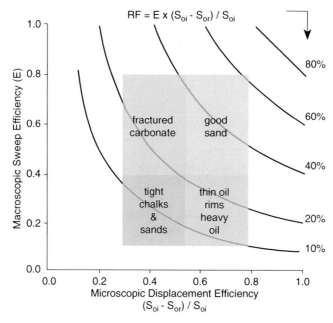

Figure 9.19   Estimating recovery factor by analogue.

and would be sufficient for most exploration and early appraisal decisions. However, when the development planning stage is reached, it is becoming common practice to build a reservoir simulation model, which allows more sensitivities to be considered in a shorter time frame. The typical sorts of questions addressed by reservoir simulations are listed in Section 9.5.

When estimating the RF, it is important to remember that a range of estimates should be provided as input to the calculation of UR, to reflect the uncertainty in the value.

9.7. ESTIMATING THE PRODUCTION PROFILE

The production profile for oil or gas is the only *source of revenue* for most projects, and making a production forecast is of key importance for the economic analysis of a proposal (e.g. FDP, incremental project). Typical shapes of production profile for the main drive mechanisms were discussed in Section 9.2, but this section will provide some guidelines on how to derive the rate of build-up, the magnitude and duration of the plateau, the rate of decline and the abandonment rate.

Figure 9.20 shows the same UR (area under the curve), produced in three different production profiles.

In the *build-up period*, profile A illustrates a gradual increase of production as the producing wells are drilled and brought on steam; the duration of the build-up period is directly related to the drilling schedule. Profile B, in which some wells have been *pre-drilled* starts production at plateau rate. The advantage of *pre-drilling* is

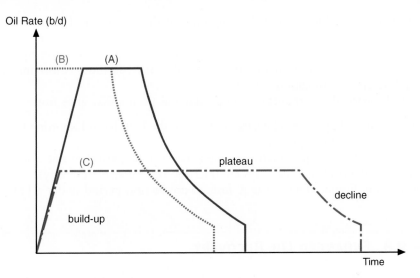

Figure 9.20   Various production profiles for the same UR.

to advance the production of oil, which improves the project cashflow, but the disadvantages are the that the cost of drilling has been advanced, and that the opportunity has been lost to gather early production information from the first few wells, which may influence the location of subsequent wells. Economic criteria (the impact on the profitability of the project) are used to decide whether to pre-drill.

The *plateau production* rates for cases A and B differ significantly from that in case C, which has a lower but longer plateau. The advantage of profile C is that it requires smaller facilities and probably less wells to produce the same UR. This advantage in reduced costs must be considered using economic criteria against the delayed production of oil (which is bad for the cashflow). One additional advantage of profile C is that the lower production rate, and therefore slower displacement in the reservoir, may improve the UR. This would be more likely in the case of unfavourable mobility ratios and low dip reservoirs where the gravity effects are smaller, as discussed in Section 9.4. The choice of plateau production rate is again an economic one, with the factors influencing the profitability being the timing of the oil production, the size and therefore cost of the facilities required and the potential for higher ultimate recoveries at lower offtake rates.

As a guideline, the plateau rate is usually between 2 and 5% of the STOIIP per year. The lower end of the range would apply to shallow dip reservoirs with an unfavourable mobility ratio, creating a rate-dependent displacement process.

Once the production potential of the producing wells is insufficient to maintain the plateau rate, the *decline period* begins. For an individual well in depletion drive, this commences as soon as production starts, and a plateau for the field can only be maintained by drilling more wells. Well performance during the decline period can be estimated by *decline curve analysis* which assumes that the decline can be described by a mathematical formula. Examples of this would be to assume an exponential decline with 10% decline per annum, or a straight line relationship between the cumulative oil production and the logarithm of the water cut. These assumptions become more robust when based on a fit to measured production data.

The most reliable way of generating production profiles, and investigating the sensitivity to well location, perforation interval, surface facilities constraints, etc., is through reservoir simulation.

Finally, *external* constraints on the production profile may arise from

- production ceilings (e.g. The Organization of the Petroleum Exporting Countries (OPEC) production quotas)
- host government requirements (e.g. generating long period of stable income)
- customer demand (e.g. gas sales contract for 10 year stable delivery)
- production licence duration (e.g. limited production period under a PSC).

9.8. Enhanced Oil Recovery

EOR techniques seek to produce oil which would not be recovered using the primary or secondary recovery methods discussed so far. Three categories of EOR exist

- *thermal techniques*
- *chemical techniques*
- *miscible processes.*

Thermal techniques are used to reduce the viscosity of heavy crudes, thereby improving the mobility, and allowing the oil to be displaced to the producers. This is the most common of the EOR techniques, and the most widely used method of heat generation is by injecting hot water or steam into the reservoir. This can be done in dedicated injectors (*hot water or steam drive*), or by alternately injecting into, and then producing from the same well (*steam soak*). A more ambitious method of generating heat in the reservoir is by igniting a mixture of the hydrocarbon gases and oxygen, and is called *in-situ combustion*.

Chemical techniques change the physical properties of either the displacing fluid, or of the oil, and comprise of *polymer flooding* and *surfactant flooding*.

Polymer flooding aims at reducing the amount of by-passed oil by increasing the viscosity of the displacing fluid, say water, and thereby improving the mobility ratio (*M*).

Recall that

$$\text{Mobility ratio } (M) = \frac{k_{rw}/\mu_w}{k_{ro}/\mu_o}$$

This technique is suitable where the natural mobility ratio is greater than 1.0. Polymer chemicals such as polysaccharides are added to the injection water.

Surfactant flooding is targeted at reducing the amount of residual oil left in the pore space, by reducing the interfacial tension between oil and water and allowing the oil droplets to break down into small enough droplets to be displaced through

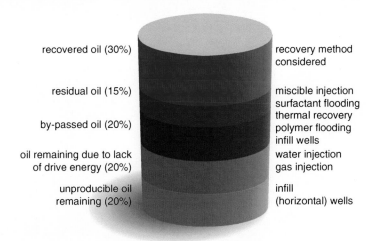

Figure 9.21   Recovering the remaining oil.

the pore throats. Very low residual oil saturations (around 5%) can be achieved. Surfactants such as soaps and detergents are added to the injection water.

Miscible processes are aimed at recovering oil which would normally be left behind as residual oil, by using a displacing fluid which actually mixes with the oil. Because the miscible drive fluid is usually more mobile than oil, it tends to by-pass the oil giving rise to a low macroscopic sweep efficiency. The method is therefore best suited to high dip reservoirs. Typical miscible drive fluids include hydrocarbon solvents, hydrocarbon gases, carbon dioxide and nitrogen.

When considering secondary recovery or EOR, it is important to establish where the remaining oil lies. Figure 9.21 shows an example of where the remaining oil may be, and the appropriate method of trying to recover it. The proportions are only an example, but such a diagram should be constructed for a specific case study to identify the 'target oil'.

One category of remaining oil shown in the above diagram is unproducible oil in *thin oil rims* (typically less than 40 ft thick), which cannot be produced without coning in unwanted oil and/or gas. Horizontal wells are an ideal form of infill well in this situation, and will be discussed in Section 10.3, Chapter 10.

WELL DYNAMIC BEHAVIOUR

Introduction and Commercial Application: Chapter 9 considered the dynamic behaviour in the reservoir, away from the influence of the wells. However, when the fluid flow comes under the influence of the pressure drop near the wellbore, the displacement may be altered by the local pressure distribution, giving rise to coning or cusping. These effects may encourage the production of unwanted fluids (e.g. water or gas instead of oil), and must be understood so that their negative impact can be minimised.

The wells provide the conduit for production from the reservoir to the surface, and are the key link between the reservoir and surface facilities. This conduit therefore needs to have the flow capacity for production (or injection) and reliability in the face of problems such as sand production, corrosion, high pressures or temperatures, mechanical failure and various production chemistry issues such as waxes, scales and hydrates. The type and number of wells required for development will dictate the drilling facilities needed, and the operating pressures, temperatures and rates of the wells and artificial lift requirements will influence the design of the production facilities. The application of horizontal or multilateral wells may, where appropriate, greatly reduce the number of wells required, which will have an impact on the cost of development. In recent years, the use of remote downhole monitoring and control through techniques such as smart wells has further extended the link between the facilities, the completion and the reservoir.

10.1. ESTIMATING THE NUMBER OF DEVELOPMENT WELLS

The type and number of wells required for development will influence the surface facilities design and have a significant impact on the cost of development. Typically the drilling expenditure for a project is between 20 and 40% of the total capex, although for a subsea development this may be higher still. A reasonable estimate of the number of wells required is therefore important.

When preparing *feasibility studies*, it is often sufficient to estimate the number of wells by considering

- the type of development (e.g. gas cap drive, water injection, natural depletion)
- the production/injection potential of individual wells.

For a particular type of development, the production profile can be estimated using the guidelines given in Section 9.6, Chapter 9. The *number of producing wells* needed to attain this profile can then be estimated from the plateau production rate and the stabilised production rates (well initial) achieved during the production tests on the exploration and appraisal wells.

$$\text{Number of production wells} = \frac{\text{Plateau production rate}}{\text{Assumed well initial}} \quad \frac{(\text{stb/d})}{(\text{stb/d})}$$

There will be some uncertainty as to the well initial rates, since the exploration and appraisal wells may not have been completed optimally, and their locations may not be representative of the whole of the field. A range of well initial rates should therefore be used to generate a range of the number of wells required. The individual well performance will depend on the fluid flow near the wellbore, the type of well (vertical, deviated or horizontal), the completion type and any artificial lift techniques used. These factors will be considered in this section. For many land and subsea developments in particular, these uncertainties can be somewhat mitigated through phased developments.

The *number of injectors* required may be estimated in a similar manner, but it is unlikely that the exploration and appraisal activities would have included injectivity tests, of say water into the water column of the reservoir. In this case, an estimate must be made of the injection potential, based on an assessment of reservoir quality in the water column, which may be reduced by the effects of compaction and diagenesis. Development plans based on water injection or natural aquifer drive often suffer from lack of data from the water bearing part of the reservoir, since appraisal activity to establish the reservoir properties in the water column is frequently overlooked. In the absence of any data, a range of assumptions of injectivity should be generated, to yield a range of number of wells required. If this range introduces large uncertainties into the development plan, then appraisal effort to reduce this uncertainty may be justified.

The presence of faults is another element that may change the number of injection/production wells required.

The type of development, type and number of development wells, recovery factor and production profile are all inter-linked. Their dependency may be estimated using the above approach, but lends itself to the techniques of reservoir simulation introduced in Section 9.4, Chapter 9. There is never an obvious single development plan for a field, and the optimum plan also involves the cost of the surface facilities required and environmental considerations. The decision as to which development plan is the best is usually based on the economic criterion of profitability. Figure 10.1

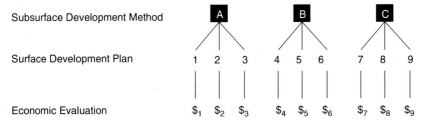

Figure 10.1   Determining the optimum development plan.

represents a range of scenarios, aimed at determining the optimum development plan (the one with the highest NPV, as defined in Chapter 14).

At the stage of *field development planning*, reservoir simulation would normally be used to generate production profiles and well requirements for a number of subsurface development options, for each of which different surface development options would be evaluated and costs estimated.

10.2. Fluid Flow Near the Wellbore

The pressure drop around the wellbore of a vertical producing well is described in the simplest case by the following profile of fluid pressure against radial distance from the well (Figure 10.2).

The difference between the flowing wellbore pressure (P_{wf}) and the average reservoir pressure ($\bar{P}$) is the pressure drawdown (ΔP_{DD})

$$\text{Pressure drawdown } \Delta P_{DD} = \bar{P} - P_{wf} \quad \text{(psi) or (bar)}$$

The relationship between the flowrate (Q) towards the well and the pressure drawdown is approximately linear for an undersaturated fluid (i.e. a fluid above the

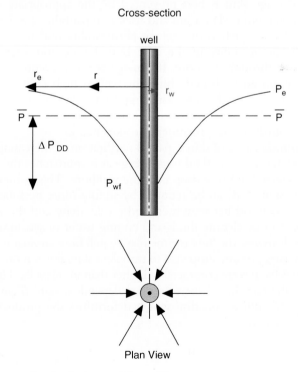

Figure 10.2 Pressure distribution around the wellbore.

bubble point), and is defined by the *productivity index* (PI).

$$\text{Productivity index (PI)} = \frac{Q}{\Delta P_{\text{DD}}} \quad \text{(bbl/d/psi) or (m}^3\text{/d/bar)}$$

For an oil reservoir a PI of 1 bbl/d/psi would be low for a vertical well, and a PI of 50 bbl/d/psi would be high.

The flowrate of oil into the wellbore is also influenced by the reservoir properties of permeability (k) and reservoir thickness (h), by the oil properties viscosity (μ) and formation volume factor (B_o) and by any change in the resistance to flow near the wellbore which is represented by the dimensionless term called *skin* (S). For *semi-steady state flow* behaviour (when the effect of the producing well is seen at all boundaries of the reservoir) the radial inflow for oil into a vertical wellbore is represented by the equation

$$Q = \frac{\Delta P_{\text{DD}} k h}{141.2 \mu B_o \{\ln((r_e/r_w) - 3/4) + S\}} \quad \text{(stb/d)}$$

The skin term represents a pressure drop which can arise due to formation damage around the wellbore. The damage is primarily caused by the invasion of solids from the drilling mud. The solid particles plug the pore throats and cause a resistance to flow, giving rise to an undesirable pressure drop near the wellbore. This so-called damage skin is best prevented by the appropriate choice of mud and completion technique. The mud and rock compatibility can be tested by core flood tests. These tests, such as the 'return permeability test' use a small core plug and measure the permeability in the core both before and after mud has been pumped against or through the core. If damage is not prevented, occasionally the damage can be removed by backflushing the well at high rates, or by pumping a limited amount of acid into the well (acidising) to dissolve the solids – assuming that they are acid soluble. Alternatively the damage will have to be by-passed by perforations or a small fracture treatment (a 'skin frac').

Another common cause of skin is partial perforation of the casing or liner across the reservoir which causes the fluid to converge as it approaches the wellbore, again giving rise to increased pressure drop near the wellbore. This component of skin is called geometric skin, and can be reduced by adding more perforations. However there is usually a trade-off between increased productivity and the very real risk of perforating close to unwelcome fluids and coning water or gas into the well.

At very high flowrates, the flow regime may switch from laminar to turbulent flow, giving rise to an extra pressure drop, due to turbulent skin; this is more common in gas wells, where the velocities are considerably higher than in oil wells. This pressure drop is dependent on the rate, hence the term 'rate dependent skin' (Figure 10.3).

In gas wells, the inflow equation which determines the production rate of gas (Q) can be expressed as

$$Q = \frac{(\bar{P}^2 - P_{\text{wf}}^2) k h}{141.2 \mu Z T \{\ln((r_e/r_w) - 3/4) + S\}} \quad \text{(Mscf/d)}$$

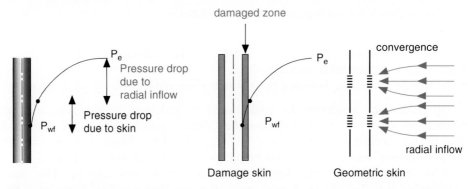

Figure 10.3 Pressure drop due to skin.

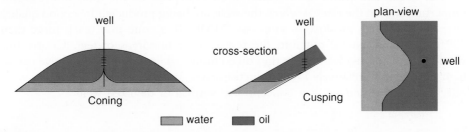

Figure 10.4 Coning and cusping of water.

where the gas flowrate Q is now measured in 1000s of standard cubic feet/day (Mscf/d); Z is the gas compressibility factor – a correction for the non 'ideal' nature of real gases. T is the reservoir temperature.

The different form of the inflow equation for gas is due to the expansion of the gas as the pressure reduces. This expansion will increase the gas velocity and therefore cause increased pressure drops.

The corresponding PI for gas becomes

$$ PI = \frac{Q}{\bar{P}^2 - P_{wf}^2} \quad (Mscf/d/psi^2) \text{ or } (m^3/d/bar^2) $$

When the radial flow of fluid towards the wellbore comes under the localised influence of the well, the shape of the interface between two fluids may be altered. The following diagrams show the phenomena of coning and cusping of water, as water is displacing oil towards the well (Figure 10.4).

Coning occurs in the vertical plane, and only when the otherwise stable oil–water contact lies directly below the producing well. Water is 'pulled up' towards the perforations, and once it reaches the perforations, the well will produce at increased water cuts.

Cusping occurs in the horizontal plane that is the stabilised OWC does not lie directly beneath the producing well. The unwanted fluid, in this case water, is pulled towards the producing well along the dip of the formation.

The tendency for coning and cusping increases if

- the flowrate in the well increases
- the distance between the stabilised OWC and the perforations reduces
- the vertical permeability increases
- the density difference between the oil and water reduces.

To reduce this tendency the well should be produced at low rate, and the perforations should be as far away from the OWC as possible. Once the unwanted fluid breaks through to a well, it may be recompleted by changing the position of the perforations during a workover, or the production rate may be reduced.

The above examples are shown for water coning and cusping. The same phenomena may be observed with overlying gas being pulled down into the producing oil well. This would be called *gas coning* or *cusping*.

The height and width of the cones or cusps depend on the fluid and reservoir properties, and on the rates at which the wells are being produced. In a good quality reservoir with high production rates (say 20 Mb/d), a cone may reach more than 200 ft high, and extend out into the reservoir by hundreds of feet. Clearly this would be a major disadvantage in thin oil columns, where coning would give rise to high water cuts at relatively low production rates. In this instance, horizontal wells offer a distinct advantage over conventional vertical or deviated wells.

10.3. Horizontal Wells

Horizontal wells were drilled as far back as the 1950s, but gained great popularity from the 1980s onwards as directional drilling technology progressed and cost pressure mounted. Horizontal wells have potential advantages over vertical or deviated wells for three main reasons

- increased exposure to the reservoir giving higher productivity indices (PIs)
- ability to connect laterally discontinuous features, for example fractures, fault blocks
- changing the geometry of drainage, for example being parallel to fluid contacts.

The *increased exposure to the reservoir* results from the long horizontal sections which can be attained (sections many kilometres in length are now routine in many fields). Because the PI is a function of the length of reservoir drained by a well, horizontal wells can give higher productivities in laterally extensive reservoirs. As an estimate of the initial potential benefit of horizontal wells, one can use a rough rule of thumb, the *productivity improvement factor* (PIF) which compares the initial productivity of a horizontal well to that of a vertical well in the same reservoir, during early time radial flow

$$\text{PIF} = \frac{L}{h}\sqrt{\frac{k_v}{k_h}}$$

where L is the length of the reservoir; h the height of the reservoir; k_h the horizontal permeability of the reservoir; k_v the vertical permeability of the reservoir.

The geometry and reservoir quality have a very important influence on whether horizontal wells will realise a benefit compared to a vertical well, as demonstrated by the following example (Figure 10.5).

In the case of the very low vertical permeability, the horizontal well actually produces at a lower rate than the vertical well. Each of these examples assumes that the reservoir is a block, with uniform properties. The ultimate recovery from the horizontal well in the above examples is unlikely to be different to that of the vertical well, and the major benefit is in the accelerated production achieved by the horizontal well.

The PIF estimate is only a qualitative check on the potential initial benefit of a horizontal well. The stabilised flowrate benefits of horizontal wells compared to vertical wells are more rigorously handled by relationships derived by Joshi (ref. Horizontal Well Technology, Pennwell, 1991). Also, in high permeability reservoirs there is actually a diminishing return of production rate on the length of well drilled, due to increasing friction pressure drops with increasing well length, shown schematically in Figure 10.6.

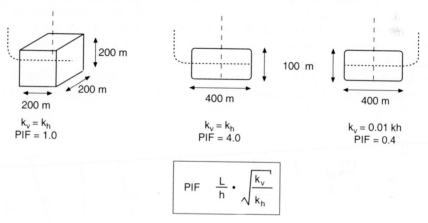

Figure 10.5 Productivity improvement factor (PIF) for horizontal wells.

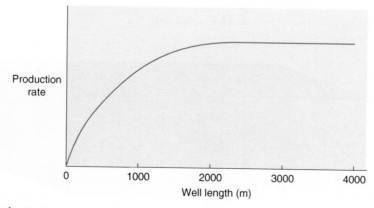

Figure 10.6 Production rate vs. horizontal well length.

The exact relationship will depend on both fluid and reservoir properties, and will be investigated during well planning. Poor completion practices may exacerbate the problem as the lower drawdown on the toe of the well compared to the heel may prevent proper clean-up of mud, filter cake and completion fluids.

Horizontal wells have a large potential to *connect laterally discontinuous* features in heterogeneous or discontinuous reservoirs. If the reservoir quality is locally poor, the subsequent section of the reservoir may be of better quality, providing a healthy productivity for the well. If the reservoir is faulted or fractured a horizontal well may connect a series of fault blocks or natural fractures in a manner which would require many vertical wells. The ultimate recovery of a horizontal well is likely to be significantly greater than for a single vertical well (Figure 10.7).

The third main application of horizontal wells is to reduce the effects of coning and cusping by *changing the geometry of drainage* close to the well. For example, a horizontal producing well may be placed along the crest of a tilted fault block to remain as far away from the advancing oil–water contact as possible during water drive. An additional advantage is that if the PI for the horizontal well is larger, then the same oil production can be achieved at much lower drawdown, therefore minimising the effect of coning or cusping. The result is that oil production is achieved with significantly less water production, which reduces processing costs and assists in maintaining reservoir pressure. Horizontal wells have a particularly strong advantage in *thin oil columns* (say, less than 40 m thick), which would be prone to coning if developed using conventional wells. The unwanted fluid in oil rim development may be water or gas,

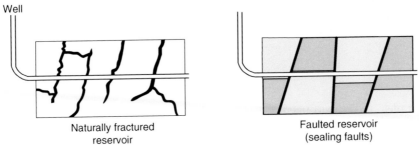

Figure 10.7   Increased recovery from a horizontal well.

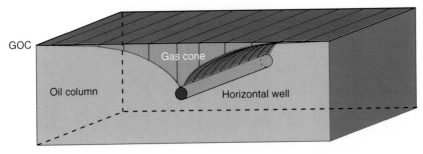

Figure 10.8   Gas cresting in oil rim development with horizontal wells.

or both. The distortion of the fluid interface near the horizontal well is referred to as *cresting* rather than coning, due to the shape of the interface. Figure 10.8 shows a schematic view of gas cresting from an overlying gas cap in an oil reservoir.

10.4. PRODUCTION TESTING AND BOTTOM HOLE PRESSURE TESTING

Routine production tests are performed, ideally at least once per month on each producing well, by diverting the production through the test separator on surface to measure the liquid flowrate, water cut and gas production rate. The tubing head pressure (also called the FTHP) is recorded at the time of the production test, and a plot of production rate against FTHP is made. The FTHP is also recorded at least once per day and used to estimate the well's production rate on a daily basis by reference to the FTHP vs. production rate plot for the well.

It is important to know how much each well produces or injects in order to identify productivity or injectivity changes in the wells, the cause of which may then be investigated. For example, the well might be scaling up. Also, for reservoir management purposes (Chapter 14) it is necessary to understand the distribution of volumes of fluids produced from and injected into the field. This data is input to the reservoir simulation model, and is used to check whether the actual performance agrees with the prediction, and to update the historical data in the model. Where actual and predicted results do not agree, an explanation is sought, and may lead to an adjustment of the model (e.g. re-defining pressure boundaries, or volumes of fluid in place).

The production testing through the surface separator gathers information at surface. Another important set of information collected during *bottom hole pressure testing* is downhole pressure data, which is used to determine the reservoir properties such as permeability and skin. In a production well, which will have been completed with production tubing, the downhole pressure measurement is typically taken by running a pressure gauge, on wireline (either electric line or with memory gauges by slickline), to the depth of the reservoir interval. The downhole pressure gauge is then able to record the pressure whilst the well is flowing or when the well is shut-in.

A *static bottom hole pressure survey* (SBHP) is useful for determining the reservoir pressure near the well, undisturbed by the effects of production. This often cannot be achieved by simply correcting a surface pressure measurement, because the tubing contents may be unknown, or the tubing contains a compressible fluid whose density varies with pressure (which itself has an unknown profile).

A *flowing bottom hole pressure survey* (FBHP) is useful in determining the pressure drawdown in a well (the difference between the average reservoir pressure and the FBHP, P_{wf}) from which the PI is calculated. By measuring the FBHP with time for a constant production rate, it is possible to determine the parameters of permeability and skin, and possibly the presence of a nearby fault, by using the radial inflow equation introduced in Section 10.2. Also, by measuring the response of the bottom hole pressure against time when the well is then shut-in, these parameters can be calculated (Figure 10.9).

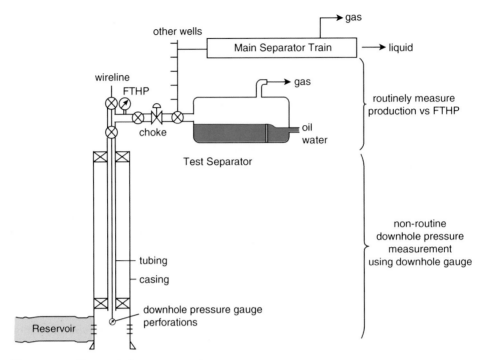

Figure 10.9   Bottom hole pressure testing.

It is common practice to record the bottom hole pressure firstly during a flowing period (pressure drawdown test), and then during a shut–in period (pressure build–up test). During the flowing period, the FBHP is drawn down from the initial pressure, and when the well is subsequently shut-in, the bottom hole pressure builds up (Figure 10.10).

In the simplest case, for a pressure drawdown survey, the radial inflow equation indicates that the bottom hole flowing pressure is proportional to the logarithm of time. From the straight line plot of pressure against the log (time), the reservoir permeability can be determined, and subsequently the total skin of the well. For a build–up survey, a similar plot (the so-called Horner plot) may be used to determine the same parameters, whose values act as an independent quality check on those derived from the drawdown survey.

Drawdown and build-up surveys are typically performed once a production well has been completed, to establish the reservoir property of permeability (k), the well completion efficiency as denoted by its skin factor (S) and the well PI. Unless the routine production tests indicate some unexpected change in the well's productivity, only SBHP surveys may be run, say once a year. A full drawdown and build-up survey would be run to establish the cause of unexplained changes in the well's productivity. In addition to pressure and temperature gauges, a whole suite of further data may be acquired at the same time in one production log. These production logging techniques (PLTs) may include spinners to measure flowrates, density meters to measure water, gas and oil contents and other more sophisticated measurements for

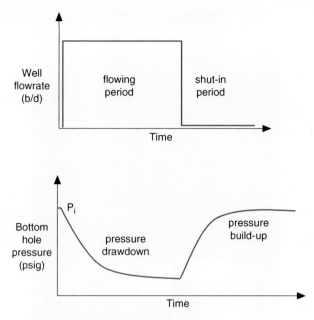

Figure 10.10 Pressure drawdown and build-up survey.

determining the type of fluids behind the casing for example. This is covered in more detail in Chapter 16.

Permanent surface read-out downhole gauges are now common for critical wells. This is especially the case for subsea wells, where running a single production log could cost in excess of $2 million as a rig or intervention vessel has to be mobilised.

Permanent downhole gauges are run with the completion. The gauges (electronic quartz crystal, or sometimes fibre optic) are normally positioned as deep as possible in order to be close to the reservoir. In practice, to avoid too many complications they are usually run above a packer. The gauge is out of the flow stream and protected, but exposed to tubing pressures and temperatures. A cable is then run beside the tubing and typically clamped at every joint of tubing. The cable will eventually pass through the Christmas tree and be connected to the facilities instrumentation system (Figure 10.11).

Gauges typically measure both pressure and temperature, although venturi effect flowmeters and densitometers can also be deployed.

In exploration wells which show hydrocarbon indications, it is often useful to test the productivity of the well, and to capture a fluid sample. This can be used as proof of whether further exploration and appraisal is justified. If the well is unlikely to be used as a production well, a method of well testing is needed which eliminates the cost of running casing across the prospective interval and installing a production tubing, packer and wellhead. In such a case, a *drill stem test* (DST) may be performed using a dedicated string, called a test string, which has gas-tight seals at the joints (Figure 10.12).

In the *open hole* DST, inflatable packers are set against the openhole section to straddle the prospective interval. Migration of hydrocarbons into the annulus is

Figure 10.11 Gauge cable and clamp being installed beside tubing (photograph copyright H. Crumpton).

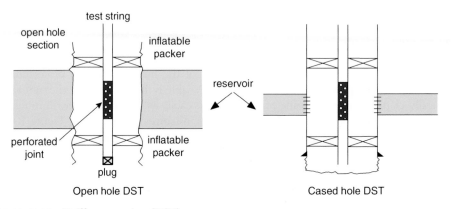

Figure 10.12 Drill stem testing (DST).

prevented by the upper packer, and a good seal is required to ensure safety. Therefore the openhole DST can only be run where the openhole section is in gauge. The safe length of the openhole test section would be determined by the strength of the casing shoe. If several intervals are to be tested independently, then a *cased hole* DST may be considered. Only the interval of interest is perforated and allowed to flow. All other intervals remain isolated behind casing. Each interval is sealed off prior to testing another. In both types of DST it is possible to run a downhole pressure gauge, and therefore to perform a drawdown and build-up survey.

10.5. Tubing Performance

The previous sections have considered the flow of fluid into the wellbore. This is commonly referred to as the 'inflow performance'. The PI indicates that as the flowing wellbore pressure (P_{wf}) reduces, so the drawdown increases and the rate of fluid flow to the well increases. Recall for an oil well (Figure 10.13)

$$\text{Drawdown pressure } \Delta P_{DD} = P - P_{wf} \quad \text{(psi) or (bar)}$$

$$\text{Productivity index (PI)} = \frac{Q}{\Delta P_{DD}} \quad \text{(bbl/d/psi) or (m}^3\text{/d/bar)}$$

Having reached the wellbore, the fluid must now flow up the tubing to the wellhead, through the choke, flowline, separator facilities and then to the export or storage point; each step involves overcoming some pressure drop.

The pressure drops can be split into three parts; the reservoir or inflow, the tubing and the surface facilities, with the linking pressures being the flowing wellbore pressure (P_{wf}) and the tubing head pressure (P_{th}). To overcome the choke and facilities pressure drops a certain tubing head pressure is required. To overcome

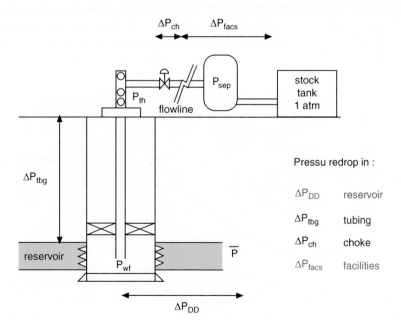

Figure 10.13 Pressure drops in the production process.

the vertical pressure drop in the tubing due to the hydrostatic pressure of the fluid in the tubing and friction pressure drops, a certain flowing wellbore pressure is required. For a single phase fluid (e.g. just water or just gas), this combination of hydrostatic and friction is relatively straightforward to calculate even accounting for the variation of gas density with pressure. For most production wells however, the complication is that there is more than one phase present and each phase has a different density and viscosity. This is *multiphase flow*. At high velocities the phases are chaotically mixed and essentially travel at the same velocity. The density and viscosities are then averages, and the friction and hydrostatic pressures can be calculated. Under these circumstances, the hydrostatic pressure does not change with rate, but the frictional pressure drop will be dependent on velocity or rate squared. At lower velocities, the phases tend to travel at different velocities. This effect is called slippage. As the lighter fluids (such as gas) travel faster than the denser phases such as water, the gas will spend less time in the tubing compared to the liquids. The gas will therefore occupy less space and have correspondingly less effect on the overall density. As the rate reduces, the overall density will therefore increase. The combined effect of friction and density is shown in Figure 10.14 where the overall pressure required (P_{wf}) to lift the fluids to a given surface pressure (P_{th}) is shown as it overcomes both the hydrostatic and friction pressure drops.

The calculation of the precise slippage and hence the friction and density is complex with no precise solution. A number of empirical correlations are normally used. The choice of correlation will depend on the fluid and rates of gas and liquid. The correct correlation can be confirmed by comparing the correlation's prediction with flowing data from downhole gauges or production logs. The overall *tubing performance relationship* (TPR) will also be significantly influenced by changes in water cuts, gas–oil ratios and the tubing size as well as the effects of artificial lift. As can be observed there is a minima in the overall TPR. This minima represents the lowest pressure required to the lift the fluids to surface; it is therefore the most efficient condition. As a rule of thumb, flow performance to the right hand side of

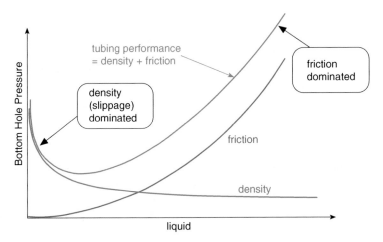

Figure 10.14   Tubing performance.

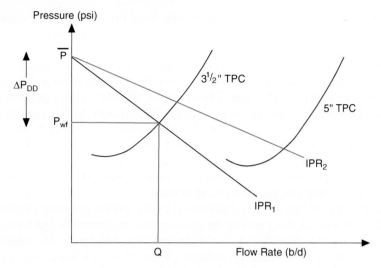

Figure 10.15   Reservoir performance and tubing performance.

the minima in the TPR is stable, whilst conditions to the left hand side of the minima, as well as being inefficient, can suffer flow instability effects such as severe slugging and are best avoided.

We now have predictions for the pressure drops in the tubing and in the reservoir. They share a common point or node. The inflow performance relationship (IPR) predicts the wellbore flowing pressure for a given reservoir and reservoir completion, whilst the TPR predicts the wellbore flowing pressure required to lift these fluids to surface through the tubing. At the (wellbore) node, the pressure and the rate must be the same and therefore the point of intersection of the IPR and the TPR is the predicted well rate and the wellbore flowing pressure. This technique is commonly called NODAL[1] analysis. The same technique can be used for the intersection of the TPR with the surface facilities pressure drops where the node is now the surface pressure.

Ignoring the surface facilities pressure drop, the following diagram shows an example of the equilibrium between the IPR and the TPR for two tubing sizes (Figure 10.15).

For the reservoir with IPR_1, the larger tubing does not achieve an equilibrium, and the well would not flow if the $5\frac{1}{2}$ in. tubing was installed. However, a different reservoir with IPR_2 would benefit from a larger tubing size which would allow greater production, and the correct selection of tubing size would be the $5\frac{1}{2}$ in. tubing if one wanted to maximise the early production from the well. An understanding of the tubing performance and the reservoir performance (which requires reservoir data gathering) is important for the correct *selection of tubing size*. Note that most of the variables (e.g. water cut, reservoir pressure) change substantially with time so designing for the life of the well will require some compromises.

Returning to the surface pressure drops across the choke and the facilities, these will also vary over the producing lifetime of the field. The choke is used to isolate the

---

[1]NODAL analysis is a trademark of Flopetrol Johnston, a division of Schlumberger Technology Corporation.

surface facilities from the variations in tubing head pressure, and the choke size is selected to create critical flow which maintains a constant downstream pressure. Initially, a small orifice will be required to control production when the reservoir pressure is high. As the reservoir pressure drops during the producing lifetime of the field, the choke size will be adjusted to reduce the pressure drop across the choke, thus helping to sustain production. The operating pressure of the separators may also be reduced over the lifetime of the field for the same reason. In fact, the linkage from the reservoir to the facilities continues down the pipeline – especially for gas fields. A high separator pressure will put a backpressure on the tubing and hence restrict production. However it will also make it easier to pump or flow the fluids through the pipeline. There will be an optimum separator pressure that balances these issues and this balance will change as the field matures.

The end of field life is often determined by the lowest reservoir pressure which can still overcome all the pressure drops described and provide production to the stock tank. As the reservoir pressure approaches this level, the abandonment conditions may be postponed by reducing some of the pressure drops, either by changing the choke and separator pressure drops as mentioned, or by introducing some form of artificial lift mechanism, as discussed in Section 10.8.

In a gas field development, the recovery factor is largely determined by how low a reservoir pressure can be achieved before finally reaching the abandonment pressure. As the reservoir pressure declines, it is therefore common to install compression facilities at the surface to pump the gas from the wellhead through the surface facilities to the delivery point. This compression may be installed in stages through the field lifetime. As gas rates decline, it might also be necessary to alter the tubing size to avoid unstable flow and liquid loading problems – the consequence of operating with too large a tubing size for the gas rates, that is essentially operating to the left hand side of the TPR minima.

10.6. Well Completions

The conduit for production or injection between the reservoir and the surface is the completion. This is commonly split into the 'lower completion' or 'reservoir completion' for the section across the reservoir interval and the 'upper completion' or 'tubing completion' for the section above the reservoir through to the wellhead.

There are a number of options for both the lower and upper completion. Options for the lower completion are shown in Figure 10.16, whilst upper completion options are shown in Figure 10.20.

Each of these five main reservoir completion options has its advantages and disadvantages, but all are in common use in various locations around the world. The barefoot completion is the simplest and cheapest. The drilled reservoir section is left as openhole and nothing is installed across the reservoir. Although cheap and simple, future reservoir access – for logging or for shutting off unwelcome fluids will be tricky. Care must therefore be taken to ensure that the drill bit does not enter into a water interval. In addition any weak intervals present might collapse and either

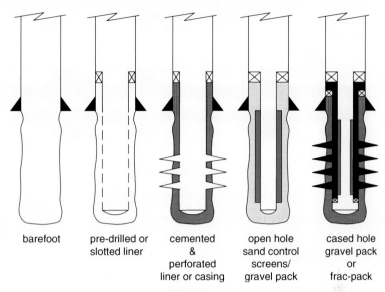

| barefoot | pre-drilled or slotted liner | cemented & perforated liner or casing | open hole sand control screens/ gravel pack | cased hole gravel pack or frac-pack |

Figure 10.16 Reservoir completion summary.

block production or produce solids to surface. However, on the upside, there is a large flow area and some future interventions such as sidetracking to a new reservoir location are relatively simple as there is no equipment in the way. Barefoot completions are common in land locations – especially those producing from competent limestones and dolomites.

The pre-drilled or pre-slotted liner is slightly more complex. The liner has holes or slots milled in it before it is installed. It is still an openhole completion in that whatever is drilled is open to production. The liner however will stop the hole from totally collapsing and aid in getting intervention or logging tools down. It is however usually impossible to make the slots small enough to stop individual grains of sand from being produced. These types of completions can use openhole packers in order to isolate water or gas intervals. An application of these in a horizontal well is shown in Figure 10.17.

These openhole packers are either inflated or can be designed to swell once in contact with reservoir fluids. They then provide a location for setting plugs or straddles when needed.

The cemented and perforated liner is more complex still, but has distinct advantages. The casing or liner is run across the reservoir section and cemented in place. Once the cement has set the well can then be perforated – typically running the perforation guns on drill pipe. Alternatively the upper completion can first be run and then the perforations run through the completion – typically with electric line. The *perforation guns* contain many shaped charges. In each shaped charge (Figure 10.18), there is a cone of explosive. When detonated, this sends out a high-pressure unidirectional jet which punches through the casing, the cement, and several feet into the formation.

Once at the correct depth the gun (as shown in Figure 10.19) is fired from surface usually either by electric cable or by hydraulic pressure. The big advantage of a

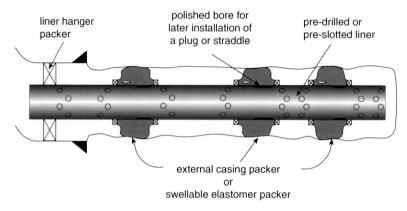

Figure 10.17 Openhole completion with zonal isolation.

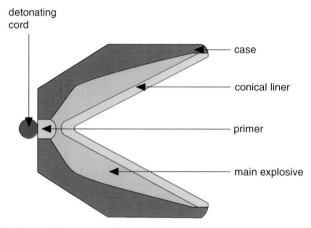

Figure 10.18 The shaped charge.

perforated completion is that the engineer can select where to perforate. The decision will be made on the basis of log data, and water, gas or weak intervals can be avoided. Perforations can also be later isolated if they produce too much water or gas.

None of the techniques mentioned already can cope with very weak sand production prone intervals, although if these intervals are few in number a perforated completion can at least avoid them. Sand control completions can either be openhole or cased hole. They employ screens (simple wire wrap or more complex mesh type) which are installed across the openhole or across a cased and perforated interval. To increase reliability, the annulus between the screen and the formation is often packed with gravel (a gravel pack). In some openhole completions, the formation is left to collapse directly onto the screen. More recently, another method of increasing reliability is to use an expandable screen in an openhole section. This screen is run across the reservoir section and then expanded against the borehole wall. In a cased hole, sand control completion, the annulus and the perforations are always packed with gravel. Cased hole gravel packs

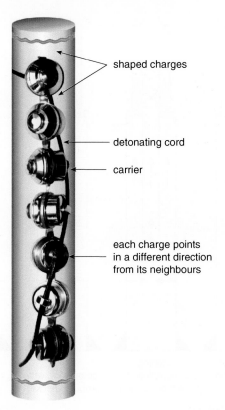

Figure 10.19   Perforating gun.

can therefore be selective, but often suffer from poor productivity. In order to increase productivity, the gravel may be pumped at high pressure and the reservoir fractured – the so-called *frac and pack* or frac-pack. This aids in packing the perforations with gravel and provides some stimulation benefit. This technique can be complex, but is used to particularly good effect offshore in the GoM.

The upper completion also covers a variety of techniques. Figure 10.20 shows four common methods. The inflow shown here is for a cased and perforated completion, but similar techniques exist for the other reservoir completion options.

The simplest method consists of producing straight up the casing with no tubing. It is cheap, can have a large flow area, but it is difficult to control, may corrode the casing and in most situations is considered unsafe due to a lack of barriers in the event of a problem. The second option, although at first glance has the same problems as the tubingless completion, has some distinct advantages. The flow may be either up the tubing, up the tubing-casing annulus or up both. It is therefore very useful in low-pressure gas wells where the flow area can be switched to overcome liquid loading problems. The other main application is with pumped wells. When a downhole pump is installed, the pumps perform best if they only produce liquid; gas production can cause problems. The gas is therefore vented up the annulus (as shown in Figure 10.21) and out of the wellhead. The liquid enters

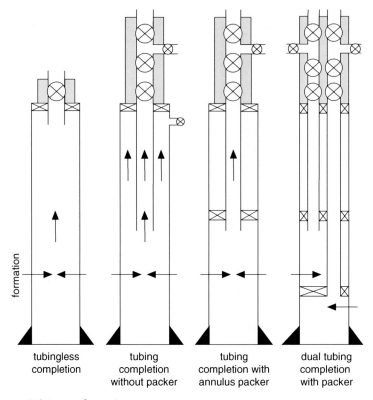

Figure 10.20   Tubing configuration summary.

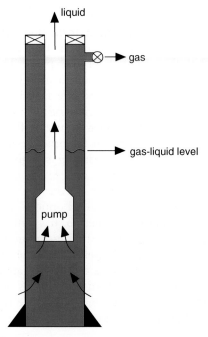

Figure 10.21   Pumped well.

the pump and is pumped up the tubing. In the event of an emergency, the pump can be stopped and tubing flow will stop. The liquid level in the annulus will rise and the well will kill itself. Pumped wells will be considered further in Section 10.8.

For naturally flowing wells, especially those flowing at moderate or high pressures, additional barriers to hydrocarbon escape are employed. In the third example, the tubing is sealed with a packer or other sealing system. Therefore in the event that the tubing develops a leak (e.g. through corrosion), the casing can withstand the pressure. Annulus pressure will be detected, the well shut-in and the tubing replaced. Replacing tubing is a much easier operation than replacing casing. Such a completion is very common offshore, where the consequences of a leak are more severe due to the proximity of people to the well.

The final option shown in Figure 10.20 is a dual string. Clearly more complex than the other options, there are however some useful advantages. This option is used in low-to-moderate rate wells where there are multiple stacked reservoirs. Flow from the two intervals is separately produced, controlled and measured and any problems with incompatible fluids are avoided. These completions can be very useful if the reservoir intervals are very different in productivity, pressure or fluids. Rates are however usually lower than the equivalent single bore commingled producer due to the size limit for two parallel strings inside the casing. In extreme examples, three strings or even four strings may be run in parallel.

10.7. COMPLETION TECHNOLOGY AND INTELLIGENT WELLS

In the last section, we considered a range of completion types for both the reservoir section and the upper completion. Let us now deal with some of the equipment you may encounter in a completion. In the example in Figure 10.22, the completion is a horizontal offshore sand control completion with many optional pieces of equipment. It does however demonstrate the types of equipment in common use and the often confusing abbreviations used.

Starting from the top of the well, we have the Christmas tree sitting on top of the wellhead. The tree is designed to control production or injection. It is the primary means of shutting in the well. Vertical access through the tree is possible for logging or other interventions. These operations can be performed on a live well (i.e. pressurised and capable of flowing) through temporary pressure control equipment installed above the swab valve (SV). Most wells will use a Christmas tree of some form, including subsea wells. Rod pumped wells however will replace the tree with a single valve and a stuffing box to allow the rods to move up and down the well whilst the well flows.

The tubing hanger is a solid piece of metal that supports the tubing. It is either installed inside the *wellhead* (as shown) or for certain types of tree it can sit inside the tree. The tubing hanger connects to the tree via seals and to the tubing below via a screwed thread. The tubing hanger will usually have penetrations for control lines, downhole gauge lines and chemical injection lines. Below the hanger comes the tubing. The tubing has to be designed to withstand high pressure (and sometimes high temperatures). The production fluids are often corrosive and the tubing

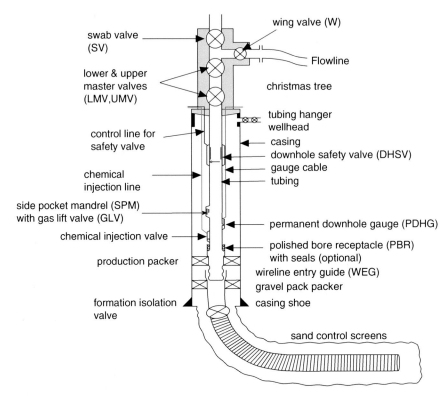

Figure 10.22 Completion schematic.

(and other completion equipment) is often made out of *corrosion resistant alloys* (CRAs) such as stainless steel – especially for critical, high-rate wells. The tubing, like casing, comes in joints typically 40 ft long and is screwed together on the rig.

Most offshore and some onshore completions employ a downhole safety valve (DHSV), as indicated in Figure 10.23. This valve is held open by pressure down a control line. If the pressure is bled off either deliberately or because the control line is damaged, a spring in the valve releases a flapper valve into the flow stream and the well is closed. For example, if there was a major incident on a platform and the tree was badly damaged, the control line would break and the well would automatically shut-in. The safety valve will also be tied into the facility shut-down control system.

Below the safety valve, comes a range of optional equipment. Shown in Figure 10.22 are three types of *mandrel*. One for the injection of gas, one for a pressure gauge and the other for chemical injection. The types of chemicals injected could include scale inhibitors, methanol for hydrate inhibition and corrosion inhibitors. Another optional piece of equipment (not shown) is a sliding side door (SSD) for the circulation of fluids into and out of the tubing.

In the example shown, the tubing is anchored and sealed to the casing with a production packer. Sometimes above the packer, an expansion device (such as the polished bore receptacle (PBR) shown) is used to allow for thermal expansion or contraction of the tubing.

flapper open   flapper closed

Figure 10.23   Downhole safety valve (DHSV).

Below the *packer*, there is usually a section (the tailpipe) with a nipple profile for the setting of plugs, temporary gauges or downhole chokes. Nipple profiles are also often found inside the tubing hanger, immediately above a safety valve and sometimes (in older completions) above the packer. The nipple profile (Figure 10.24) is a permanent part of the completion, but the lock is installed and removed by slickline. The lock can have a variety of components screwed to it, including a blank plug, check valve or a gauge.

In this example, the screen is connected (and sealed to) the casing via a further packer. Without this seal, sand could flow into the tubing. Below this gravel pack packer is another valve (formation isolation valve) that crops up in many modern completions (often with a confusing array of proprietary names!) for a variety of purposes. In this case it is used to safely isolate the reservoir whilst the upper completion is run. In other applications it can, for example, be used as a downhole barrier to run long perforating guns. They are usually mechanically closed (e.g. by the pipe used to run the screens) and then opened by a series of pressure cycles.

A further type of completion is the well with remote downhole flow control. These are often called *smart wells* or *intelligent wells*, although as there is rarely any

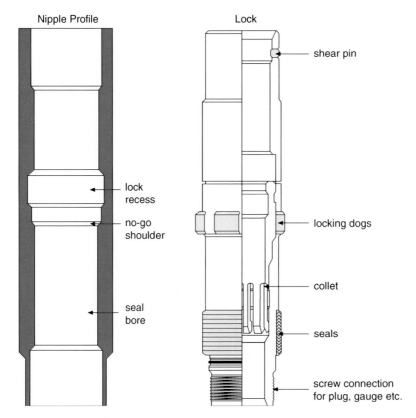

Figure 10.24 Nipple profile and associated lock.

direct decision-making capability incorporated into the flow control, the term intelligent is a misnomer.

The configuration shown in Figure 10.25 is for a cased and perforated completion, although (more complex) variations exist for sand control completions. The valves are either electrically or hydraulically controlled from surface via the control cables. Packers isolate each section of the reservoir or different reservoirs. The valves can be the on-off type or for added control, and complexity, can be variable. Most smart wells will also deploy gauges for pressure and temperature measurements both inside and outside of the tubing at each control valve. The gauge signals are usually multiplexed into the valve control signals. Smart completions are particularly useful where reservoir access by traditional means (through tubing intervention) is limited. This could be because the well is remote, for example an unmanned offshore platform or difficult or expensive to access as in the case of a subsea well.

Multilaterals are a further, relatively recent, complex completion. If a wellbore is partially abandoned and a hole milled above the abandoned section to the side of the main bore, this is a *sidetrack*. It is a common operation when the original wellbore is no longer usable, but there are nearby reserves. It is usually considerably cheaper

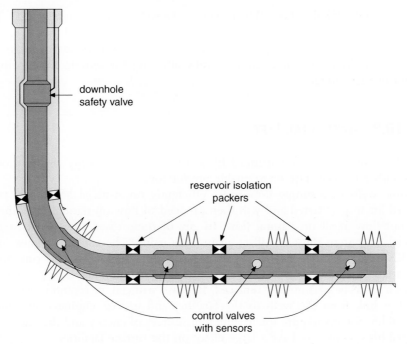

Figure 10.25   Smart well example.

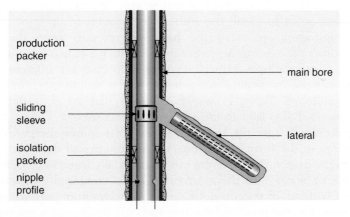

Figure 10.26   Multilateral example.

than drilling a new well. If the original wellbore is not abandoned, but only temporarily isolated and a sidetrack drilled, production or injection from both the sidetrack and original wellbore can be achieved. This is a multilateral. There is no limit to how many laterals can be drilled from one bore, though more than five laterals are rare. Figure 10.26 shows a two branch multilateral. In this case, the completion has been designed to maintain access into the original bore and to be able to shut-off either bore. The lateral can be isolated by closing the SSD. This is achieved by *slickline intervention*. The original bore can be isolated by setting a plug (by slickline again) in the tailpipe. It is also possible, and increasingly common, to mix smart wells with multilaterals, thereby allowing for remote monitoring and control of each lateral.

10.8. ARTIFICIAL LIFT

The objective of any artificial lift system is to add energy to the produced fluids, either to accelerate or to enable production.

Some wells may simply flow more efficiently on artificial lift, others require artificial lift to get started and will then proceed to flow on natural lift, others yet may not flow at all on natural flow. In any of these cases, the total cost of the artificial lift system must be offset against the gains. The total cost must include CAPEX on the wells and facilities and the operating costs for running and maintaining the equipment. Operating costs may be considerable, especially when pumps have to be periodically replaced. Because artificial lift requires energy, there is a clear link from the wells to the facilities and process engineering. Different artificial lift systems require totally different sources of energy and the choice of an artificial lift system may have a large effect on the surface facilities.

Artificial lift systems are mostly required later in a field's life, when reservoir pressures decline and therefore well productivities drop. If a situation is anticipated where artificial lift will be required or will be cost-effective later in a field's life, it may be advantageous to install the artificial lift equipment up front and use it to accelerate production, provided the increased revenues from the accelerated production offset the cost of the earlier investment. In some other cases it may be beneficial to install multiple artificial lift systems to cater for different wells, or to change the artificial lift system during the life of the well to cater for the different operating conditions. Typical examples are wells that are converted to *electrical submersible pumps* (ESPs) later in life as water cut increases.

Lifting the fluids from the reservoir to surface requires energy. All reservoirs contain energy in the form of pressure, in the compressed fluid itself and in the rock, due to the overburden. Pressure can be artificially maintained or enhanced by injecting gas or water into the reservoir. This is commonly known as *pressure maintenance*. Artificial lift systems distinguish themselves from pressure maintenance by adding energy to the produced fluids in the well; the energy is not transferred to the reservoir.

The following types of artificial lift are commonly available today

- Beam pump
- Progressive cavity pump
- Electric submersible pump (ESP)
- Hydraulic submersible pump (HSP)
- Jet pump
- Continuous flow gas lift

- Intermittent gas lift
- Plungers.

The first five on the list are all pumps, literally squeezing, pushing or pulling the fluids to surface, thus transferring mechanical energy to the fluids, albeit in different ways. The gas lift systems add energy by adding light gas and thus lowering the overall density of the produced fluids. A brief introduction to each of the systems follows. Their schematics are shown in Figure 10.27.

10.8.1. Beam pump

These pumps are also commonly known as 'rod pumps' or 'nodding donkeys'. The beam pump has a subsurface plunger. The plunger is rocked up and down by the movement of the walking beam on surface. The plunger has a check valve (the travelling valve), whilst underneath the plunger there is a second valve (the standing valve). As the plunger moves up liquid is sucked from the reservoir. On the downstroke, the plunger refills, thus there is only pumping on the upstroke. The walking beam is driven by an electric or reciprocating motor. The downhole plunger and walking beam are mechanically connected by sucker rods. Different plunger sizes (both area and length) allow for a large range of possible flowrates. For a given plunger size, the flow rate can be further adjusted by altering stroke length and pump speed. Even lower flow rates can easily be accommodated by cycling the pump on and off. Finding the right balance between stroke length and pump speed is the art of beam pump design. Sub-optimal designs lead to poor efficiencies and excessive rod and pump wear. A 'dynamometer' is used to monitor the system. The dynamometer shows the relationship between pump travel and load. Beam pumps are very common on land wells, but are usually limited to a few hundred barrels per day. Occasionally, designs can accommodate larger rates, especially if the surface beam system is replaced with a hydraulic piston.

10.8.2. Progressive cavity pump

The progressive cavity pump consists of a rotating corkscrew like subsurface assembly which is driven by a surface mounted motor – usually electrical. Beam pump rods are used to connect the two. The flowrate achieved is mainly a function of the rotational speed of the subsurface assembly. There is in principle very little that can go wrong with progressive cavity pumps, although energy is lost by rod wear in transferring the torque down the well – especially in deviated and deep wells. For this reason, deeper pumps can employ a downhole motor similar to that used in an ESP. Progressive cavity pumps excel in low productivity shallow wells with viscous crude oils and can also handle significant quantities of produced solids.

10.8.3. Electric submersible pump

The ESP is an advanced multistage centrifugal pump, driven directly by a downhole electric motor. The ESP's output is more or less pre-determined by the type and

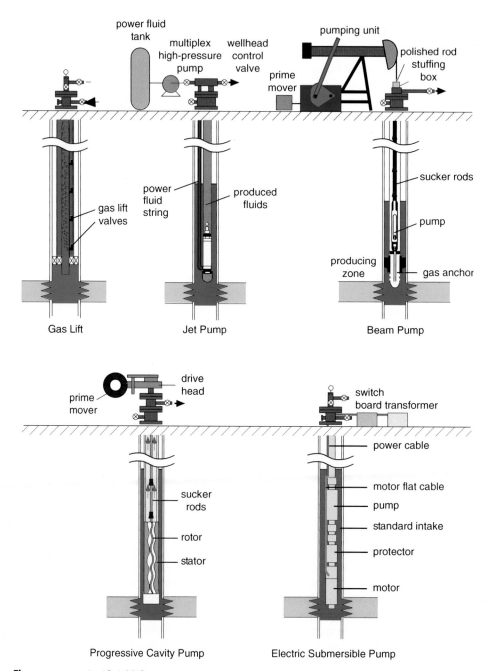

Figure 10.27   Artificial lift systems.

number of pump stages. At additional cost, a variable speed drive can be installed to allow the motor speed, and thus the flow rate, to be changed.

As can be seen from Figure 10.28, a centrifugal pump performs work by lifting the fluid a certain 'head'. The head is the vertical distance the fluid is lifted. The head is independent of the fluid, but the resulting pressure increase will depend significantly on the fluid – lifting water will generate a bigger pressure increase (and require more energy from the motor) than lifting gas. The diagram shows only the output of one stage. The total required head determines the number of stages – and the power requirement of motor. ESP design concerns itself primarily with choosing the right type of pump, the optimum number of stages and the corresponding motor and cable size to ensure the smooth functioning of the system. As can be seen from the diagram, changes in well productivity are hard to accommodate.

The performance of the system is monitored primarily by the use of a current and voltage meter, measuring the motor load and by analysing the fluid throughput against the hydraulic head. If the rates are too high or too low for the pump, then significant thrust loads will develop on the pump stages and the pump may self-destruct. Other problems include electrical short circuits especially when cable penetrators are required (e.g. through packers). Runs lives vary enormously from many years in shallow, low temperature, solids free wells to less than a year in more extreme environments or where there are errors in the design, installation or operation of the pumps.

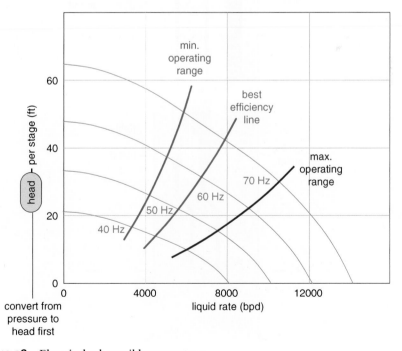

Figure 10.28 Electrical submersible pump stage.

10.8.4. Hydraulic submersible pump

These pumps are similar to ESPs in that a multistage centrifugal pump lifts the
fluids. The electrical motor is replaced with a turbine. The turbine is powered by
high-pressure fluid from surface. The turbine will rotate much quicker than an ESP
motor and thus HSPs require fewer pump stages and are therefore much more
compact. They can be deployed with the tubing or through the completion (if large
enough). One of the main factors limiting their use is the problem with circulating
the power to and from the pump. The power fluid can be mixed with the produced
fluid and then separated and pressurised at surface. A second method using a dual
bore completion is shown (Figure 10.29).

A third alternative is to dispose of the power fluid downhole. This requires a
suitably placed aquifer or an injection zone. The power fluid in these open systems will
typically be water – probably boosted to a high enough pressure to drive the turbine.
Because of the lack of electrical components, HSPs are generally more reliable than
ESPs, though they are still susceptible to damage by solids. Power fluids may be
corrosive to the casing and their cooling action may help precipitate wax or hydrates.

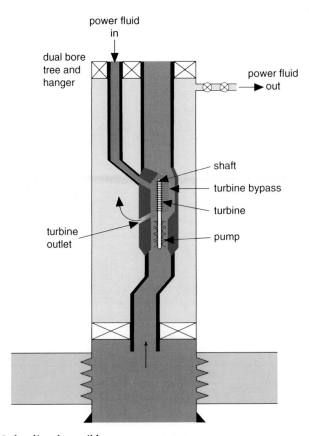

Figure 10.29 Hydraulic submersible pump.

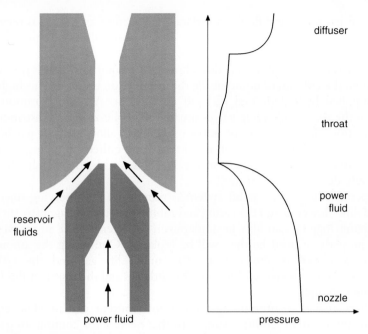

Figure 10.30   Jet pump.

10.8.5. Jet pump

Like the HSP, the jet pump relies on the hydraulic power being delivered subsurface. There the similarity ends. The high-pressure power fluid is accelerated through a nozzle. The high velocity creates a low pressure through the Bernoulli or venturi principle. This low pressure sucks up reservoir fluids (Figure 10.30).

The low-pressure mix of power fluid and reservoir fluid then enters the diffuser where the velocity is converted to pressure. The pressure is then sufficient to deliver the fluids to surface. The jet pump has no moving parts and can be made very compact and is typically installed on wireline. It is applicable to low-to-moderate rate wells, but the efficiency is poor. The power fluid is either water injection quality water or high-pressure crude oil. Unlike an HSP, the power fluid always mixes with the reservoir fluid and must be separated at surface and either reused or disposed of and a fresh supply found.

10.8.6. Gas lift

Gas lift systems aim at lightening the liquid column by injecting gas into it, essentially stimulating natural flow. In its simplest form, hydrocarbon gas is dried and compressed at surface and enters the well via the wellhead. It travels down inside the production casing and, as deep as possible, it enters the completion via a *gas lift valve* (GLV, comprising a check valve and restriction or orifice). The GLV is replaceable by wireline as it sits inside a side pocket mandrel (SPM). The reservoir fluid and lift gas mixes and flows to surface, where some of the gas is recycled. No gas is 'consumed' by gas lift, although the compressor will require power – usually from fuel gas.

Such a system, although simple downhole, does require a high gas pressure to get started when the column of liquid inside the completion is dense. This high pressure can be supplied by a dedicated kick-off compressor. A more common solution however is the use of unloading valves situated in gas lift mandrels distributed down the completion. These valves are initially open and allow shallow gas lift. As the column of liquid in the completion is lightened and either the tubing or the casing pressure reduces, the valves automatically close and thus push the lift gas progressively deeper down the well.

The performance of a gas lift system is monitored by observing injection and produced fluid flowrates and the casing and tubing pressures. Diagnostic surveys such as production logging can also be undertaken to determine if any valves are not working properly; if need be they will be replaced. Optimising the amount of gas going to each well is also critical. Too much lift gas and the system will underperform due to increased friction. Too little gas and the reservoir fluids will not be lightened enough.

An alternative to continuous gas lift is intermittent gas lift. The equipment needed for intermittent gas lift is similar to that needed for continuous gas lift, but the operating principle is different. Whereas in a smoothly operating continuous gas lift system the gas is dispersed in the liquid, intermittent gas lift relies on a finite volume of gas lifting a liquid column to surface at regular intervals, as piston-like as possible. The lift gas can be separated from the oil by a plunger; *plunger assisted gas lift*. This has proven more efficient in viscous crude oils or in crude oils prone to emulsions. The performance of the system is again monitored by observing the casing and tubing pressures.

Figure 10.31 provides an overview of the application envelope and the respective advantages and disadvantages of the various artificial lift techniques. As can be seen, only a few methods are suited for high rate environments: gas lift, ESPs and HSPs. Beam pumps are generally unsuited to offshore applications because of the low rate and bulk of the required surface equipment. Whereas the vast majority of the world's artificially lifted wells are beam pumped, the majority of these are *stripper wells* producing less than 10 bpd.

10.9. SUBSEA VS. PLATFORM TREES

With an offshore development, it is often possible to develop the field using subsea technology and put the Christmas trees on the seabed – hence the term '*wet*' *trees*. Alternatively a platform can be constructed with the trees on the platform – '*dry*' *trees*. In any offshore development, this is a big decision. It is sometimes viewed as simply a capital cost analysis. For a small development in deep water, subsea development to either a host platform or a floating production vessel will be cheaper. For larger well numbers, or for shallower water, the lower *individual well* cost of a platform will promote the use of fixed platform. However, capital costs are

	ESP	JET	GL	HSP	BP	PC
CAPEX*	$$	$	$$$	$$	$	$
OPEX	$$$	$$	$$	$	$	$$

TYPICAL LIFT CAPABILITIES

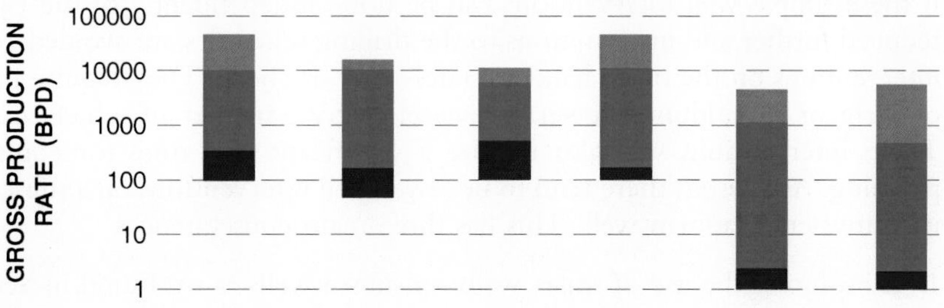

	ESP	JET	GL	HSP	BP	PC
FLEXIBILITY	-	+	++	+	++	+
RELIABILITY		+	++	+	+	+

HYDRAULIC EFFICIENCY

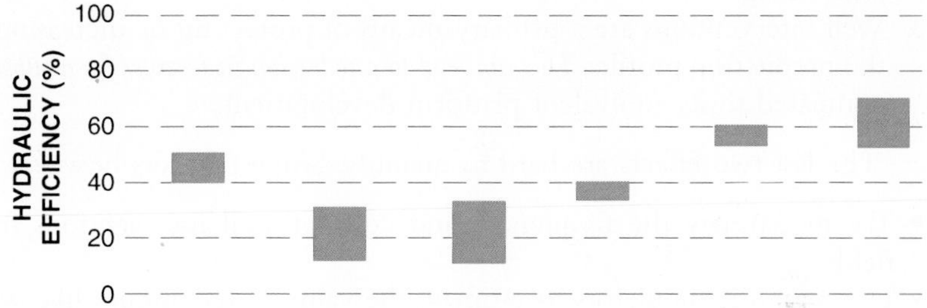

CONSTRAINTS

	ESP	JET	GL	HSP	BP	PC
DRAWDOWN		—	—		++	+
VISCOUS OIL	+	+	—		+	++
SOLIDS	—	+	+	—	—	++
GAS	—	+	++	—	—	—
HIGH TEMP	—	+	++	+	+	—
DEPTH		+	+	+		—
DEVIATED		++	—	++	—	—
DOGLEGGED	+	+	++	+	—	—
OPERATNG & TESTING	—	+	+		+	+
OFFSHORE	+	++	++	++	—	—

* ASSUMING THAT NO INFRASTRUCTURE ALREADY EXISTS

Figure 10.31   Overview of artificial lift techniques.

not the only issue. Other issues include well interventions, reserves and production profiles, operating costs, flexibility and decommissioning.

Well interventions might be for value *protection* (e.g. repairing or preventing corrosion or scale, maintaining artificial lift systems) or value *creation* (e.g. shutting off water or adding artificial lift). If a platform has been properly designed, well intervention from a fixed platform should be relatively straightforward and cheap. If these simple well interventions can be done independently of the rig, costs are reduced further and interruptions to the drilling schedules are avoided. Subsea well interventions on the other hand are rarely easy or cheap. They require mobilisation of a rig or a multipurpose service vessel (MSV); neither of which are low cost. These interventions will also require a significant lead time for equipment and planning. As a result, there tend to be fewer well interventions on a subsea well than its equivalent platform well. This has three main consequences

1. It promotes the use of smart wells on subsea wells as we found in Section 10.7. Smart wells are more expensive.
2. Partly because well interventions are less common subsea, data on the state of the well or the reservoir can be missing. This can reduce the probability of success or increase the cost for the intervention. This loop can be self-reinforcing.
3. Well interventions are a primary means of protecting or increasing reserves and the production profile. *This can lead to a reduction in reserves for a subsea development* compared to its equivalent platform development.

The last two effects are hard to quantify. Some pointers however

- Try to estimate the frequency (and cost) of well interventions from analogue fields.
- Use reservoir simulators to estimate the value of techniques like water shut-off.

Reservoir simulators can automate the decision-making process in a plan and scripts can be used to decide whether these interventions can be done economically. The effect on production profiles and reserves can therefore be quantified.

Subsea developments do however have a few advantages over platform wells when it comes to phasing and sequencing of a field development

1. A platform, by virtue of its fixed size and large cost requires a relative low level of uncertainty in reserves and productivity before the project is sanctioned. A subsea development is much more flexible; the subsea wells and architecture are typically phased in over many years, often with multiple drill sites. The cost of a floating production vessel will typically be lower than its equivalent platform. As a result, a subsea development can be sanctioned with a relative high level of reserves uncertainty.
2. The ability to place the seabed locations of subsea wells remote from the host platform or floating production facility, gives huge flexibility for drilling and construction activities. For example, multiple drilling rigs can be used, or drilling on one location can progress whilst the subsea infrastructure is installed.

3. A very high proportion of a subsea development cost is the flowlines, manifolds and wells. These can all be phased over many years. This delay in CAPEX will have a beneficial effect on the present value (PV) of the project.

Figure 10.32 Subsea horizontal tree (copyright D. Thomas, 2006).

Operating costs for a subsea development will often be higher than for a platform. For example, replacing a choke on a subsea tree will be at least an order of magnitude more expensive than it would be on a platform. Flow assurance (making sure the flowlines do not plug up with scale, wax etc.) may require energy, or chemicals. The colder sea floor temperatures often make these problems more severe (Figure 10.32).

Finally, the decommissioning of a subsea development will be different from that of a fixed platform. Removing a floating production facility will be cheaper than removing a fixed platform, but decommissioning the wells and subsea equipment might be substantially higher.

CHAPTER 11

SURFACE FACILITIES

Introduction and Commercial Application: This section covers the processes applied to fluids produced at the wellhead in preparation for transportation or storage. Oil and gas are rarely produced from a reservoir already at an export quality. More commonly, the process engineer is faced with a mixture of oil, gas and water, as well as small volumes of undesirable substances, which have to be separated and treated for export or disposal. Oil and gas processing facilities also have to be designed to cope with produced volumes which change quite considerably over the field lifetime, whilst the specifications for the end product, for example export crude, generally remain constant. The consequences of a badly designed process can be, for example, reduced throughput or expensive plant modifications after production start-up (i.e. costs in terms of capital spending and loss of income). However, building in overcapacity or unnecessary process flexibility can also be very costly.

Though the type of processing required is largely dependent upon fluid composition at the wellhead, the equipment employed is significantly influenced by location, whether, for example, the facilities are based on land or offshore, in tropical or arctic environments. Sometimes conditions are such that a process which is difficult or expensive to perform offshore can be 'exported' to the coast and handled much more easily on land.

As well as meeting transport or storage specifications, consideration must also be given to legislation covering levels of emission to the environment. Standards in most countries are becoming increasingly rigorous and upgrading in order to reduce emissions can be much more costly once production has started. Engineering skills should be focused on adding greatest value to the product at least cost, whilst working within a consistent set of health, safety and environmental policies.

Most projects can be sub-divided into four parts: wells, gathering system, processing plant and export facilities. Some or all of these components need to be supported on a platform, which can be a land site, the seabed, a fixed steel jacket or a floating structure. Though projects are often characterised by platform type, the design of a project usually starts by consideration of the process required to handle the reservoir fluids. The selection of a platform type can come quite late in the project design and will generally be influenced mainly by the physical environment in which the process plant has to be located. The following sections are laid out with this logic in mind. Process facilities will be discussed first, followed by a description of platform type and selection.

11.1. OIL AND GAS PROCESSING

This section will consider the physical processes which oil and gas (and unwanted fluids) from the wellhead must go through to reach product specifications. These processes will include gas–liquid separation, liquid–liquid separation, drying of gas, treatment of produced water, and others. The *process engineer* is typically concerned with determining the sequence of processes required, and will work largely with chemical engineering principles, and the phase envelopes for hydrocarbons presented in Section 6.2, Chapter 6. The design of the hardware to achieve the processes is the concern of the *facilities engineer*, and will be covered in Section 11.2.

11.1.1. Process design

Before designing a process scheme, it is necessary to know the specification of the raw material input (or *feedstock*) and the specification of the *end product* desired. Designing a process to convert fluids produced at a wellhead into oil and gas products fit for evacuation and storage is no different. The characteristics of the well stream or streams must be known and specifications for the products agreed. At the simplest level the majority of process facilities are designed to split commingled wellhead fluids into three streams of mainly gas, oil and water, as soon as possible (Figure 11.1). Thereafter, each of these streams is further treated to achieve a defined product specification by passing through one or more processes arranged in series. The process streams are described in more detail in Sections 11.1.2 and 11.1.3.

11.1.1.1. Description of wellhead fluids

The quality and quantity of fluids produced at the wellhead is determined by hydrocarbon composition, reservoir character and the field development scheme. Whilst the first two are dictated by nature, the last can be manipulated within technological and market constraints.

The main hydrocarbon properties which will influence process design are

- *PVT characteristics:* which describe whether a production stream will be in the gas or liquid phase at a particular temperature and pressure

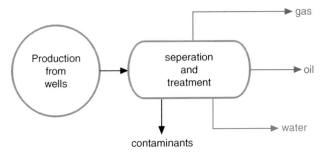

Figure 11.1   Oil and gas process schematic.

- *Composition:* which describes the proportion of hydrocarbon components (C_1–C_{7+}) and non–hydrocarbon substances (e.g. nitrogen, carbon dioxide and hydrogen sulphide) present
- *Emulsion behaviour:* which describes how difficult it will be to separate the liquid phases
- *Viscosity and density:* which help determine how easily the fluids will move through the process facility.

If *formation water* production is expected, a chemical analysis of the water will also be required. It is good practice to record the details of the methods used for sampling and analysis in each case so that measurement uncertainties can be assessed.

In addition to fluid properties, it is important to know how *volumes and rates* will change at the wellhead over the life of the well or field. *Production profiles* are required for oil, water and gas in order to size facilities, and estimates of wellhead temperatures and pressures (over time) are used to determine how the character of the production stream will change. If reservoir pressure support is planned, details of *injected water or gas* which may ultimately appear in the well stream are required.

It is important to put a realistic range of uncertainty on all the information supplied and, at the feasibility study stage, to include all production scenarios under consideration. Favoured options are identified during the field development planning stage as project design becomes firmer. Whilst designing a process for continuous throughput, engineers must also consider the implications of starting up and shutting down the process, and whether special precautions and procedures will be required.

11.1.1.2. Product specification

The end product specification of a process may be defined by a customer (e.g. gas quality), by transport requirements (e.g. pipeline corrosion protection) or by storage considerations (e.g. pour point). Product specifications normally do not change, and one may be expected to deliver within narrow tolerances, though specification can be subject to negotiation with the customer, for example in gas contracts.

Typical product specification for the oil, gas and water would include value for the following parameters:

Oil	True vapour pressure (TVP), BS&W content, temperature, salinity, hydrogen sulphide content
Gas	Water and hydrocarbon dew point, hydrocarbon composition, contaminants content, heating value
Water	Oil and solids content

Table 11.1 provides some quantitative values for typical product specifications.

Table 11.1 Typical product specifications

Oil	True vapour pressure (TVP)	$<83\,kPa$ @ $15°$
	Base sediment and water (BS&W)	<0.5 vol%
	Temperature	$>$ Pour point
	Salinity (NaCl)	$<70\,g/m^3$
	Hydrogen sulphide (H_2S)	$<70\,g/m^3$
Gas	Liquid content	$<100\,mg/m^3$
	Water dew point at $-5°C$	$<7\,Pa$
	Heating value	$>25\,MJ/m^3$
	Composition, CO_2, N_2, H_2S	
	Delivery pressure and temperature	
Water	Dispersed oil content	$<40\,ppm$
	Suspended solids content	$<50\,g/m^3$

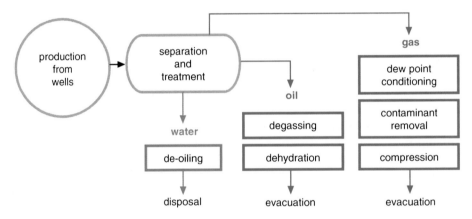

Figure 11.2 A process flow schematic.

11.1.1.3. The process model

Once specifications for the input stream and end product are known, the process engineer must determine the minimum number of steps required to achieve the transformation.

For each step of the process a number of factors must be considered:

- product yield (the volumes of gas and liquids from each stage)
- inter-stage pressure and temperatures
- compression power required (for gas)
- cooling and heating requirements
- flowrates for equipment sizes
- implications of changing production profile.

A schematic diagram describing the process steps required for a mixed well stream is shown in Figure 11.2.

When an oil or gas field has just been discovered, the quality of the information available about the well stream may be sparse, and the amount of detail put into the process design should reflect this. However, early models of the process along with broad cost estimates are needed to progress, and both design detail and cost ranges narrow as projects develop through the feasibility study and field development planning phases (see Chapter 12 for a description of project phases).

11.1.1.4. Process flow schemes

To give some structure to the process design it is common to present information and ideas in the form of *process flow schemes* (PFS). These can take a number of forms and be prepared in various levels of detail. A typical approach is to divide the process into a hierarchy, differentiating the main process from both utility and safety processes.

For example, a PFS for crude oil stabilisation might contain details of equipment, lines, valves, controls and mass and heat balance information where appropriate. This would be the typical level of detail used in the project definition and preliminary design phase described in Chapter 13.

Equipment				
	V-101 (Low Pressure Production Separator)	V-102 (Crude Oil Stabiliser Vessel)		P-101 (Stabilised Crude Oil Pumps)
ID × length (cm)	250 × 750	180 × 720	Capacity (m³/h)	150
Volume (m³)	39.5	19.8	Head (metres of liquid)	23
Type/Make	B.S & B	Kunzel	Type/Make	BS-50F

Operation Stream								
	1		2	3	4[a]	5	6	7
Phase	Vapour	Liquid	Vapour	Liquid	Liquid	Vapour	Liquid	Liquid	Vapour
tons/d	67	2840	67	2840	1996	9	2830	2820	67
kg/sec	0.8	33	0.8	33	23	0.1	33	32.5	0.8
MW or SG	44	0.9	44	0.9	1.04	44	0.9	0.9	43
Density (kg/m³)	5.8	880	5.8	880	1035	4.1	880	875	5.6
Viscosity (mm²/sec)	–	16	–	16	–	–	16	15	–
Pressure (barg)	2.5		2.5	2.5	2.45	1.4		0.05	2.45
Temperature (°C)	41		41	41	43	41		45	34

[a]Normally no flow, design only, for line sizing based on 60% water cut at 3000 m³/d.

A PFS such as the one shown in Figure 11.3 would typically be used as a basis for

- preparing preliminary equipment lists
- advanced ordering of long lead time equipment

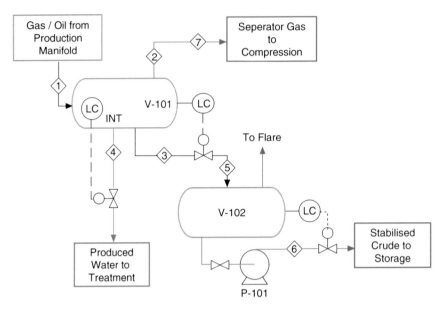

Figure 11.3   Main process flow scheme (PFS).

- preparing a preliminary plant layout
- supporting early cost estimates (25–40% accuracy)
- preparing engineering design sheets
- basic risk analysis.

Detailed engineering design work and preparation of utilities and safety flow schemes will often require input from specialist engineering disciplines such as rotating equipment engineers, and instrument and control engineers. It is common for oil and gas companies to contract out the detailed engineering design and construction work once preliminary designs have been accepted. Utilities refer to support systems such as power, instrumentation, water and safety systems.

Once the main components of the PFS have been selected, the process engineer will perform a preliminary simulation to see whether the process works 'on paper'. This identifies the points at which the flow stream needs, for example, temperature or pressure elevation and second-order pieces of equipment, such as heaters and pumps can be added where appropriate. A process simulation is usually run under a range of operating conditions from start to end of the field life cycle. A check will also be made to ensure that flow will start up again after a plant shutdown.

The diagram given in Figure 11.4 will be discussed in more detail in Chapter 13, but is included here to introduce the different phases of a project, and the corresponding levels of design detail.

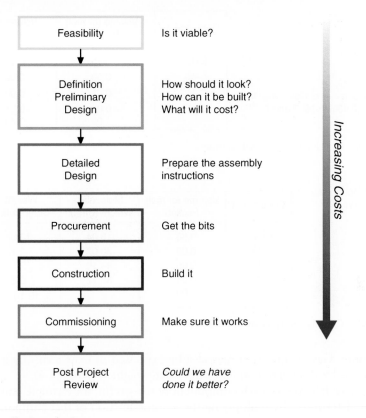

Figure 11.4 Project phasing.

11.1.1.5. Describing hydrocarbon composition

Before oil and gas processing are described in detail in the following sections, it is useful to consider how oil and gas volumes and compositions are reported.

A container full of hydrocarbons can be described in a number of ways, from a simple measurement of the dimensions of the container to a detailed compositional analysis. The most appropriate method is usually determined by what you want to do with the hydrocarbons. If, for example, hydrocarbon products are stored in a warehouse prior to sale, the dimensions of the container are very important, and the hydrocarbon quality may be completely irrelevant for the store keeper. However, a process engineer calculating yields of oil and gas from a reservoir oil sample will require a detailed breakdown of hydrocarbon composition, that is what components are present and in what quantities.

Compositional data are expressed in two main ways: components are shown as a *volume fraction* or as *weight fraction* of the total (Figure 11.5).

The volume fraction would typically be used to represent the make up of a gas at a particular stage in a process and describes gas composition, for example 70% methane and 30% ethane (also known as mol fractions) at a particular temperature

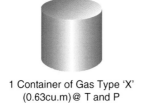

Component	Volume (or mol) fraction	Volume in cu.m.
CH_4	0.85	0.53
C_2H_6	0.09	0.06
C_3H_8	0.05	0.03
C_4H_{10}	0.01	0.01
	1.00	0.63

1 Container of Gas Type 'X'
(0.63cu.m)@ T and P

Figure 11.5   Fractional and actual volumes.

1 Mol of Gas Type 'X'

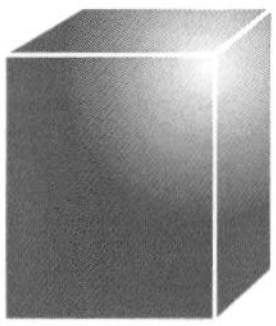

Component	Volume (or mol) fraction	Molecular Weight (g/mol)	Weight Composition
CH_4	0.85	16.04	13.6
C_2H_6	0.09	30.07	2.7
C_3H_8	0.05	44.10	2.2
C_4H_{10}	0.01	58.12	0.6
	1.00		19.1g/mol

Figure 11.6   Calculating (relative) molar mass.

and pressure. Gas composition may also be expressed in mass terms by multiplying the fractions by the corresponding molecular weight (Figure 11.6).

The actual flowrate of each component of the gas (in, for example, cubic metres) would be determined by multiplying the volume fraction of that component by the total flowrate.

For a further description of the chemistry and physics of hydrocarbons, refer back to Section 6.2, Chapter 6.

11.1.2. Oil processing

In this section, we describe hydrocarbon processing in preparation for evacuation, either from a production platform or land-based facilities. In simple terms, this means splitting the hydrocarbon well stream into liquid and vapour phases and treating each phase so that they remain as liquid or vapour throughout the evacuation route. For example, crude must be stabilised to a *TVP* specification to minimise gas evolution during transportation by tanker, and gas must be dew point conditioned to prevent liquid dropout during evacuation to a gas plant.

11.1.2.1. Separation

When oil and gas are produced simultaneously into a separator a certain amount (mass fraction) of each component (e.g. butane) will be in the vapour phase and the rest in the liquid phase. This can be understood using phase diagrams (such as those described in Section 6.2, Chapter 6) which describe the behaviour of multicomponent mixtures at various temperatures and pressures. However, to determine how much of each component goes into the gas or liquid phase, the *equilibrium constants* (or equilibrium vapour liquid ratios) K must be known (Figure 11.7).

These constants are dependent upon pressure, temperature and also the composition of the hydrocarbon fluid, as the various components within the system will interact with each other. K values can be found in gas engineering data books. The basic separation process is similar for oil and gas production, though the relative amounts of each phase differ.

For a *single-stage separator*, that is only one separator vessel, there is an optimum pressure which yields the maximum amount of oil and minimises the carry over of heavy components into the gas phase (a phenomenon called *stripping*). By adding additional separators to the process train the yield of oil can be increased, but with each additional separator the incremental oil yield will decrease (Figure 11.8).

Capital and operating costs will increase as more separator stages are added to the process train, so a balance has to be struck between increased oil yield and cost. It is uncommon to find that economics support more than three stages of separation and one- or two-stage separation is more typical. The increased risk of separation shutdown is also a contributing factor in limiting numbers.

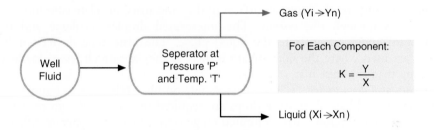

X - Mol fraction of each component in the vapour phase
Y - Mol fraction of each component in the liquid phase

Figure 11.7   Equilibrium constant (K value).

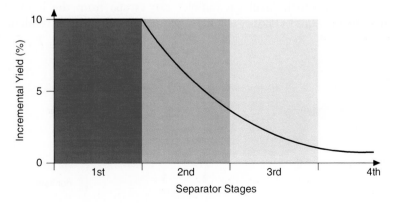

Figure 11.8   Incremental oil yield vs. separator stages.

Multistage separation may also be constrained by low wellhead pressures. The separation process involves a pressure drop, therefore the lower the wellhead pressure the less scope there is for separation.

11.1.2.2. Separation design

Although there are many variations in separator design, certain components are common.

The *inlet section* is designed to separate out most of the liquid phase such as large *slugs* or *droplets* in a two-phase stream. These simple devices redirect the inlet flow towards the liquid at the bottom of the vessel, separating the stream without generating a mist (Figure 11.9).

As small droplets of liquid are usually still present in the gas phase, *demisting* sections are required to recover the liquid mist before it is 'carried over' in the gas stream leaving the separator. The largest liquid droplets fall out of the gas quickly under the action of gravity but smaller droplets (less than 200 μm) require more sophisticated extraction systems.

Impingement demister systems are designed to intercept liquid particles before the gas outlet. They are usually constructed from wire mesh or metal plates and liquid droplets impinge on the internal surfaces of the 'mist mats' or 'plate labyrinth' as the gas weaves through the system. The intercepted droplets coalesce and move downward under gravity into the liquid phase. The plate type devices or 'vane packs' are used where the inlet stream is dirty as they are much less vulnerable to clogging than the mist mat.

Centrifugal demister (or cyclone) devices rely on high velocities to remove liquid particles and substantial pressure drops are required in cyclone design to generate these velocities. Cyclones have a limited range over which they operate efficiently; this is a disadvantage if the input stream flowrate is very variable.

In addition to preventing liquid 'carry over' in the gas phase, gas 'carry under' must also be prevented in the liquid phase. Gas bubbles entrained in the liquid phase must be given the opportunity (or *residence time*) to escape to the gas phase under buoyancy forces.

The ease with which small gas bubbles can escape from the liquid phase is determined by the liquid viscosity; higher viscosities require longer residence times.

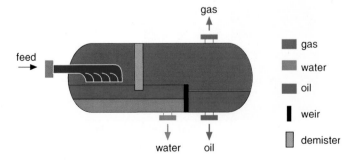

Figure 11.9 A basic three-phase separator.

Typical residence times vary from some 3 minutes for a light crude to up to 20 minutes for very heavy crudes.

In summary, separator sizing is determined by three main factors:

- gas velocity (to minimise mist carry over)
- viscosity (residence time)
- surge volume allowances (up to 50% over normal operating rates).

11.1.2.3. Separator types

Basic separator types can be characterised in two ways: firstly, by main function (bulk or mist separation), and secondly, by *orientation* (either vertical or horizontal).

Knockout vessels are the most common form of basic separator. The vessel contains no internals and demisting efficiency is poor. However, they perform well in dirty service conditions (i.e. where sand, water and corrosive products are carried in the well stream).

Demister separators are employed where liquid carry over is a problem. The recovery of liquids is sometimes less important than the elimination of particles prior to feeding a compression system.

Both separators can be built vertically or horizontally. Vertical separators are often favoured when high oil capacity and ample surge volume is required, though degassing can become a problem if liquid viscosity is high (gas bubbles have to escape against the fluid flow). Horizontal separators can handle high gas volumes and foaming crudes (as degassing occurs during cross flow rather than counter flow). In general, horizontal separators are used for high flowrates and high gas–liquid ratios (GLRs) (Figure 11.10).

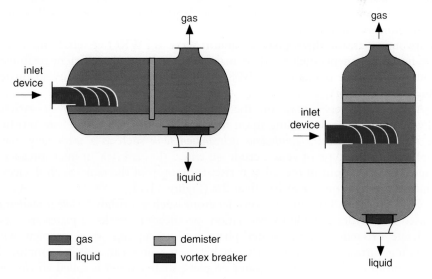

Figure 11.10   Horizontal and vertical demister separators.

11.1.2.4. Dehydration and water treatment

Produced water has to be separated from oil for two main reasons: firstly, because the customer is buying oil not water, and secondly, to minimise costs associated with evacuation (e.g. volume pumped, corrosion protection for pipelines). A water content of less than 0.5% is a typical specification for sales crude.

Water separated from oil usually contains small amounts of oil which have to be removed before the water can be released to the environment. Specifications are getting tighter but standards ranging from 10 to 100 ppm (parts per million) *oil in water* before disposal are currently common. In most areas, 40 ppm of oil in water is the legal requirement, that is 40 mg/l.

The simplest way to dehydrate or de-oil an oil–water mixture is to use *settling* or *skimming tanks*, respectively. Over time the relative density differences will separate the two liquids. On land in situations where weight and space considerations are not an issue, this can often be accomplished by using combined settling/storage tanks in a tank farm, either within a field or at an export facility. Unfortunately, this process takes time and space, both of which are often a constraint in offshore operations. Highly efficient equipment which combines both mechanical and chemical separation processes are now more commonly used.

11.1.2.5. Dehydration

The choice of dehydration equipment is primarily a function of how much water the well stream contains. Where water cut is high it is common to find a 'free water knockout vessel' (FWKO) employed for primary separation. The vessel is similar to the separators described earlier for oil and gas. It is used where large quantities of water need to be separated from the oil/oil emulsion. The incoming fluid flows against a diverter plate which causes an initial separation of gas and liquid. Within the liquid column an oil layer will form at the top, a water layer at the bottom and dispersed oil and oil emulsion in the middle. With time, coalescence will occur and the amount of emulsion will reduce. If considerable amounts of gas are present in the incoming stream, three-phase separators serve as FWKO vessels. Some FWKOs have heating elements built in, their main purpose being the efficient treatment of emulsions. In some operations, a FWKO will produce oil of a quality adequate for subsequent transport.

A knockout vessel may, on the other hand, be followed by a variety of dehydrating systems depending upon the space available and the characteristics of the mixture. On land a *continuous dehydration tank* such as a *wash tank* may be employed. In this type of vessel, crude oil enters the tank via an inlet spreader and water droplets fall out of the oil as it rises to the top of the tank. Such devices can reduce the water content to less than 2% (Figure 11.11).

Where space and weight are considerations (such as offshore) *plate separators* may be used to dehydrate crude to evacuation specification. Packs of plates are used to accelerate extraction of the water phase by intercepting water droplets with a coalescing surface. An electric charge between plates can also be employed to further promote the capture of water droplets, such a vessel is called an *electrostatic coalescer*. Plate separators have space requirements similar to that of knockout vessels.

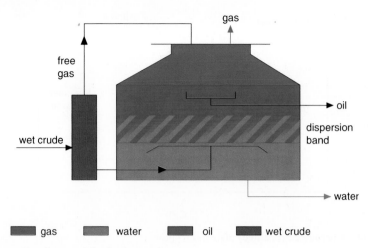

Figure 11.11 Continuous dehydration tank.

For dehydration of very high viscosity crudes, *heaters* can be used in combination with dehydration tanks. The temperature to which the crude is heated is a function of the viscosity required for effective separation.

If oil and water are mixed as an *emulsion,* dehydration becomes much more difficult. Emulsions can form as oil-in-water or water-in-oil if mixed production streams are subjected to severe turbulence, as might occur in front of perforations in the borehole or in a pump. Emulsions can be encouraged to break (or destabilise) using chemicals, heat or just gentle agitation. *Chemical destabilisation* is the most common method and laboratory tests would normally be conducted to determine the most suitable combination of chemicals.

11.1.2.6. De-oiling
Skimming tanks have already been described as the simplest form of de-oiling facility; such tanks can reduce oil concentrations down to less than 200 ppm but are not suitable for offshore operations. Plate coalescer vessels or hydrocyclones (shown in Figure 11.13) are generally used offshore and maybe used in tandem with other water-cleaning devices such as gas flotation units. The choice of de-oiling unit is influenced by throughput, variability of the feed (in terms of oil content), space and weight considerations.

Another type of gravity separator used for small amounts of oily water, the *oil interceptor,* is widely used both offshore and onshore. These devices work by encouraging oil particles to coalesce on the surface of plates. Once bigger oil droplets are formed they tend to float to the surface of the water faster and can be skimmed off. A *corrugated plate interceptor* (CPI) is shown in Figure 11.12 and demonstrates the principle involved. Plate interceptors can typically reduce oil-in-water content to 50–150 ppm.

To reduce oil content to levels which meet disposal standards it is often necessary to employ rather more sophisticated methods. Two such techniques which can reduce oil in water to less than 40 ppm are *gas flotation* and *hydrocyclone* processes.

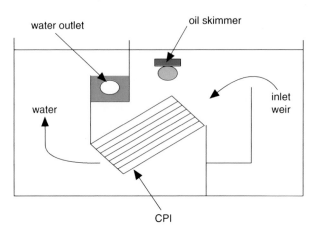

Figure 11.12 Corrugated (or tilted) plate interceptor.

In a gas flotation unit, air is bubbled through oily water to capture oil particles which then rise with the bubble to form a scum at the surface of the flotation unit. The scum can be removed by rotating paddles. Chemicals are often added to destabilise the inlet stream and enhance performance.

Hydrocyclones (Figure 11.13) have become common on offshore facilities and rely on centrifugal force to separate light oil particles from the heavier water phase. As the inlet stream is centrifuged, the heavier water phase is 'spun' to the outside of the cyclone whilst oil particles move to the centre of the cyclone, coalesce and are drawn off upwards. The heavier water is taken out at the bottom.

To ensure disposal water quality is in line with regulatory requirements (usually 40 ppm), the oil-in-water content is monitored by solvent extraction and infrared spectroscopy. The specification of 40 ppm refers to an oil-in-water content typically averaged over a 1 month period. Oil-in-water standards are generally tightening and whilst 40 ppm remains acceptable in some areas, 10 ppm or less is becoming more common. In closed marine environments such as the Caspian Sea, partially closed environments like the Arabian Gulf and elsewhere, many companies are re-injecting produced water back into reservoirs to meet prevailing or self-imposed corporate environmental standards. This alternative is known as *produced water re-injection* (PWRI).

11.1.2.7. Multiphase pumps

Since the 1990s, multiphase pumps, capable of simultaneously pumping oil, water and gas, have gained increased acceptance in oil field production. In some cases, they have replaced conventional processing equipment at the wellhead or platform, effectively exporting processing requirements downstream to central gathering facilities. Eliminating in-field equipment such as separators, compressors, individual pumping equipment, heaters, gas flares and separate flowlines can also significantly reduce processing costs.

Additional benefits include reduced environmental impact for onshore installations as multiphase pumps require only a fraction of the space conventional

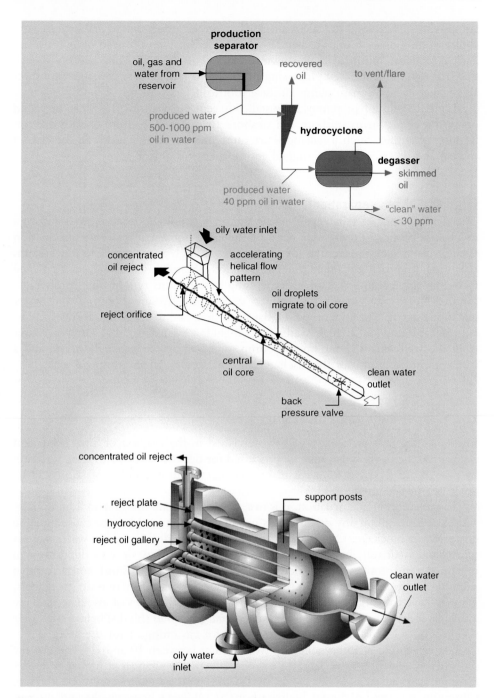

Figure 11.13 Hydrocyclone for oil-in-water removal.

equipment needs, resulting in much smaller project footprints. Furthermore, the ability of multiphase pumps to handle gas in a closed system instead of venting and flaring reduces emissions to the environment. The ability to pressure boost well flows to remote centralised processing units and to handle very low inlet pressures also make the multiphase pump an ideal tool to develop marginal fields.

Typically, multiphase pumps are installed in remote locations. They can be found on wellhead platforms offshore, on onshore fields far away from the production facilities or subsea. Such installations are often unmanned (obviously in the case of subsea) and therefore require rugged and reliable equipment.

11.1.2.8. Subsea separation
The development of reliable subsea separation technologies that allow companies to process offshore production without the need for above sea level production facilities have been progressing for some years, though as yet there are few systems operating other than on a field trial basis. In 2007, Statoil installed what was claimed to be the first commercial subsea processing station of its type in the Tordis Field (in the North Sea) which is a subsea separation, pressure boosting and water injection system. The system, which has been designed to allow the Tordis Field to continue production at high water cuts, includes a separator that removes water from the well stream, a multiphase pump for boosting the production rate and a water injection pump for the discharge of separated water in a disposal well.

11.1.3. Upstream gas processing

In this section, gas processing will be described in the context of 'site' needs and evacuation, that is how gas may be processed for disposal or prior to transportation by pipeline to a downstream gas plant. Gas fractionation and liquefaction will be described in Section 11.1.4.

To prepare gas for evacuation, it is necessary to separate the gas and liquid phases and extract or inhibit any components in the gas which are likely to cause pipeline corrosion or blockage. Components which can cause difficulties are water vapour (corrosion, hydrates), heavy hydrocarbons (two-phase flow or wax deposition in pipelines), and contaminants such as carbon dioxide (corrosion) and hydrogen sulphide (corrosion, toxicity). In the case of associated gas, if there is no gas market, gas may have to be flared or re-injected. If significant volumes of associated gas are available, it may be worthwhile to extract *natural gas liquids* (NGLs) before flaring or re-injection. Gas may also have to be treated for gas lifting or for use as a fuel.

Gas processing facilities generally work best at between 10 and 100 bar. At low pressure, vessels have to be large to operate effectively, whereas at higher pressures facilities can be smaller but vessel walls and piping systems must be thicker. Optimum recovery of heavy hydrocarbons is achieved between 20 and 40 bar. Long-distance pipeline pressures may reach 150 bar and re-injection pressure can be as high as 700 bar. The gas process train will reflect gas quality and pressure as well as delivery specifications.

11.1.3.1. Pressure reduction

Gas is sometimes produced at very high pressures which have to be reduced for efficient processing and to reduce the weight and cost of the process facilities. The first pressure reduction is normally made across a choke before the well fluid enters the primary oil–gas separator.

Note that primary separation has already been described in Section 11.1.2.

11.1.3.2. Gas dehydration

If produced gas contains water vapour, it may have to be dried (dehydrated). Water condensation in the process facilities can lead to *hydrate formation* and may cause corrosion (pipelines are particularly vulnerable) in the presence of carbon dioxide and hydrogen sulphide. Hydrates are formed by physical bonding between water and the lighter components in natural gas. They can plug pipes and process equipment. Charts such as the one given in Figure 11.14 are available to predict when hydrate formation may become a problem.

Dehydration can be performed by a number of methods: cooling, absorption and adsorption. Water removal by *cooling* is simply a condensation process; at lower temperatures the gas can hold less water vapour. This method of dehydration is often used when gas has to be cooled to recover heavy hydrocarbons. *Inhibitors* such as glycol may have to be injected upstream of the chillers to prevent hydrate formation.

One of the most common methods of dehydration is *absorption* of water vapour by *tri-ethylene glycol (TEG)* contacting. Gas is bubbled through a contact tower and water is absorbed by the glycol. Glycol can be regenerated by heating to boil off the water. In practice, glycol contacting will reduce water content sufficiently to prevent water dropout during evacuation by pipeline. Glycol absorption should not be confused with glycol (hydrate) inhibition, a process in which water is not removed (Figure 11.15).

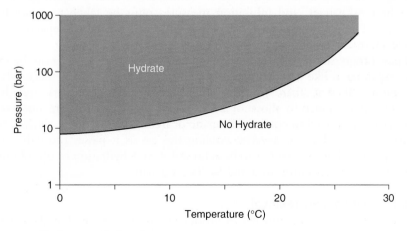

Figure 11.14 Hydrate prediction plot.

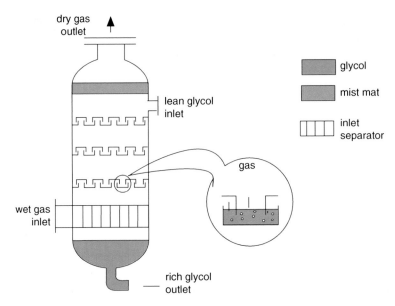

Figure 11.15   Glycol contacting tower.

11.1.3.3. Heavy hydrocarbon removal

Condensable hydrocarbon components are usually removed from gas to avoid liquid dropout in pipelines, or to recover valuable NGLs where there is no facility for gas export. Cooling to ambient conditions can be achieved by air or water heat exchange, or to sub-zero temperatures by gas expansion or refrigeration. Many other processes such as compression and absorption also work more efficiently at low temperatures.

If high wellhead pressures are available over long periods, cooling can be achieved by expanding gas through a valve, a process known as *Joule Thomson (JT) throttling*. The valve is normally used in combination with a liquid gas separator and a heat exchanger, and inhibition measures must be taken to avoid hydrate formation. The whole process is often termed 'low temperature separation' (LTS) (Figure 11.16).

If gas compression is required following cooling, a *turbo-expander* can be used. A turbo-expander is like a centrifugal compressor in reverse, and is thermodynamically more efficient than JT throttling. Pressure requirements and hydrate precautions are similar to those of LTS. When high pressures are not available, *refrigeration* can be used to cool gas. Propane or freon is compressed, allowed to cool and then expanded across a valve, cooling the gas as it passes through a chiller. Temperatures as low as $-40°C$ can be achieved. Gas dehydration or glycol injection must precede the operation to avoid hydrate formation.

11.1.3.4. Contaminant removal

The most common contaminants in produced gas are carbon dioxide (CO_2) and hydrogen sulphide (H_2S). Both can combine with free water to cause corrosion and

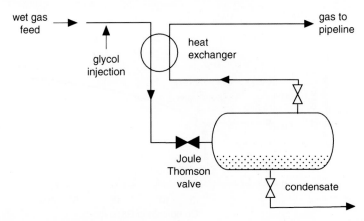

Figure 11.16   Low temperature separation (LTS).

H_2S is extremely toxic even in very small amounts (less than 0.01% volume can be fatal if inhaled). Because of the equipment required, extraction is performed onshore whenever possible, and providing gas is dehydrated, most pipeline corrosion problems can be avoided. However, if third party pipelines are used, it may be necessary to perform some extraction on site prior to evacuation to meet pipeline owner specifications. Extraction of CO_2 and H_2S is normally performed by absorption in contact towers like those used for dehydration, though other solvents are used instead of glycol.

Historically, CO_2 removed from hydrocarbon gas was vented, in most cases it still is, but since CO_2 is increasingly associated with global climate change more companies are making efforts to capture and store the gas. Amine, an organic compound derived from ammonia, can be used both as a CO_2 and H_2S absorber in contact towers. The CO_2- and/or H_2S-rich amine output stream is subsequently heated and de-pressurised so that the CO_2 and H_2S gas is boiled off and can be injected into, for example, a depleted gas reservoir or aquifer.

11.1.3.5. Pressure elevation (gas compression)

After passing through several stages of processing, gas pressure may need to be increased before it can be evacuated, used for gas lift or re-injected. If gas flowing from wells at a low wellhead pressure requires processing at higher separator pressures, inter-stage compression may be required.

The main types of compressor used in the gas industry are *reciprocating* and *centrifugal compressors*. The power requirements of a reciprocating compressor are shown in Figure 11.17. Notice that at compression ratios above 5 and 25, two-stage and three-stage compression, respectively, becomes necessary to accommodate inter-stage cooling. Apart from the need for inter-stage cooling, additional compression capacity may be installed in phases through the life of a gas field as reservoir pressure declines.

Gas turbine driven centrifugal compressors are very efficient under the right operating conditions but require careful selection and demand higher levels of

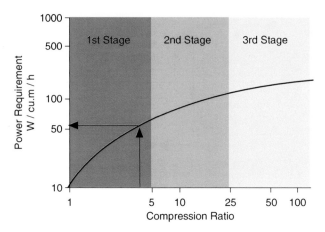

Figure 11.17 Compression power.

maintenance than reciprocating compressors. Compression facilities are generally the most expensive item in an upstream gas process facility.

11.1.4. Downstream gas processing

The gas processing options described in the previous section were designed primarily to meet on-site usage or evacuation specifications. Before delivery to the customer further processing would normally be carried out at dedicated gas processing plants, which may receive gas from many different gas and oil fields. Gas piped to such plants is normally treated to prevent liquid dropout under pipeline conditions (dew point control) but may still contain considerable volumes of NGLs and also contaminants.

The composition of natural gas varies considerably from lean non–associated gas which is predominantly methane to rich associated gas containing a significant proportion of NGLs. *NGLs* are those components remaining once methane and all non-hydrocarbon components have been removed, that is $(C_2–C_{5+})$ (Figure 11.18).

Butane (C_4H_{10}) and propane (C_5H_{12}) can be further isolated and sold as *liquefied petroleum gas* (LPG). This is commonly seen as bottled gas and is a useful method of distributing energy to remote areas. LPG is further discussed at the end of Section 11.1.4.

Sales gas, which is typically made up of methane (CH_4) and small amounts of ethane (C_2H_6), can be exported by refrigerated tanker rather than by pipeline and has to be compressed by a factor of 600 (and cooled to $-160°C$). This is then termed *liquefied natural gas* (LNG).

11.1.4.1. Contaminant removal

Although gas may have been partially dried and dew point controlled (for hydrocarbons and water) prior to evacuation from the site of production, some

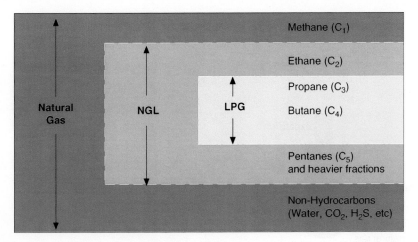

Figure 11.18 Terminology of natural gas.

heavier hydrocarbons, water and other non-hydrocarbon components can still be present. Gas arriving at the gas plant may pass through a 'slug catcher', a device which removes any slugs of liquid which have condensed and accumulated in the pipeline during the journey. Following this, gas is dehydrated, processed to remove contaminants and passed through a demethaniser to isolate most of the methane component (for 'sales' gas). Specifications for sales gas may accommodate small amounts of impurities such as CO_2 (up to 3%), but gas feed for either LPG or LNG plants must be free of practically all water and contaminants (Figure 11.19).

Gases which are high in H_2S are subject to a *de-sulphurisation* process in which H_2S is converted into elemental sulphur or a metal sulphide. There are a number of processes based on absorption in contactors, adsorption (to a surface) in molecular sieves or chemical reaction (e.g. with zinc oxide).

Carbon dioxide (CO_2) will solidify at the temperatures required to liquefy natural gas, and high quantities can make the gas unsuitable for distribution. Removal is usually achieved in contacting towers.

Water can be removed by *adsorption* in molecular sieves using solid desiccants such as silica gel. More effective desiccants are available and a typical arrangement might have four drying vessels: one in adsorbing mode, one being regenerated (heating to drive off water), one cooling and a fourth on standby for when the first becomes saturated with water.

11.1.4.2. Natural gas liquid recovery

When gases are rich in ethane, propane, butane and heavier hydrocarbons and there is a local market for such products, it may be economic to recover these condensable components. NGLs can be recovered in a number of ways, some of which have already been described in the previous section. However, to maximise recovery of the individual NGL components, gas would have to be processed in a *fractionation plant*.

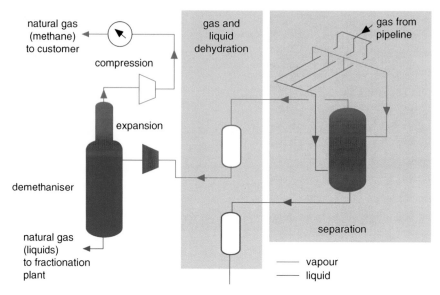

Figure 11.19   Gas separation facilities.

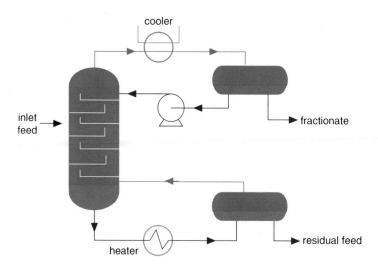

Figure 11.20   Fractionating column.

In such a plant, the gas stream passes through a series of fractionating columns in which liquids are heated at the bottom and partly vaporised, and gases are cooled and condensed at the top of the column. Gas flows up the column and liquid flows down through the column, coming into close contact at trays in the column. Lighter components are stripped to the top and heavier products stripped to the bottom of the tower (Figure 11.20).

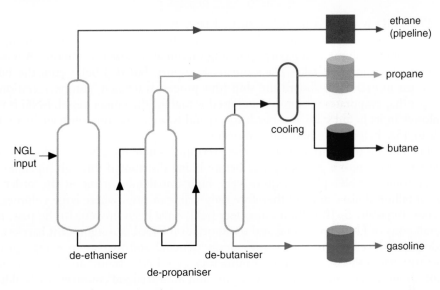

Figure 11.21   NGL fractionation plant.

In the fractionation plant, the first column (de-ethaniser) removes ethane which, after treatment for storage, may be used as feed for an ethylene plant. The heavier hydrocarbons pass to the next fractionating column (or de-propaniser) where propane is removed and so on, until butane has been separated and the remaining NGLs can be stored as natural gasoline. The lighter components can only be recovered at very low temperatures – ethane, for example, has to be reduced to $-100°C$. Propane can be stored as a liquid at about $-40°C$ and butane at $0°C$. Natural gasoline does not require cooling for storage (Figure 11.21).

Gas fractionation plants require considerable investment and in many situations would not be economic. However, less complete NGL recovery methods may still prove cost-effective.

The component factor gives the unit yield for each component and includes a volume conversion factor. The factors can be obtained from tables.

11.1.4.3. Liquefied natural gas

Where the distance to the customer is very large or where a gas pipeline would have to cross too many countries, gas may be shipped as a liquid. To condition the gas for liquefaction any CO_2, H_2S, water and heavier hydrocarbons must be removed, by the methods already discussed. The choice of how much propane and butane to leave in the LNG depends upon the heating requirements negotiated with the customer. At a LNG plant consisting of one or more 'trains', gas is condensed by chilling it to around -120 to $-170°C$ and compressing it up to 60 atm/870 psi. Once the gas has been condensed the pressure is reduced for storage and shipping. In order to keep the gas in liquid form, the LNG must be kept at temperatures below $-83°C$ independent of pressure.

The LNG is shipped in well-insulated tankers as a very cold liquid at boiling point (typically −160°C) under close to atmospheric pressure, where it occupies 1/600th of the volume of the corresponding vapour at standard conditions. A minor amount of evaporation takes place during transport, but this helps cool the bulk of the gas in each tank aboard the ship (this process is termed 'autorefrigeration'). The gas that evaporates is captured and used as fuel for the tanker vessel. LNG is not explosive in its liquid state, but needs to expand to a gas and mix with air in a ratio of 5 to 15% before an explosion can occur.

The LNG is transported to a receiving terminal where the LNG is unloaded and converted back into a gaseous phase before being distributed through pipelines to the customers. LNG plants require very high initial investment in the order of several billion dollars, and are therefore only viable in cases where large volumes of reserves (typically 5–10 trillion cubic feet [tcf]) have been proven. In the past, the overall costs of LNG processing and transportation have been significant barriers to widespread development, but rising energy demand and increasing energy prices have stimulated the construction of numerous new LNG 'trains' and regasification plants in several countries. Both developing and industrialised countries are building LNG regasification plants in order to diversify their energy supply and increase energy security.

Overall LNG demand is forecast to grow three-fold from 2006 to 2020, from 160 million to around 500 million tons/year. The main suppliers are Indonesia, Malaysia, Brunei, Australia, Trinidad, Algeria and Qatar, with new liquefaction plants under construction or recently on stream in, for example Egypt, Nigeria, Oman, Norway and Russia. The number of LNG reception terminals is also increasing, with several new regasification plants already built or under construction in, for example Great Britain, Canada, USA, Spain, China and India, with the result that there is now an emerging global spot market in LNG. A further development is floating LNG facilities, whereby a complete LNG plant is built on a vessel in order to bring the facilities to the gas deposits, which would be a viable scenario for stranded gas accumulations, or regions where there is an absence of gas infrastructure and/or no local gas market (Figure 11.22).

11.1.4.4. Gas to liquids

Gas to liquids (GTL) technology has been available for a long time, but has challenging economic hurdles. GTL provides an alternative method for commercialising stranded or remote gas resources, in addition to well-established technologies like LNG plants and gas pipelines. Shell has had a GTL plant in operation at Bintulu, Malaysia since 1993.

The term GTL conversion refers to a group of processes that convert natural GTL fuels, which are easier and cheaper to transport, market and distribute. Another advantage of GTL is that the products contain less pollutants and fine particulates than conventional liquid fuels, and have a better energy yield (higher cetane number) resulting in improved engine performance.

In theory virtually any hydrocarbon can be synthesised from any other, and since the 1920s several processes have emerged that can be used to produce synthetic

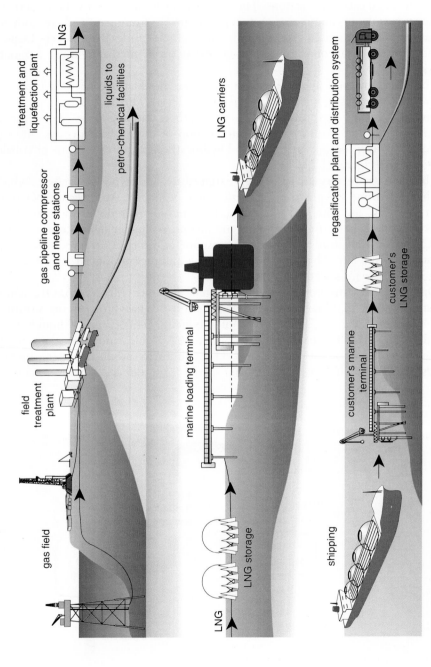

Figure 11.22 Schematic diagram of full chain of LNG facilities.

liquid fuels from natural gas. Essentially, all the processes chemically bind short-chain (gas) hydrocarbon molecules together to form long-chain (liquid) hydrocarbons through catalytic reactions. The two main technologies for GTL to derive a synthetic petroleum liquid are a direct conversion from gas, and an indirect conversion via synthesised gas (syngas), for example using the Fischer–Tropsch (F-T) synthesis with cobalt or iron catalysts, or from methanol using a zeolite catalyst.

The direct method avoids the cost of syngas production, but is difficult to control and requires a high activation energy level. The F-T syngas processes convert the natural gas to hydrogen and carbon monoxide by either steam reforming or partial oxidation, or a combination of these two processes, and the subsequent conversion of syngas to liquid hydrocarbons requires an iron- or cobalt-based catalyst (Figure 11.23).

The key parameters are pressure, temperature and type of catalyst, which together determine the grade of synthetic crude produced. At lower temperatures (e.g. 180–250°C and with a cobalt catalyst) the syncrude is predominantly diesel and waxes which are almost free of sulphur and olefins. At higher temperatures around 330°C and with an iron catalyst mainly gasoline and olefins are produced, again almost free of sulphur. The range and grade of liquid fuels produced can be further optimised by careful selection of the catalysts.

One of the leading companies in this arena is South Africa's SASOL, which in partnership with Chevron and Qatar Petroleum has brought on stream a GTL plant in Qatar in 2006 capable of producing 34,000 bpd of liquids from a feed of 330 MMscf/d of gas. These plants are scalable, and firm plans are in progress for expansion of the Qatar GTL facilities to produce 450,000 bpd of liquids by 2015. The overall costs per barrel of produced liquids, including

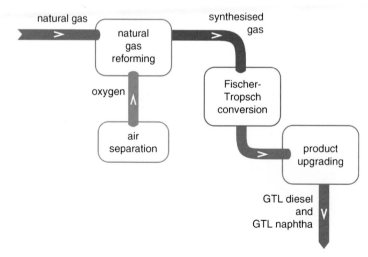

Figure 11.23 Schematic flowchart for Fischer–Tropsch GTL conversion process (source: www.sasolchevron.com/technology.htm).

gas feedstock, CAPEX, and operating expenditure (OPEX), are in the range US$20–25 per barrel.

11.1.4.5. Liquefied petroleum gas

Liquefied petroleum gas (LPG) is a mixture of propane and butane used as fuel for heating appliances, vehicles and also as a replacement for environmentally damaging gases previously used as refrigerants and aerosol propellants (e.g. chlorofluorocarbons (CFCs)). The ratio of propane and butane depends on the intended use of the LPG, and also indirectly on the season (summer vs. winter).

LPG is generated as a by-product during crude oil refining, and is also recovered from oil and gas production during the standard processing at the surface. As LPG has a significantly higher calorific value than typical natural gas, it cannot be substituted directly for methane, but has to be diluted with air to produce a synthetic natural gas (SNG) that can be used in emergency back-up systems for civilian and military installations, and more widely in emerging markets before a natural gas distribution system has been fully developed. Where LPG is used as fuel in motor vehicles it is referred to as autogas, and is stored in a separate tank within the vehicle and blended in with the flow of petrol to the engine.

11.2. FACILITIES

The hardware items with which the processes described in Section 11.1 are achieved are called facilities, and are designed by the facilities engineer. The previous section described the equipment items used for the main processes such as separation, drying, fractionation and compression. This section will describe some of the facilities required for the systems which support production from the reservoir, such as gas injection, gas lift and water injection, and also the transportation facilities used for both offshore and land operations.

11.2.1. Production support systems

Although the type of production support systems required depend upon reservoir type the most common include

- water injection
- gas injection
- artificial lift.

11.2.1.1. Water injection

Water may be injected into the reservoir to supplement oil recovery or to dispose of produced water. In some cases, these options may be complementary. Water will generally need to be treated before it can be injected into a reservoir, whether it is 'cleaned' seawater or produced water. Once treated it is injected into the reservoir, often at high pressures. Therefore, to design a PFS for water injection one needs specifications of the source water and injected water.

Possible *water sources* for injection are seawater, fresh surface water, produced water or aquifer water (not from the producing reservoir). Once it has been established that there is enough water to meet demand (not an issue in the case of seawater), it is important to determine what type of treatment is required to make the water suitable for injection. This is investigated by performing laboratory tests on representative water samples.

The principle parameters studied in an analysis are

(1) *Dissolved solids* to determine whether precipitates (such as calcium carbonate) will form under injection conditions or due to mixing with formation waters.
(2) *Suspended solids* (such as clays or living organisms) which may reduce injection potential or reservoir permeability.
(3) *Suspended oil* content where produced water is considered for re-injection. Oil particles can behave like suspended solids.
(4) *Bacteria* which may contribute to formation impairment or lead to reservoir souring (generation of H_2S).
(5) *Dissolved gases* which may encourage corrosion and lead to reservoir impairment by corrosion products.

The likely impact of each of these parameters on injection rates or formation damage can be simulated in the laboratory, tested in a pilot scheme or predicted by analogy with similar field conditions (Table 11.2).

Once injection water treatment requirements have been established, process equipment must be sized to deal with the anticipated throughput. In a situation where water injection is the primary source of reservoir energy it is common to apply a *voidage replacement* policy, that is produced volumes are replaced by injected volumes. An allowance above this capacity would be specified to cover equipment downtime (Figure 11.24).

Table 11.2 Water treatment considerations

Problems	Possible Effect	Solution
Suspended solids	Formation plugging	Filtration
Suspended oil	Formation plugging	Flotation or filtration
Dissolved precipitates	Scaling and plugging	Scale inhibitors
Bacteria	Loss of injectivity (corrosion products) and reservoir souring	Biocides and selection of sour service materials
Dissolved gas	Facilities corrosion and loss of injectivity	Degasification

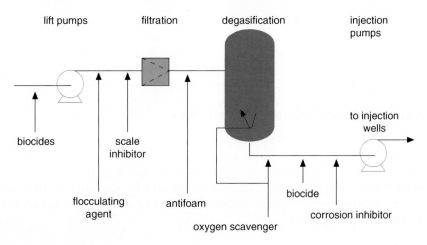

Figure 11.24 Injection water treatment scheme.

11.2.1.2. Gas injection

Gas can be injected into reservoirs to supplement recovery by maintaining reservoir pressure or as a means of disposing of gas which cannot be flared under environmental legislation, and for which no market exists.

Facilities for the treatment and compression of gas have already been described in earlier sections. However, there are a number of differences in the specifications for injected gas that differ from those of export gas. Generally, there is no need to control hydrocarbon dew point as injected gas will get hotter not cooler, but it may be attractive to remove heavy hydrocarbons for economic reasons. Basic liquid separation will normally be performed, and due to the high pressures involved it will nearly always be necessary to dehydrate the gas to avoid water dropout.

Injection gas pressures are usually much higher than lift-gas or gas pipeline pressures and special care has to be taken to select compressor lubricants that will not dissolve in high-pressure gas. Such a situation could lead to inadequate lubrication and may impair well injectivity.

11.2.1.3. Artificial lift

The most common types of artificial lift are gas lift, beam pumping and downhole pumping, and the mechanics of these systems are described in Section 10.8, Chapter 10. Gas-lifting systems require a suitable gas source though at a lower pressure than injection gas. Gas treatment considerations are similar except that heavy ends are not normally stripped out of the gas, as a lean gas would only resaturate with NGLs from the producing crude in the lifting operation. Gas compression can be avoided if a gas source of suitable pressure exists nearby, for example an adjacent gas field. Since gas lift is essentially a closed-loop system, little gas is consumed in gas-lifting operations but gas must be available for starting up operations after a shutdown ('kicking off' production). Alternatively, nitrogen pumped through coiled tubing could be used for kick off, though this is expensive and may be subject to availability restrictions.

Beam pumping and ESPs require a source of power. On land it may be convenient to tap into the local electricity network, or in the case of the beam pump to use a diesel powered engine. Offshore (ESP only) provision for power generation must be made to drive downhole electric pumps.

11.2.2. Land-based production facilities

We have so far discussed process design and processing equipment rather than the *layout* of production facilities. Once a process scheme has been defined, the fashion in which equipment and plant is located is determined partly by transportation considerations (e.g. pipeline specifications) and also by the surface environment. However, regardless of the surface location, a number of issues will always have to be addressed. These include

- how to gather well fluids
- how and where to treat produced fluids
- how to evacuate or store products.

Providing the land surface above a reservoir is relatively flat, it is generally cheaper to drill and maintain a vertical well than to access a reservoir from a location that requires a deviated borehole. In unpopulated areas such as desert or jungle locations, it is common to find that the pattern of wellheads at surface closely reflects the pattern in which wells penetrate the reservoir. However, in many cases constraints will be placed on drill site availability as a result of housing, environmental concerns or topography. In such conditions, wells may be drilled in *clusters* from one or a number of sites as close as possible to the surface location of the reservoir.

Figure 11.25 shows a typical arrangement of land-based facilities in a situation where there are some constraints on the location of wellsites and processing plant.

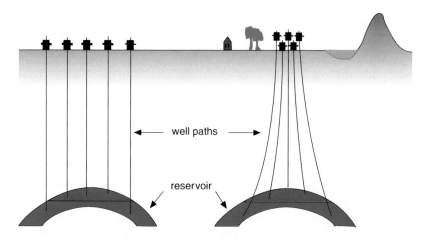

Figure 11.25   Impact of surface constraints on drilling.

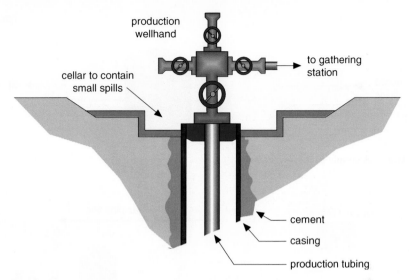

Figure 11.26   Single wellhead arrangement.

11.2.2.1. Wellsites

The first function of a wellsite is to accommodate drilling operations. However, a wellsite must be designed to allow access for future operations and maintenance activity, and in many cases provide containment in the event of accidental emission. Production from a single wellhead or wellhead cluster is routed by pipeline to a gathering station, often without any treatment. In this case, the pipeline effectively becomes an extension of the production tubing. If a well is producing naturally or with assistance from a downhole pump, there may be little equipment on site during normal operations, apart from a wellhead and associated pipework (Figure 11.26).

11.2.2.2. Gathering stations

The term 'gathering station' may describe anything from a very simple gathering and pumping station to a complex processing centre in which produced fluids are treated and separated into gas, NGLs and stabilised crude.

If several widely spaced fields are feeding a single gathering and treatment centre, it is common to perform primary separation of gas and oil (and possibly water) in the field. A *field station* may include a simple slug catcher, temporary storage tanks and pumps for getting the separated fluids to the main gathering and treatment centre (Figure 11.27).

A complex gathering station may include facilities to separate produced fluids, stabilise crude for storage, dehydrate and treat sales gas, and recover and fractionate NGLs. Such a plant would also handle the treatment of waste products for disposal (Figure 11.28).

On a land site where space and weight are not normally constraints, advantage can be taken of tank type separation equipment such as wash tanks and settling tanks, and

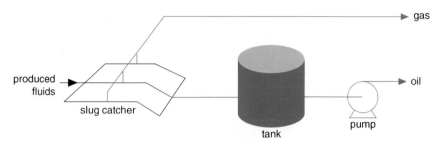

Figure 11.27 A simple gathering station.

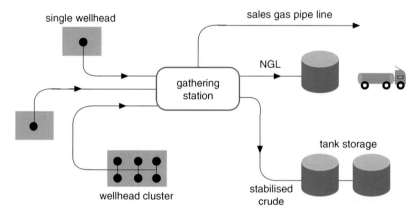

Figure 11.28 Land-based production facilities.

batch processing methods. Such equipment is generally cheaper to maintain than continuous throughput vessels, though a combination of both may be required.

11.2.2.3. Evacuation and storage

Once oil and gas have been processed the products have to be evacuated from the site. Stabilised crude is normally stored in tank farms at a distribution terminal which may involve an extended journey by pipeline. At a distribution terminal, crude is stored prior to further pipeline distribution or loading for shipment by sea (Figure 11.29).

Sales gas is piped directly into the national gas distribution network (assuming one exists) and NGL products such as propane and butane can be stored locally in pressurised tanks. NGL products are often distributed by road or rail directly from the gathering station, although if ethane is recovered it is normally delivered by pipeline.

Two basic types of oil *storage tank* are in common use: fixed roof tanks and floating roof tanks. Floating roof tanks are generally used when large diameters are required and there are no restrictions on vapour venting. Such tanks only operate at atmospheric pressures, and the roof floats up and down as the volume of crude

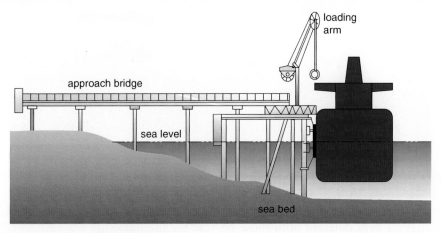

Figure 11.29   Tanker loading terminal.

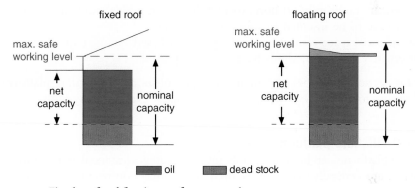

Figure 11.30   Fixed roof and floating roof storage tanks.

increases or decreases. There are a variety of fixed roof tanks for storing hydrocarbons at atmospheric pressure without vapour loss and for storage at elevated pressures (Figure 11.30).

Storage tanks should always be closely surrounded by *bund walls* to contain crude in the event of a spillage incident, such as a ruptured pipe or tank, and to allow fire-fighting personnel and equipment to be positioned reasonably close to the tanks by providing protected access (Figure 11.31).

Drainage systems inside the bund wall should only be open when the outlet can be monitored to avoid hydrocarbon liquids run-off in the event of an unforeseen release of crude.

11.2.3. Offshore production facilities

The functions of offshore production facilities are very much the same as those described for land operations. An offshore production platform is rather like a gathering station; hydrocarbons have to be collected, processed and evacuated for

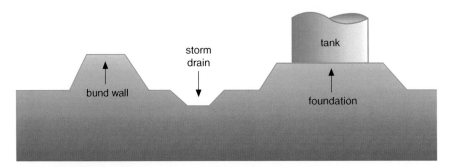

Figure 11.31   Bund wall and drainage arrangements.

further treatment or storage. However, the design and layout of the offshore facilities are very different from those on land for the following reasons:

(1) A platform has to be installed above sea level before drilling and process facilities can be placed offshore.
(2) There are no utilities offshore, so all light, water, power and living quarters, etc. also have to be installed to support operations.
(3) Weight and space restrictions make platform-based storage tanks non-viable, so alternative storage methods have to be employed.

This section describes the main types of offshore production platform and satellite development facilities, as well as associated evacuation systems.

11.2.3.1. Offshore platforms

Offshore platforms can be split broadly into two categories: fixed and floating. Fixed platforms are generally classified by their mechanical construction. There are four main types:

- steel jacket platforms
- gravity-based platforms
- tension leg platforms (TLPs)
- minimum facility systems.

Floating platforms can also be categorised into three main types:

- semi-submersible vessels
- ship-shaped monohull vessels (such as floating production, storage and offloading (FPSO))
- SPAR platforms.

TLPs may also be considered as a type of floating platform.

Artificial islands could be regarded as platforms but fall somewhere between land and offshore facilities.

Steel piled jackets are the most common type of platform and are employed in a wide range of sea conditions, from the comparative calm of the South China Sea to the hostile Northern North Sea. Steel jackets are used in water depths of up to

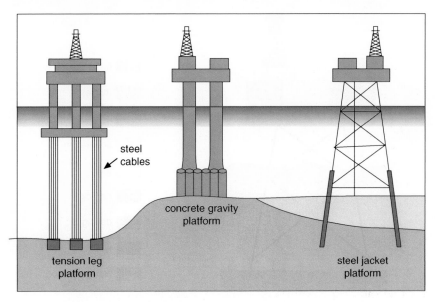

Figure 11.32   Fixed production platforms.

150 m and may support production facilities a further 50 m above mean sea level (MSL). In deepwater, all the process and support facilities are normally supported on a single jacket, but in shallow seas it may be cheaper and safer to support drilling, production and accommodation modules on different jackets. In some areas, single well jackets are common, connected by subsea pipelines to a central processing platform (Figure 11.32).

Steel jackets are constructed from welded steel pipe. The jacket is fabricated onshore and then floated out horizontally on a barge and set upright on location. Once in position a jacket is pinned to the seafloor with steel piles. Prefabricated units or modules containing processing equipment, drilling and other equipment (see Figure 11.33) are installed by lift barges on to the top of the jacket, and the whole assembly is connected and tested by commissioning teams. Steel jackets can weigh 20,000 tons or more and support a similar weight of equipment.

Concrete or steel gravity-based structures can be deployed in similar water depths to steel jacket platforms. Gravity-based platforms rely on weight to secure them to the seabed, which eliminates the need for piling in hard seabeds. Concrete gravity-based structures (which are by far the most common) are built with huge ballast tanks surrounding hollow concrete legs. They can be floated into position without a barge and are sunk once on site by flooding the ballast tanks. For example, the Mobil Hibernia Platform (offshore Canada) weighs around 450,000 tons and is designed and constructed to resist iceberg impact!

The *legs* of the platform can be used as settling tanks or temporary storage facilities for crude oil where oil is exported via tankers, or to allow production to continue in the event of a pipeline shutdown. The Brent D platform in the North Sea weighs more than 200,000 tons and can store over a million barrels of oil.

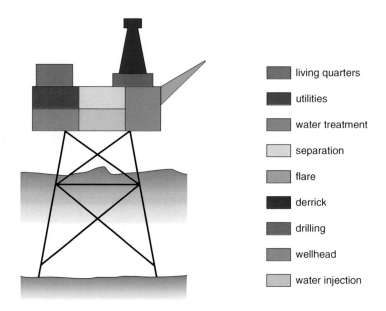

living quarters

utilities

water treatment

separation

flare

derrick

drilling

wellhead

water injection

Figure 11.33   A steel jacket platform.

Topside modules are either installed offshore by lift barges, or can be positioned before the platform is floated out.

TLPs are used mainly in deepwater where rigid platforms would be both vulnerable to bending stresses and very expensive to construct. A TLP is rather like a semi-submersible rig, tethered to the seabed by jointed legs kept in tension. Tension is maintained by pulling the floating platform down into the sea below its normal displacement level. The 'legs' are secured to a template or anchor points installed on the seabed.

Floating production systems are becoming much more common as a means of developing smaller fields which cannot support the cost of a permanent platform and for deepwater development. Ships and semi-submersible rigs have been converted or custom built to support production facilities which can be moved from field to field as reserves are depleted. Production facilities were initially more limited compared to the fixed platforms, though the later generation of floating production systems have the capacity to deal with much more variable production streams and additionally provide for storage and offloading of crude, and hence are referred to as *FPSOs (floating production, storage and offloading)*. The newer vessels can provide all services which are available on integrated platforms, in particular three-phase separation, gas lift, water treatment and injection.

Ship-shaped FPSOs must be designed to 'weather vane', that is must have the ability to rotate in the direction of wind or current. This requires complex mooring systems and the connections with the wellheads must be able to accommodate the movement. The mooring systems can be via a single buoy or, in newer vessels designed for the harsh environments of the North Sea, via an internal or external turret (Figure 11.34).

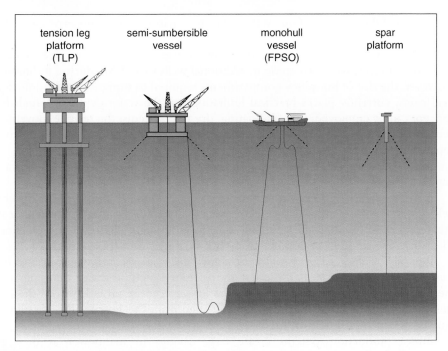

Figure 11.34 Floating production systems.

The typical process capability for FPSOs is around 100,000 barrels per day, with storage capacity up to 800,000 bbls. However, in the recent deepwater developments in West Africa some FPSOs exist which are over double this capacity.

Semi-submersible vessels have been used for many years as floating production systems, examples include Argyll (now abandoned), Buchan, Balmoral and Veslefrikk in the North Sea, and more recently Mad Dog and Thunder Horse (in more than 1800 m of water) in the GoM. In Brazil, the Campos and Roncador Fields use semi-subs in 930 and 1360 m, respectively.

A semi-sub may either be a new build, for example Balmoral, or a converted drilling rig, for example Argyll and Buchan. Conversion offers the potential for short lead times but may limit topsides weight capacity compared to recent new builds like Thunder Horse. New semi-subs can be built very large and designed to accommodate very large deck loads and support heavy steel risers (one reason why this concept was chosen in favour of a SPAR for Thunder Horse).

SPAR platforms were first employed as a concept by Shell, when it was used as a storage facility for the Brent Field in the North Sea. It had no production facilities but was installed simply for storage and offshore loading (see Figure 11.43). More recently, SPAR structures have incorporated drilling, production, storage and offshore loading facilities as an integrated development option. Examples of SPAR developments include Genesis, Neptune, Hoover and the Devils Tower Fields in the GoM, the latter of which was installed in 1710 m water depth. SPARs are more common in the GoM than anywhere else as until 2006 the US authorities would not licence FPSOs in the Gulf waters due to environmental concerns associated with such systems.

SPARs can support dry trees and allow conventional access to development wells. It is also relatively easy to tie in additional wells to a SPAR during its lifetime. However, the size of buoyancy compartments required to support and tension rigid steel risers, currently places practical limitations on the water depth in which this concept can be employed. Like all floating 'dry tree' systems, the relative movement between the risers and SPAR deck also requires flexible 'jumpers' that can limit application in high-temperature reservoir environments. In future, it may be possible, with subsea trees and flexible riser technology, to deploy SPARs in water depths as much as 3000 m.

Subsea production systems are an alternative development option for an offshore field. They are often a very cost-effective means of exploiting small fields which are situated close to existing infrastructure, such as production platforms and pipelines. They may also be used in combination with floating production systems.

Typically, a *subsea field development* or *subsea satellite development* would consist of a cluster of special *subsea trees* positioned on the seabed with produced fluids piped to the host facility. Water injection, as well as lift gas, can be provided from the host facility. Control of subsea facilities is maintained from the host facility via control umbilicals and subsea control modules.

Subsea production systems create large savings in manpower as they are unmanned facilities. However, these systems can be subject to very high OPEX from the well servicing and subsea intervention point of view as expensive vessels have to be mobilised to perform the work. As subsea systems become more reliable this OPEX will be reduced.

In 1986 when the oil price crashed to $10 a barrel, operators began to look very hard at the requirements for offshore developments and novel slimline, reduced facilities platforms began to be considered. The reduced capital outlay and early production start-up capability, coupled with the added flexibility, ensured that all companies now consider subsea systems as an important field development option. Although the interest and investment in subsea systems increased dramatically, subsea systems still had to compete with the new generation of platforms, which were becoming lighter and cheaper.

In mature regions, the focus has turned towards developing much smaller fields, making use of the existing field infrastructure. This, in combination with advances in subsea completion technology and the introduction of new production equipment, has further stimulated the application of subsea technology.

11.2.4. Satellite wells, templates and manifolds

Various types of subsea production systems are being used and their versatility and practicality is being demonstrated in both major and marginal fields throughout the world. These are illustrated in Figure 11.35.

The most basic subsea satellite is a single *subsea wellhead* with *subsea tree*, connected to a production facility by a series of pipelines and umbilicals. A control

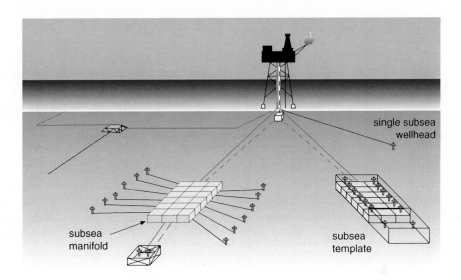

Figure 11.35   Typical subsea field development options – tied back to a host facility.

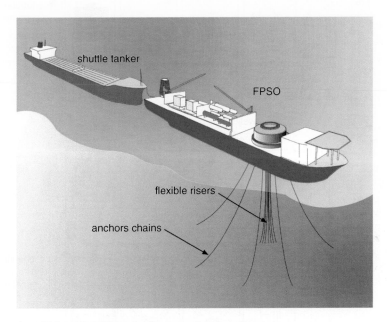

Figure 11.36   FPSO with offshore loading to a shuttle tanker.

module, usually situated on the subsea tree, allows the production platform to remotely operate the subsea facility via its valves and chokes (Figure 11.36).

These single satellites are commonly used to develop small reservoirs near to a large field. They are also used to provide additional production from, or peripheral water injection support to, a field which could not adequately be covered by drilling extended reach wells from the platform.

An exploration or appraisal well, if successful, can be converted to a subsea producer if hydrocarbons are discovered. In this case, the initial well design would have to allow for any proposed conversion.

The *subsea production template* is generally recommended for use with six or more wells. It is commonly used when an operator has a firm idea of the number of wells that will be drilled in a certain location. All subsea facilities are contained within one protective structure (Figure 11.37).

The templates are fabricated from large tubular members and incorporate a receptacle for each well and a three- or four-point levelling system. Drilling equipment guidance is achieved through the use of integral guide posts or retrievable guide structures. It is possible to place part of the template over existing exploration wells and tie them individually into the template production facilities.

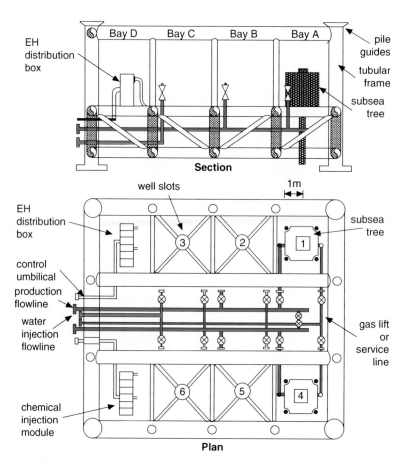

Figure 11.37   Simple subsea production template.

The template will be constructed and fitted out at a fabrication yard and then transported offshore to the drilling location. The template is lowered to the seabed using a crane barge or, if small enough, lowered beneath a semi-submersible rig. Prior to drilling the first well, piles are driven into the seabed to hold the template in place.

As the first well is being drilled the template is connected to the host facility with flowlines, umbilicals and risers. A *chemical injection umbilical* will also typically be laid to the template or subsea facility and connected to a distribution manifold.

As soon as the subsea tree on the first well has been commissioned, production can commence. The rig will then move to another template slot and start drilling the next well.

If one or more clusters of single wells are required then an *underwater manifold system* can be deployed and used as a subsea focal point to connect each well. The subsea trees sit on the seabed and are tied back to the main manifold through subsea flowlines.

Only one set of pipelines and umbilicals (as with the template) are required from the manifold back to the host facility, saving unnecessary expense. Underwater manifolds offer a great deal of flexibility in field development and can be very cost-effective.

The manifold is typically a tubular steel structure (similar to a template) which is host to a series of remotely operated valves and chokes. It is common for subsea tree control systems to be mounted on the manifold and not on the individual trees. A complex manifold will generally have its own set of dedicated subsea control modules (for controlling manifold valves and monitoring flowline sensors).

11.2.5. Control systems

As subsea production systems are remote from the host production facility there must be some type of system in place which allows personnel on the host facility to control and monitor the operation of the unmanned subsea system.

Modern subsea trees and manifolds are commonly controlled via a complex *electro-hydraulic system*. Electricity is used to power the control system and to allow for communication or command signalling between surface and subsea. Signals sent back to surface will include, for example, subsea valve status and pressure/temperature sensor outputs. Hydraulics are used to operate valves on the subsea facilities (e.g. subsea tree and manifold valves). The majority of the subsea valves are operated by hydraulically powered actuator units mounted on the valve bodies.

With the electro-hydraulic system the signals, power and hydraulic supplies are sent from a *master control station* (or MCS) on the host facility down *control umbilicals* (Figure 11.38) to individual *junction boxes* on the seabed or subsea structure.

The MCS allows the operator to open and close all the systems remotely operated valves, including tree and manifold valves and downhole safety valve.

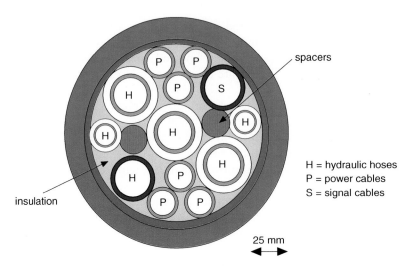

Figure 11.38 Electro-hydraulic control umbilical bundle.

Sensors on the tree allow the control module to transmit data such as tubing head pressure, tubing head temperature, annulus pressure and production choke setting. Data from the downhole gauge is also received by the control module. With current subsea systems more and more data are being recorded and transmitted to the host facility. This allows operations staff to continuously monitor the performance of the subsea system.

The pressure to accelerate first oil and to decrease development costs is continuously demanding innovations in offshore engineering. This has led to the development of minimum facilities, such as *monopods* and related systems. Instead of multilegged jackets, monopods consist of a single central column, the well protector caisson. The caisson is driven into the seafloor and held in place by an arrangement of cables anchored to the seafloor. The system can accommodate up to six wells internally and externally of the caisson and it can carry basic production facilities. The main advantage of this innovative concept is the low construction costs and the fact that installation can be carried out by the drilling rig, a diver support vessel (DSV) and a tug boat in some 3 months. There has been some concern regarding the stability of monopods, but a number of installations in the GoM and offshore Australia have been exposed to hurricanes without sustaining major damage. Water depth is a limiting factor of the system (currently about 300 ft) (Figure 11.39).

11.2.5.1. Offshore evacuation systems

Crude oil and gas from offshore platforms are evacuated by *pipeline* or alternatively, in the case of oil, by *tanker*. Pipeline transport is the most common means of evacuating hydrocarbons, particularly where large volumes are concerned. Although a pipeline may seem a fairly basic piece of equipment, failure to design

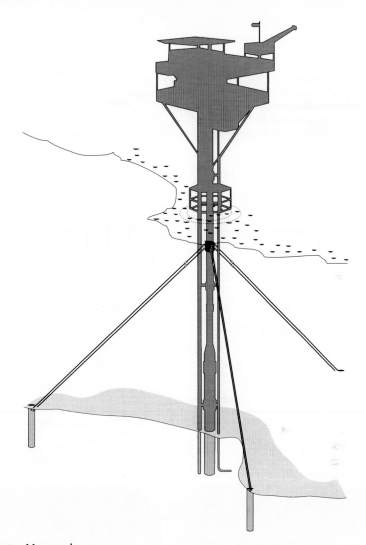

Figure 11.39 Monopod system.

a line for the appropriate capacity, or to withstand operating conditions over the field lifetime, can prove very costly in terms of deferred oil production.

Long pipelines are normally installed using a lay barge on which welded connections are made one at a time as the pipe is lowered into the sea. Pipelines are often buried for protection as a large proportion of pipeline failures result from external impact. For shorter lengths, particularly in-field lines, the pipeline may be constructed onshore either as a single line or as a bundle. Once constructed the pipeline is towed offshore and positioned as required. It has become common practice to integrate pipeline connectors into the towing head, both for protection and easier tie in (Figure 11.40).

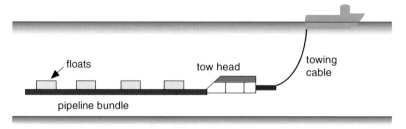

Figure 11.40   Towing a pipeline.

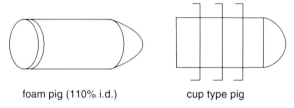

Figure 11.41   Foam and cup type pigs.

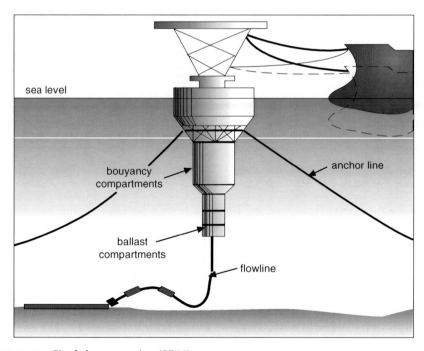

Figure 11.42   Single buoy mooring (SBM).

Pipelines are cleaned and inspected using 'pigs'. *Pigs* usually have a steel body fitted with rubber cups and brushes or scrapers to remove wax and rust deposits on the pipe wall, as the pig is pumped along the pipe. Sometimes spherical pigs are sed for product separation or controlling liquid hold-up. In-field lines handling untreated crude may have to be insulated to prevent wax formation (Figure 11.41).

In recent years much more attention has been given to pipeline isolation, after instances in which the contents of export pipelines fed platform fires, adding significantly to damage and loss of life. Many export and in-field pipelines are now fitted with *emergency shutdown valves* (ESDV) close to the production platform, to isolate the pipeline in the event of an emergency.

11.2.5.2. Offshore loading

In areas where seabed relief makes pipelines vulnerable or where pipelines cannot be justified on economic grounds, tankers are used to store and transport crude from production centres. The simplest method for evacuation is to pump stabilised crude from a processing facility directly to a tanker (Figure 11.42).

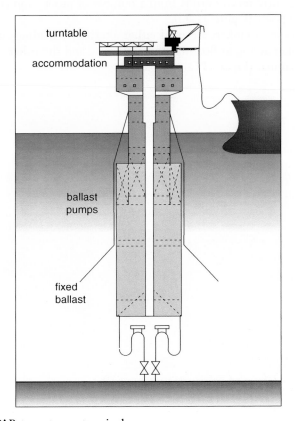

Figure 11.43   SPAR type storage terminal.

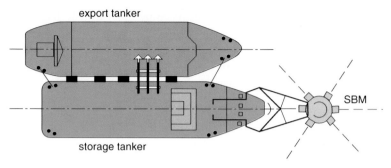

Figure 11.44 Tanker storage and export.

Loading is carried out through a *single buoy mooring* (SBM) to which the tanker can tie up and rotate around to accommodate the prevailing weather conditions. The SBM has no storage facility, but if a production facility has storage capacity sufficient to continue production whilst the tanker makes a round trip to off-load, then only a single tanker may be required. In some areas, the SBM option has been developed to include storage facilities such as the 'SPAR' type storage terminals used in the North Sea. Such systems may receive crude from a number of production centres and act as a central loading point (Figure 11.43).

In some cases, two tankers are used either alternately loading and transporting, or with one tanker acting as floating storage facility and the other shuttling to and from a shore terminal (Figure 11.44).

CHAPTER 12

PRODUCTION OPERATIONS AND MAINTENANCE

Introduction and Commercial Application: During the development planning phase of a project, it is important to define how the field will be produced and operated and how the facilities are to be maintained. The answers to these questions will influence the design of the facilities. The typical development planning and project execution period may be 5 or 6 years, but the typical producing lifetime of the field may be 25 years. Because the facilities will need to be operated and will incur OPEX for this long period, the production and maintenance modes should be an integrated part of the facilities design. Figure 12.1 puts the operating period into perspective.

The disciplines dealing with the development planning, design and construction phases are typically petroleum and well engineering and facilities engineering, whilst production operations and maintenance are run by a separate group. Early input into the FDP from the production operations and maintenance group is essential to ensure that the mode of production and maintenance is considered in the design of the facilities.

Over the lifetime of the field, the total undiscounted OPEX is likely to exceed the CAPEX. It is therefore important to control and reduce OPEX at the project design stage as well as during the production period.

The operations group will develop *general operating and maintenance objectives* for the facilities which will address product quality, cost, safety and environmental issues. At a more detailed level, the *mode of operations and maintenance* for a particular project will be specified in the FDP. Both specifications will be discussed in this chapter which will focus on the input of the production operations and maintenance departments to a *FDP.* The management of the field during the producing period is discussed in Chapter 16.

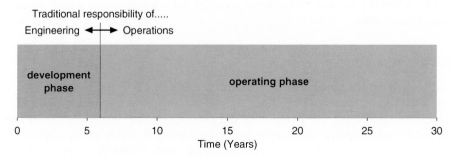

Figure 12.1 The operating phase in perspective.

12.1. OPERATING AND MAINTENANCE OBJECTIVES

The production operations and maintenance group will develop a set of *operating and maintenance objectives* for the project. This will be a guideline when specifying the mode of operation and maintenance of the equipment items and systems, and will incorporate elements of

- business objectives
- responsibilities to the customer
- health safety and environmental management systems
- reservoir management
- product quality and availability
- cost control.

An example of the operating and maintenance objectives for a project might include statements which cover technical, business and environmental principles, such as

- meeting the company objectives of, say, maximising the economic recovery of hydrocarbons
- ensuring that the agreed quantities of hydrocarbons are delivered to the customer on time, to specification and in a safe manner
- ensuring an uptime of offshore facilities of, say, 98%
- minimising manpower offshore
- providing a safe working environment for all staff and contractors
- complying with all local legislation
- measuring hydrocarbon delivery to a specified accuracy
- providing certain levels of employment within a local community.

12.2. PRODUCTION OPERATIONS INPUT TO THE FDP

When preparing a FDP, the production operations department will become involved in determining how the field will be operated, with specific reference to areas such as those shown in Table 12.1.

The following section will indicate some of the considerations which would be made in each area.

12.2.1. Production

One of the primary objectives of production operations is to deliver product at the required rate and quality. Therefore, the *product quality specification* and any *agreed contract terms* will drive the activities of the production operations department, and will be a starting point for determining the preferred mode of operation. The specifications, such as delivery of stabilised crude with a BS&W of less than 0.5%, and a salinity of $70 \, \text{g/m}^3$, and contractually agreed *fiscalisation points* (where the

Table 12.1 Operations and maintenance in the FDP

Production	Product quality specification
	Contractual agreements
	Capacity and availability
	Concurrent operations (e.g. drilling and production)
	Monitoring and control
	Testing and metering
	Standardisation
	Flaring and venting
	Waste disposal
	Utilities systems
Manning	Manned/unmanned operations
	Accommodation
Logistics	Transport
	Supplies of materials
	Storage
Communications	Requirements for operations
	Evacuation routes in emergency
Cost control	Measurement and control of OPEX

crude will be metered for fiscal purposes) should be clearly stated in the FDP. In gas sales contracts, the quantity of gas sales is specified, and any shortfall often incurs a severe penalty to the supplier. In this situation, it is imperative that the selected mode of operation aims to guarantee that the contract is met.

Product quality is not limited to oil and gas quality; certain *effluent streams* will also have to meet a legal specification. For example, in disposal of oil in water, the legislation in many offshore areas demands less than 40 ppm (parts per million) of oil in water for disposal into the sea, or in some cases zero discharge of oily water to sea. In the UK, oil production platforms are allowed to flare gas up to a legal limit.

The *capacity and availability* of the equipment items in the process need to be addressed by both the process engineers and the production operations group during the design phase of the project. Sufficient capacity and availability (as defined in Section 16.2, Chapter 16) must be provided to achieve the production targets and to satisfy contracts. The process and facilities engineers will design the equipment for a range of capacities (maximum throughputs), but the mode of operation and maintenance, as well as the performance of the equipment will determine the availability (the fraction of the time which the item operates). Consultation with the production operators is essential to design the right mode of operation, and to include previous experience when estimating availability.

Concurrent operations refers to performing the simultaneous activities of production and drilling, or sometimes production, drilling and maintenance. In some areas simultaneous production and drilling is abbreviated to *SIPROD*. Clearly, the issues which drive the operator's decision on whether to carry out SIPROD are safety and cost. Shutting in production whilst drilling will reduce the consequences of a drilling incident such as a blowout, but will incur a loss of revenue.

Risk analysis techniques may be used to help in this decision, and if SIPROD is adopted, then procedures will be written specifying how to operate in this mode. It is common practice in production operations to close in production from a well when another nearby operation is rigging up or rigging down, to avoid the more serious consequences of a load being dropped during equipment movements. Another more general term used is *SIMOPS*.

Monitoring and control of the production process will be performed by a combination of instrumentation and control equipment plus manual involvement. The level of sophistication of the systems can vary considerably. For example, monitoring well performance can be done in a simple fashion by sending an operator to write down and report the tubing head pressures of producing wells on a daily basis, or at the other extreme, by using CAO. This uses a remote computer-based system to record and control production on a well by well basis with no physical presence at the wellhead.

CAO involves the use of computer technology to support operations, with functions ranging from collection of data using simple calculators and PCs to integrated computer networks for automatic operation of a field. In the extreme case, CAO can be used for totally unmanned offshore production operations with remote monitoring and control from shore-based locations. In considering the requirements for operations at FDP stage, the inclusion of CAO would have a great impact on the mode of operations. CAO may also be applied to reporting, design and simulation of possible situations, leading to performance optimisation, improved safety and better environmental protection.

By providing more accurate monitoring and control of the production operations, CAO is proven to provide benefits such as

- *increased production rates*: through controlling the system to produce closer to its design limits, reducing downtime and giving early notice of problems
- *reduced OPEX*: less manpower costs, reduced maintenance costs due to better surveillance and faster response and reduced fuel costs
- *reduced CAPEX*: by increasing throughput, less facilities capacity required, less accommodation and office space and reduced instrumentation
- *increased safety*: less people in hazardous areas, less driving, better monitoring of toxic gases and better alarm systems
- *improved environmental protection*: control of effluent streams and better leak detection
- *improved database*: more and better organised historical data, simulation capability, better reporting and use as training for operators.

The cost of implementing CAO depends of course on the system installed, but for a new field development is likely to be in the order of 1–5% of the project CAPEX, plus 1–5% of the annual OPEX.

An example of an application of CAO is its use in optimising the distribution of gas in a gas–lift system (Figure 12.2). Each well will have a particular optimum GLR, which would maximise the oil production from that well. A CAO system may be used to determine the optimum distribution of a fixed amount of compressed gas between the gas–lifted wells, with the objective of maximising the overall oil production from the field. Measurement of the production rate of each well and its

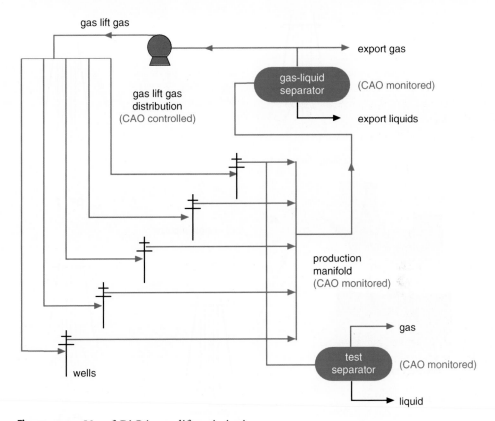

Figure 12.2   Use of CAO in gas-lift optimisation.

producing GOR (using the test separator) provides a CAO system with the information to calculate the optimum gas-lift gas required by each well, and then distributes the available gas–lift gas (a limited resource) between the producing wells.

   Testing of the production rate of each well on a routine basis can be performed at the drilling platform or at the centralised production facility (Figure 12.3). Consider an offshore development with four eight-well drilling platforms and one centralised production platform. If the production from each drilling platform is manifolded together for transfer to the production platform then the there are two principal ways of testing the production from each well on a routine basis (required for reservoir management described in Section 16.1, Chapter 16):

1. A test separator is provided on each drilling platform and is used to test the wells sequentially. The capacity of the test separator would have to be equal to the production from the highest rate well.
2. A test separator is provided on the production platform, and is large enough to handle the production from any one of the drilling platforms. An individual well would be tested by passing the production from its drilling platform through the test separator and then shutting in the well under consideration and calculating its production from the reduction in rate. This is referred to as 'testing by difference'.

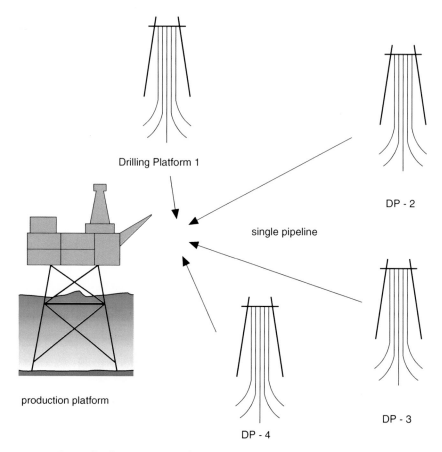

Figure 12.3   Centralised vs. remote production testing.

The benefits of having a test separator on each drilling platform are that the individual wells can be tested more frequently and with much greater accuracy since they are measured directly. However, this will require either a manned operation on the drilling platform or the installation of CAO for remotely operating the test separator. A single centralised test separator is cheaper but less accurate and can only test the wells at a quarter of the frequency. This is an example of the need for the reservoir engineers (who require the data for reservoir management) to liaise with the production operations department (who require the data for programming) and the facilities engineers who are designing the equipment. This discussion must take place whilst planning the field development.

In new developments, test separators may be substituted by *multiphase metering* devices which can quantitatively measure volumes of oil, gas and water without the need of separation. This technology is now developed and a viable alternative.

Metering of the production for fiscal (taxation), tariffing and re-allocation purposes may take place as the product leaves the production platform, or as it arrives at the delivery point such as the crude oil terminal. If the export pipeline is used by other fields (including third-party users), it would be common practice to meter the production as it leaves the platform.

Standardisation of equipment items is an area for potential cost savings, both in terms of CAPEX and OPEX, and is a decision which should be taken in consultation with the production operations department at the FDP stage. Standardisation can be applied to equipment items ranging from drilling platforms to valves. The benefits of standardisation are

- reduced design and capital costs
- reduced spares stock required and less inventory management
- less operating procedures, hence better safety and lower OPEX
- less training required.

The drawbacks of standardisation are

- less equipment available to select from (less variations possible)
- fewer vendors to select from.

Flaring and venting policies will often be driven by legislation which states maximum allowable limits for these activities. Such existing regulations must be established at the FDP stage, but it is good practice to anticipate future legislation and to determine whether it is worth designing this into the initial facilities. Even if constant flaring of excess gas is avoided by gas re-injection or export, a flare or vent system will be required to relieve the process facilities in case of shutdown. Flaring can be performed from a fixed flare boom or from a separate, more remote platform. Venting is usually from a separate vent jacket. Venting is more environmentally damaging than flaring, since methane is approximately 20 times worse as a contributor to the greenhouse effect than carbon dioxide.

Waste disposal is an aspect of the production process which must be considered at FDP stage. This should cover all effluent streams other than the useful product including

- waste to be discharged to the *sea* or *land* (drill cuttings, drilling mud, sewage, food, empty drums/crates/packaging, used lubricants, used coolants and fire-fighting fluids, drain discharges)
- effluents discharged to the air (hydrocarbon gases, coolant vapours, noise and light).

The treatment of these issues will be discussed jointly with the HSE departments within the company and with the process and facilities engineers, and their treatment should be designed in conjunction with an *EIA*. Some of the important basic principles for waste management are to

- eliminate the waste at source where possible (e.g. slim-hole drilling)
- re-use materials wherever possible (e.g. recycling of drilling mud)

- re-inject waste material into the reservoir where possible (e.g. re-injection of drilling cuttings).

Utilities systems support production operations, and should also be addressed when putting together a FDP. Some examples of these are

- power system (fuel gas and diesel)
- seawater and potable water treatment system
- chemicals and lubrication oils
- alarm and shutdown system
- fire protection and fire-fighting system
- instrument/utility air system.

12.2.2. Manning

Manning of production facilities is a key part of field development planning. Every person offshore requires accommodation, transport, administrative support, managing and at least one back-up to operate a shift system. Typically, every one person offshore requires between three and five other employees as support. If a platform is manned, then life-saving systems must be provided, along with other items like a mess, recreation room, radio and telecommunications facilities, medical and sick-bay facilities. This is one of the main reasons for the drive towards *minimum manning* or *unmanned operations*; it is not only safer, but also cheaper. Along with the introduction of CAO, unmanned operations are now a reality.

If it is decided that an operation does require to be manned, then it may need to be manned on a 24-h basis, or a 12-h basis, or only for daily inspection. Accommodation may be provided on a separate living quarters platform or as part of an integrated platform, or on a floating hotel.

12.2.3. Logistics

Logistics refers to the organisation of transport of people, and supply and storage of materials. The transport of people is linked to the mode of manning the operation, and is clearly much simplified for an unmanned operation.

For a typical operation in the North Sea, the transport of personnel to and from the facilities is by helicopter. The transport of materials is normally by supply boat.

The *storage of chemicals*, lubricants, aviation fuel and diesel fuel is normally on the platforms, with chemicals kept in bulk storage or in drums depending on the quantities. A typical diesel storage would be adequate to run back-up power generators for around a week, but the appropriate storage for each item would need to be specified in the FDP.

12.2.4. Communications

Telecommunications systems will include internal communications within the platforms (telephone, radio, walkie-talkie, air-ground-air, navigation and public address) and external systems (telephone and internet, telex, fax, telemetry, VHF radio and satellite links). These systems are designed to handle the day-to-day communications as well as emergency situations.

If the development is so far from shore that direct *line of sight* communication is not possible, then *satellite* communications will be installed, with one platform acting as a satellite link for the area.

In case of a major disaster, one platform in a region will be equipped to act as a control centre from which rescue operations are co-ordinated. Evacuation routes will be provided, and where large complexes are clustered together, a standby vessel will be available in the region to supply emergency services such as fire fighting and rescue.

12.2.5. Measurement and control of operating costs

As discussed in Chapters 14 and 16, the management of OPEX is a major issue, since initial estimates of OPEX are often far exceeded in reality, and may threaten the overall profitability of a project. Within the FDP, it is therefore useful to specify the system which will be used to measure the OPEX. Without measuring OPEX, there is no chance of managing it. This will involve the joint effort of production operations, finance and accounting and the development managers.

The projection of OPEX should be budgeted on an annual basis, to reflect the *annual work programme* for the following year. Maintaining good records of actual operating costs simplifies the process of budgeting for the future, as well as comparing actual expenditure with budget. These statements sound obvious, but require a considerable amount of integrated effort to perform effectively.

12.3. Maintenance Engineering Input to the FDP

In conjunction with the production operations input into the FDP, describing how the process will be operated, maintenance engineering will outline how the equipment will be maintained. Maintenance is required to ensure that equipment is capable of safely performing the tasks for which it was designed. This is often stated as maintaining the 'technical integrity' of the equipment.

The mechanical performance of equipment is likely to deteriorate with use due to wear, corrosion, erosion, vibration, contamination and fracture, which may lead to failure. Since this would threaten a typical production objective of meeting quality and quantity specifications, the maintenance engineering department provides a service which helps to safely achieve the production objective.

The service provided by maintenance engineering was traditionally that of repairing equipment items when they failed. This is no longer the case, and a maintenance department is now proactive rather than reactive in its approach. Maintenance of equipment items will be an important consideration in the FDP, because the mode and cost of maintaining equipment play an important part in the facilities design and in the mode of operation.

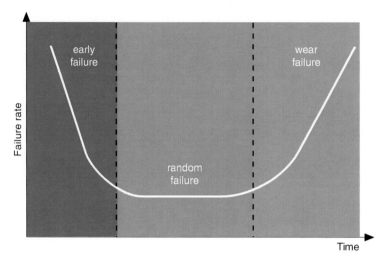

Figure 12.4   The bathtub curve for failure frequency.

Increasingly, maintenance engineers think in terms of the performance and maintenance of equipment over the whole life of the field. This is often at the centre of the decision on *CAPEX–OPEX trade-offs*; for example spending higher CAPEX on a more reliable piece of equipment in anticipation of less maintenance costs over the life of the equipment.

Statistical analysis of *failures of equipment* shows a characteristic trend with time, often described as the 'bathtub curve' (Figure 12.4).

Early failures may occur almost immediately, and the failure rate is determined by manufacturing faults or poor repairs. *Random failures* are due to mechanical or human failure, whilst *wear failure* occurs mainly due to mechanical faults as the equipment becomes old. One of the techniques used by maintenance engineers is to record the *mean time to failure* (MTF) of equipment items to find out in which period a piece of equipment is likely to fail. This provides some of the information required to determine an appropriate *maintenance strategy* for each equipment item.

Equipment items will be maintained in different ways, depending upon their

- *criticality* which is associated with the consequence of failure
- failure mode.

Criticality refers to how important an equipment item is to the process. Consider the role of the export pump in the situation given in Figure 12.5.

The choice of the size of the export pump will involve both maintenance and production operations. If a single export pump with a capacity of 12 Mb/d is selected, then this item becomes critical to the continuous export of oil, though not to the production of oil, since the storage tank is sufficient to hold 4 days of production. If continuous export is important, then the pump should be maintained in a way which gives very high reliability. If, however, two 12 Mb/d were provided for export as part of the production operations 'sparing' philosophy, then the pumps

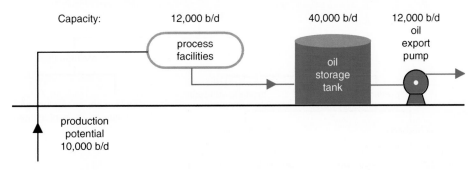

Figure 12.5 Criticality of equipment.

could be maintained in a different way, such as allowing one to run to failure and then switching to the spare pump whilst repairing the failed one.

Criticality in the above example is set within the context of guaranteeing production. However, a similar analysis will be performed with respect to the criticality of guaranteeing safety and minimum impact on the environment.

The *failure mode* of an equipment item describes the reason for the failure, and is often determined by analysing what causes historic failures in the particular item. This is another good reason for keeping records of the performance of equipment. For example, if it is recognised that a pump typically fails due to worn bearings after 8000 h in operation, a maintenance strategy may be adopted which replaces the bearings after 7000 h if that pump is a critical item. If a spare pump is available as a back-up, then the policy may be to allow the pump to run to failure, but keep a stock of spare parts to allow a quick repair.

12.3.1. Maintenance strategies

For some cheap, easily replaceable equipment, it may be more economic to do *no maintenance* at all, and in this case the item may be replaced on failure or at planned intervals. If the equipment is more highly critical, availability of spares and rapid replacement must be planned for.

If maintenance is performed, there are two principal maintenance strategies: *preventive* and *breakdown* maintenance. These are not mutually exclusive, and may be combined even in the same piece of equipment. Take for example a private motor car. The owner performs a mixture of preventive maintenance (by adding lubricating oil, topping up the battery fluid, hydraulic fluid and coolant) with breakdown maintenance (e.g. only replacing the starter motor when it fails, rather than at regular intervals).

Figure 12.6 summarises the alternative forms of maintenance.

Breakdown maintenance is suitable for equipment whose failure does not threaten production, safety or the environment, and where the cost of preventing failure would be greater than the consequence of failure. In this case, the equipment would be repaired either on location or in a workshop. Even with this policy, it is assumed that the recommended lubrication and minor servicing is performed, just as with a motor car.

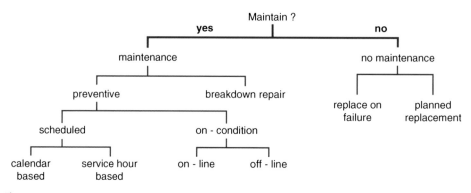

Figure 12.6   Maintenance strategies.

Preventive maintenance includes inspection, servicing and adjustment with the objective of preventing breakdown of equipment. This is appropriate for highly critical equipment where the cost of failure is high, or where failure implies a significant negative impact on safety or the environment. This form of maintenance can be *scheduled* on a calendar basis (e.g. every 6 months) or on a service hour basis (e.g. every 5000 running hours).

If the performance of the equipment is monitored on a continuous basis, then abnormal behaviour can be identified, and preventive maintenance can be performed as and when required; this is called *on-condition* preventive maintenance. The condition of equipment may be established by inspection, that is taking it *off-line*, opening it up and looking for signs of wear, corrosion, etc. This obviously takes the equipment out of service, and may be costly.

A more sophisticated and increasingly popular method of on-condition maintenance is to monitor the performance of equipment *on-line*. For example, a piece of rotating equipment such as a turbine may be monitored for vibration and mechanical performance (speed, inlet and outlet pressure, throughput). If a baseline performance is established, then deviations from this may indicate that the turbine has a mechanical problem which will reduce its performance or lead to failure. This would be used to alert the operators that some form of repair is required.

One of the most cost-effective forms of maintenance is to train the operators to visually inspect the equipment on a daily basis. Careful selection of staff, appropriate training and incentives will help to improve what is often called *first-line maintenance*.

12.3.2. Measurement and control of maintenance costs

Maintenance costs account for a large fraction of the total OPEX of a project. Because of the bathtub curve mentioned above, maintenance costs typically increase as the facilities age; just when the production and hence revenues enter into decline. The measurement and control of OPEX often becomes a key issue during the producing lifetime of the field, as discussed in Chapter 16. However, the problem should be anticipated when writing the FDP.

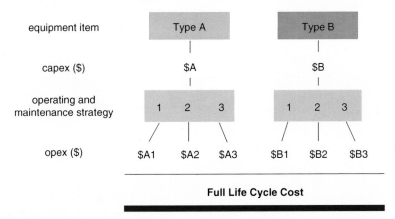

Figure 12.7 Full life cycle costing.

A suitable maintenance strategy should be developed for equipment by considering the criticality and failure mode, and then applying a mixture of the forms of maintenance described above. In particular, the long-term cost of maintenance of an item of equipment should be estimated over the whole life of the project and combined with its capital cost to select both the type of equipment and form of maintenance which gives the best *full life cycle cost* (on a discounted basis), whilst meeting the technical, safety and environmental specifications (Figure 12.7).

Although Figure 12.7 indicates a linear step-wise procedure for selecting the equipment type and the operating and maintenance strategies, the actual procedure will involve a number of loops to select the best option. This procedure will require input from the process engineers, facilities engineers, production operators and the maintenance engineers, and demonstrates the integrated approach to field development planning.

When estimating the operating and maintenance costs for various options, it is recommended that the actual activities which are anticipated are specified and costed. This will run into the detail of frequency and duration of maintenance activities such as inspection, overhaul, painting, etc. This technique allows a much more realistic estimate of OPEX to be made, rather than relying on the traditional method of estimating OPEX based on a percentage of CAPEX. The benefits of this *activity-based costing* are further discussed in Chapters 14 and 16.

By diligent operations and maintenance activity, operators are able to achieve overall uptimes on plants of around 95%, excluding planned shutdowns. This is critical in meeting production targets which will have factored in the anticipated uptime during the forecasting exercise. *Uptime* refers to the fraction of time the plant is available.

CHAPTER 13

PROJECT AND CONTRACT MANAGEMENT

Introduction and Commercial Application: Large, capital-intensive projects are characteristic of the oil and gas industry. Planning and controlling a project which may involve hundreds of personnel, millions of individual items and a significant investment, has become a discipline in its own right. This section describes how and why a typical project is organised in a number of well-defined stages, and discusses the methods used to ensure that cost and time expectations are fulfilled, and 'products' delivered to an agreed specification.

Many oil and gas companies use contract staff to perform the part of a project between preliminary design and commissioning. This is either because they do not immediately have the staff or the skills to perform these tasks, or it is cheaper and more efficient to pass the work to a contractor. Contracting out tasks is not limited to project work, and affects most departments in a company, from the drilling department through to the catering services. The fraction of a company's expenditure directed to contract services may be very significant, especially when major projects are being performed. Every contract needs to be managed, and this section outlines some of the reasons for contracting out work and the main types of contract used in the oil and gas industry.

13.1. PHASING AND ORGANISATION

A 'Project' can be defined as a task that has to be completed to a defined specification within an agreed time and for a specific price. Although simple to define, a large project requires many people bringing different skills to bear, as the task evolves from conception to completion. Large businesses, including those in the oil and gas industry, find it more manageable to divide projects into phases, which reflect changing skill requirements, levels of uncertainty and commitment of resources.

As mentioned in Section 11.1, Chapter 11, a typical project might be split into the following phases (Figure 13.1).

- Feasibility
- Definition and preliminary design
- Detailed design
- Procurement
- Construction
- Commissioning
- Review.

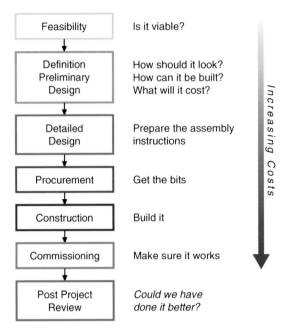

Figure 13.1   Project phases.

Also note that an alternative set of nomenclature for the same phased approach is discussed in Section 15.3, Chapter 15.

13.1.1. Project phasing

The first three phases listed above are sometimes defined collectively as the pre-project stage. This is the stage in which ideas are developed and tested, but before large funding commitments are made.

In the *feasibility* phase the project is tested as a concept. Is it technically feasible and is it economically viable? There may be a number of ways to perform a particular task (such as develop an oil field) and these have to be judged against economic criteria, availability of resources and risk. At this stage, estimates of cost and income (production) profiles will carry a considerable uncertainty range, but are used to filter out unrealistic options. Several options may remain under consideration at the end of a feasibility study.

In the *definition* phase options are narrowed down and a preferred solution is proposed. The project becomes better defined in terms of what should be built and how it should be operated, and an assessment of how the project may be affected by changes beyond the control of the company (e.g. the oil price) should be made. Normally a clear statement should be prepared, describing why the option is preferred and what project specifications must be met, to be used as a basis for further work.

Providing a project is viable, resources are available and risk levels acceptable, work can continue on *preliminary design* and tighter cost estimates. The object of the preliminary design phase is to prepare a document that will support an application for funds. The level of detail must be sufficient to give fund holders confidence that the project is technically sound and commercially robust, and may also have to be used to gain a licence to proceed from government bodies. Tried and tested engineering issues may not need a great deal of elaboration, but issues with a high novelty value have to be identified and clearly explained. If work is subsequently contracted out the document can form the basis for a tender. This phase is also referred to as front end engineering design (FEED).

Once a project has been given approval then *detailed design* can begin. This phase often signals a significant increase in spending as teams of design engineers are mobilised to prepare detailed engineering drawings. It is also quite common for oil companies to contract out the work from this stage, though some company staff may continue to work with the contractor in a liaison role. The detailed engineering drawings are used to initiate procurement activities and construction planning. By this stage the total expenditure may be 5% of the total project budget, and yet around 80% of the hardware items will have been specified. The emphasis at the detailed design stage is to achieve the appropriate design and to reduce the need for changes during subsequent stages.

Procurement is a matter of getting the right materials together at the right time and within a specified budget. For items which can be obtained from a number of sources a *tendering* process may appropriate, possibly from a list of company approved suppliers. Very exotic items, or items which are particularly critical, may be acquired through a single source contract where reliability is paramount. Complex items such as turbines will often be accompanied by test certification which has to be checked for compliance with performance and safety standards. Equipment must be inspected when the company takes delivery, to ensure that goods have not been damaged in shipment. The procurement team may also be responsible for ensuring that the supply of spare parts is secure. Spending at this stage can range anywhere from 10 to 40% of the total project cost.

The character of a project *construction* phase can vary considerably depending on the nature of the contract. The construction of a gas plant in a rural setting will raise very different issues from that of a refurbishment project on an old production platform. Construction activities will normally be carried out by specialised contractors working under the supervision of a company representative such as a construction manager (or resident engineer). The *construction manager* is responsible for delivering completed works to specification and within time and budget limits. When design problems come to light the construction manager must determine the impact of changes and co-ordinate an appropriate response with the construction contractor and design team.

As construction nears completion the *commissioning* phase will begin. The objective of the commissioning phase is to demonstrate that the facility constructed performs to the design specification. Typically a construction team will hand over a project to an operating team (which may be company staff) once the facility or equipment has been successfully tested. The receiving party will normally confirm their acceptance by signing a 'hand over' document, and responsibility for the project is passed on. The hand over document may also carry a budget to finish outstanding items if these can be handled more easily by the operator.

It is good practice to *review* a project on completion and record the reasons for differences between planned and actual performance. Where lessons can be learned, or opportunities exploited, they should be incorporated into project management guidelines. Some companies hold post project sessions with their contractors to explore better ways of handling particular issues, especially when there is an expectation of additional shared activities.

The project phasing covered so far is still the most common approach used in industry. However other concepts have also been tested. *Parallel Engineering* is a project management style aimed at significantly reducing the time span from discovery to first oil and thus *fast tracking* new developments. In the North Sea, conventional developments during the 1990s on average took some 9 years from discovery to first oil. Parallel engineering may help to half this time frame by carrying out appraisal, conceptual design and construction concurrently. The approach carries a higher risk for the parties involved and this has to be balanced with the potentially much higher rewards resulting from acceleration of first oil. For example, if conceptual design is carried out prior to appraisal results being available, considerable uncertainty will have to be managed by the engineers. The conceptual design needs to be continuously refined and possibly changed as additional information becomes available. All tendering processes for vessels, equipment and services are more difficult due to the lack of reliable data. Examples of fast track developments include the Foinaven and Schiehallion Fields in the UK West of Shetland basin. Figure 13.2 contrasts traditional and fast track development approaches.

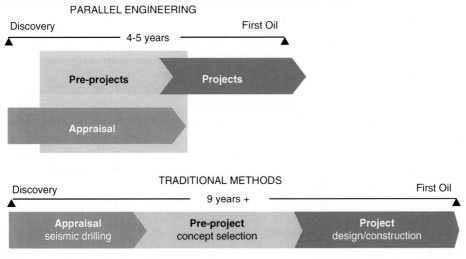

Figure 13.2   Parallel Engineering approach.

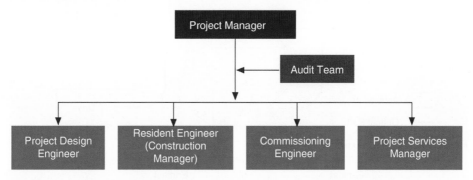

Figure 13.3   Example of a project team organisation.

13.1.2. Project organisation

Although a single project manager may direct activities throughout a project life, he or she will normally be supported by a project team whose composition should reflect the type of project and the experience levels of both company and contractor personnel. The make-up and size of the team may change over the life of a project to match the prevailing activity levels in each particular section of the project (Figure 13.3).

An organisation such as the example above includes sub groups for each of the main activities and a support (or services) group to manage information and procurement. Auditing commitments may be fulfilled by an 'independent' in-house team or by external auditors.

13.2. PLANNING AND CONTROL

In order to manage a project effectively it is important to have planning and control processes in place that are recognised and understood through all supervisory and management levels. Large projects in particular can suffer if engineering teams become isolated or lose touch with the common interests of the project group or company business objectives.

Project planning techniques are employed to prepare realistic schedules within manpower, materials and funding constraints. Realistic schedules are those that include a time allocation for delays where past experience has shown they may be likely, and where no action has been taken to prevent reoccurrence. Once agreed, schedules can be used to monitor progress against targets and highlight departure from plans.

13.2.1. Network analysis

A technique widely used by the industry is *Critical Path Analysis* (CPA or 'network analysis') which is a method for systematically analysing the schedule of large projects, so that activities within a project can be phased logically, and dependencies

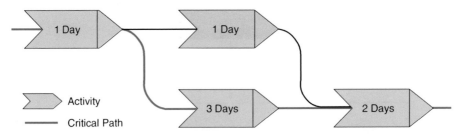

Figure 13.4   Project planning network.

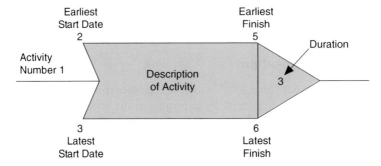

Figure 13.5   Activity symbol convention.

identified. All activities are given a duration and the longest route through the network is known as the *critical path*.

In Figure 13.4, the relationship between four activities of different duration is shown. In this case the critical path is indicated by the lower route (6 days), since the last activity cannot start until all the previous activities have been completed.

In reality all activities are listed and dependency relationships are identified. Activities are given a duration, and an earliest start and finish date is determined, based on their dependency with previous activities. Latest start and finish dates (without incurring project delays) can be calculated once the network is complete, and indicate how much 'play' there is in the system.

A typical 'activity symbol' convention is shown in Figure 13.5. Other information that may be included in a network is: milestones (e.g. first oil), weather windows and restraints (e.g. permit to continue requirements).

Once a network has been constructed it can be reviewed to determine whether the completion date and intermediate key dates are acceptable. If not, activity duration reductions have to be sought, for example, by increasing manpower or changing suppliers.

13.2.2. Bar charts

Whilst network analysis is a useful tool for estimating timing and resources, it is not a very good means for displaying schedules. Bar charts are used more commonly to illustrate planning expectations and as a means to determine resource loading.

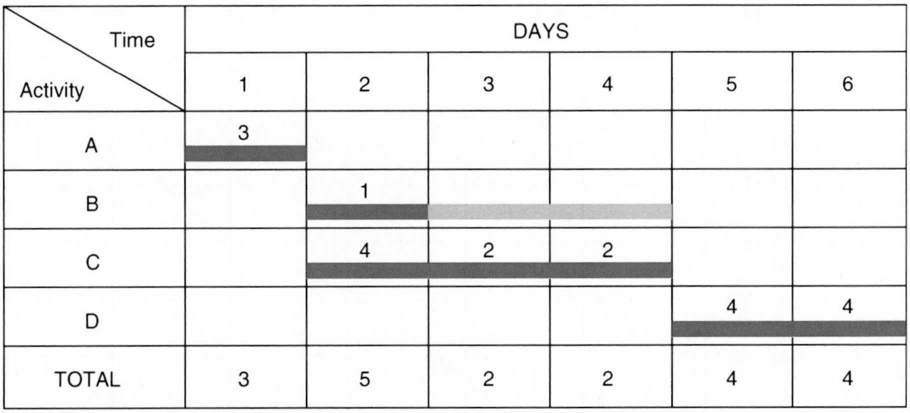

Activity \ Time	DAYS					
	1	2	3	4	5	6
A	3					
B		1				
C		4	2	2		
D					4	4
TOTAL	3	5	2	2	4	4

Figure 13.6 Bar chart with resource loading.

Table 13.1 Resource weighting matrix

Activity	Weight per Day						Weight per Activity (%)
	1	2	3	4	5	6	
A	15.0						15
B			5.0				5
C		20.0	10.0	10.0			40
D					20.0	20.0	40
Total	15%	20%	15%	10%	20%	20%	100

The bar chart below is a representation of the network shown in Figure 13.4. In addition the chart has been used to display the resource loading (Figure 13.6).

The bar chart indicates that activity 'B' can be performed at any time within days 2, 3 and 4, without delaying the project. It also shows that the resource loading can be smoothed out if activity 'B' is performed in either day 3 or 4, such that the maximum loading in any period does not exceed 4 units. Resource units may be, for example, 'man hours' or 'machine hours'.

The resource loading can be represented in percentage terms (see Table 13.1) to give an indication of the resource 'weighting' distribution on a daily basis and per activity (note that activity 'B' has been moved to day 3 to smooth resource loading).

13.2.3. 'S'-curves

By plotting the cumulative resource weighting against time, the planned progress of the project can be illustrated, as shown in Figure 13.7. This type of plot is often referred to as an 'S'-Curve, as projects often need time to gain momentum and slow down towards completion (unlike the example shown).

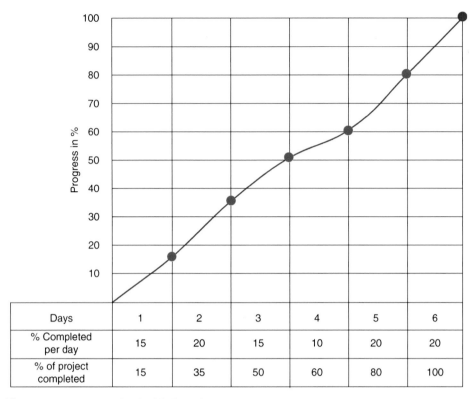

Figure 13.7   Progress plot (or 'S'-Curve).

Plots such as this can be used to compare actual to planned progress. If progress is delayed at any point, but the completion date cannot be slipped, the plot can be used to determine how many extra resource units have to be employed to complete the project on time.

 ## 13.3. Cost Estimation and Budgets

At each phase of a project, cost information is required to enable decisions to be taken. In the conceptual phase these estimates may be very approximate (e.g.+40% accuracy), reflecting the degree of uncertainty regarding both reservoir development and surface options. This is sometimes referred to as an 'order of magnitude cost estimate'. As the project becomes better defined the accuracy of estimates should improve.

An appropriate estimate of technical cost is important for economic analysis. *Underestimating* costs may lead to funding difficulties associated with cost overruns, and ultimately, lower profitability than expected. Setting estimates too high can kill a project unnecessarily. Costs are often based on suppliers' price lists and historical data. However, many recent oil and gas developments can be considered pioneering

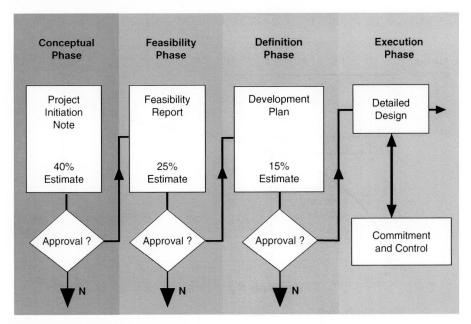

Figure 13.8 Cost estimate evolution.

ventures in terms of the technology and engineering applied. Estimating solely on the basis of historical costs can be inappropriate.

Cost estimates can usually be broken into firm items, and items which are more difficult to assess because of associated uncertainties or novelty factor. For example, the construction of a pipeline might be a firm item but its installation may be weather dependent, so an 'allowance' could be included to cover extra lay-barge charges if poor sea conditions are likely (Figure 13.8).

Firm items such as pipelines are often estimated using charts of cost vs. size and length. The total of such items and allowances may form a preliminary project estimate. In addition to allowances some *contingency* is often made for expected but undefined changes, for example to cover design and construction changes within the project scope. The objective of such an approach is to define an estimate that has as much chance of under running as over running (sometimes termed a 50/50 estimate) (Figure 13.9).

A budget containing a number of 50/50 project estimates is more likely to balance than if no allowances or contingencies are built in. However such systems should not be abused to give insurance against budget overrun; inflated estimates tend to hide inefficiency and distort project ranking. Allowances should generally be supported by statistical evidence, and contingencies clearly qualified. Contingency levels should normally reduce as planning detail increases.

Minimum risk estimates are sometimes used to quantify either maximum exposure in monetary terms or, in the case of an annual work plan containing multiple projects, to help determine the proportion of firm projects. Firm projects are those which have budget cover even if costs overrun. A minimum risk estimate is one

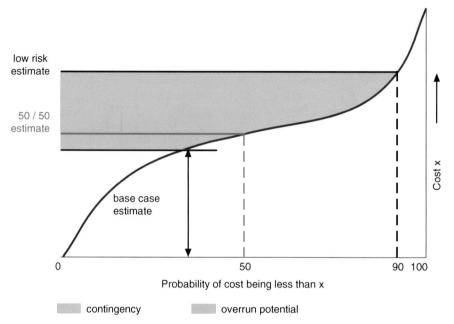

low risk estimate

50 / 50 estimate

base case estimate

Cost x

0   50   90 100

Probability of cost being less than x

contingency   overrun potential

Figure 13.9   Estimates and contingency.

with little or no probability of overrun, and can be used to reflect the risk associated with very complex or novel projects.

This may be referred to as the p90 cost estimate. Note that it is at the high end of the range, whereas to a typical subsurface engineer a p90 reserves estimate is at the low end of the range. Care must be taken when quoting p90 and p10 estimates, as they may mean different things to different disciplines or even to different companies!

13.4. REASONS FOR CONTRACTING

Many oil and gas companies do not consider the detailed design and construction of production facilities as part of their core business. This is often the stage at which work is contracted out to engineering firms and the client company will switch manpower resources elsewhere, although some degree of project management is commonly retained.

Contracts are used by an oil company where

- the services offered by a contractor can be provided more cheaply or more efficiently than using in-house resources
- the services required are of a specialist nature, and are not available in-house
- services are required for a peak of demand for a short period of time, and the oil company prefers not to recruit staff to meet this peak.

13.5. TYPES OF CONTRACT

To protect both parties in a contract arrangement it is good practice to make a contract in which the scope of work, completion time and method of reimbursement are agreed. Contracts are normally awarded through a competitive *tendering* process or after negotiation if there is only one suitable contractor.

There are many varieties of contract for many different services, but some of the more common types include

- *Lump Sum* contract; contractor manages and executes specified work to an agreed delivery date for a fixed price. Penalties may be due for late completion of the work, and this provides an incentive for timely completion. Payment may be staged when agreed milestones are reached.
- *Bills of Quantities* contract; the total work is split into components which are specified in detail, and rates are agreed for the materials and labour. The basis of handling variations to cost are agreed.
- *Schedule of Rates* contract; the cost of the labour is agreed on a rate basis, but the cost of materials and the exact hours are not specified.
- *Cost Plus Profit* contract; all costs incurred by the contractor are reimbursed in full, and the contractor then adds an agreed percentage as a profit fee.

Lump sum contracts tend to be favoured by companies awarding work (if the scope of work can be well defined) as they provide a clear incentive for the contractor to complete a project on time and within an agreed price.

The choice of contract type will depend on the type of work, and the level of control which the oil company wishes to maintain. There is a current trend for the oil company to consider the contractor as a partner in the project (*partnering arrangements*), and to work closely with the contractor at all stages of the project development. The objective of this closer involvement of the contractor is to provide a common *incentive* for the contractor *and* the oil company to improve quality, efficiency, safety and most importantly to reduce cost. This type of contract usually contains a significant element of sharing risk and reward of the project.

CHAPTER 14

PETROLEUM ECONOMICS

Introduction and Commercial Application: Investment opportunities in the exploration and production (E&P) sector of oil and gas business are abundant. Despite areas such as the North Sea, shallow water GoM and the North Slope in Alaska being mature, there are still many new fields under development in those regions, and new business interests are emerging in South America, Africa and South East Asia. Countries such as Russia and China have opened up their regions to foreign investment, driven by the requirement for capital investment and technical expertise. Some fields which have been producing for decades are being redeveloped using technologies such as multilateral wells, which were not available during their initial development. With a period of sustained high oil prices since the mid 1990s, formerly uneconomic enhanced recovery processes have created interest in further investment, even though the field is considered mature. Such ongoing opportunities all require analysis of the attractiveness of the investment.

Development of an oil or gas accumulation is characteristically a high cost venture, especially offshore, and the uncertainties are large. A typical capital investment for a medium-sized offshore oil field (say 100 MMbbl recoverable reserves) would be in the order of one billion US$, and the range of uncertainty on recoverable reserves may be plus or minus 25% prior to committing to the development. The result of technical or commercial failure when the investment is so high is very significant to most investors, who therefore expend great effort to understand and quantify the uncertainties and assess the consequent levels of risk and reward in investment proposals.

Petroleum economics provides the tools with which to quantify and assess the financial risks involved in field exploration, appraisal and development, and is the consistent basis used for comparing alternative investments. The techniques are applied to advise management on the attractiveness of investment opportunities, to assist in selecting the best options, to identify the financial exposure and to determine how to maximise the value of existing assets.

14.1. BASIC PRINCIPLES OF DEVELOPMENT ECONOMICS

Sections 14.1–14.7 will deal mainly with the economics of a new field development. Exploration economics is introduced in Section 14.8. The general approach will be to look at an investment proposal from an operator's point of view.

The economic analysis of investment opportunities requires the gathering of much information, such as capital costs, operating costs, anticipated hydrocarbon production profiles, contract terms, fiscal (tax) structures, forecast oil/gas prices, the timing of the project and the expectations of the stakeholders in the investment.

These data must be collected from a number of different departments and bodies (e.g. petroleum engineering, engineering, taxation and legal, host government) and each data set carries with it a range of uncertainty. Data gathering and establishing realistic ranges of uncertainty can be very time consuming.

The *economic model* for evaluation of investment (or divestment) opportunities is normally constructed as a spreadsheet, using the techniques to be introduced in this section. Specialised software packages are available for the analysis, but are generally costly due to the requirement for sufficient flexibility to model different fiscal systems around the world. Petroleum economists often write bespoke spreadsheet-based models, tailored to a particular fiscal system, but this in turn creates an issue of consistency within an organisation.

The uncertainties in the model's input data are handled by establishing a *base case* (often using the 'best guess' values of the input variables) and then investigating the impact of varying the values of key inputs in a *sensitivity analysis*.

It is useful to firstly consider a very simple model of an investment (Figure 14.1). From an overall economic viewpoint, any investment proposal may be considered as an activity which initially *absorbs money* and later *generates money*. The money invested may be raised as *loan capital* (*debt*) or from *shareholders' capital* (*equity*). This is invested in the project to purchase plant and equipment and pay for operating costs. The net money generated (i.e. revenues less all costs, which will include taxes) may be used to repay interest on loans and loan capital. The residual balance belongs to the shareholders, and is called *shareholders' profit*. This can either be paid out as dividends, or reinvested in the company to fund the existing venture or new ventures. The diagram indicates the overall flow of funds for a proposed project. The detailed cash movements are contained within the box labelled 'the project'. We will look inside this box in the coming sections.

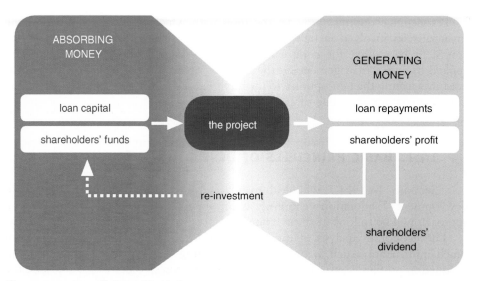

Figure 14.1   Overall flow of funds for a project.

From this overview, it is apparent that the project must generate sufficient return on the money absorbed to at least pay the interest on loans and pay the dividend expected by the shareholders. Any remaining cash generated can be used to pay off loans or reinvested in the same or alternative projects. The minimum return expected from the investment in a project will be further discussed in Section 14.4.

Take, for example, the investment opportunity to be the development of an oil field. Within the project box, the *cashflow* of the project is the forecast of the money absorbed and the money generated during the project lifetime. Initially the cashflow will be dominated by the CAPEX required to design, construct and commission the hardware for the project (e.g. platform, pipeline, wells, compression facilities).

Once production commences (possibly 3–8 years after the first CAPEX) *gross revenues* are received from the sale of the hydrocarbons. These revenues are used to recover the CAPEX of the project, to pay for the OPEX of the project (e.g. manpower, maintenance, equipment running costs, support costs), and to provide the *host government take* which may in the simplest case be in the form of taxes and royalty.

The oil company's after-tax share of the profit is then available for repayment of interest on loans, repayment of loan capital, distribution to the shareholders as dividends or reinvestment on behalf of the shareholders in this or other projects.

So, from the oil company's point of view, the balance of the money absorbed by the project (CAPEX, OPEX) and the money generated (the oil company's after-tax share of the profit) yields the *project net cashflow*, which can be calculated on an annual basis. It is often referred to simply as the project cashflow.

The project cashflow forms the basis of the economic evaluation methods which will be described. From the cashflow a number of *economic indicators* can be derived and used to judge the attractiveness of the project. Some of the techniques to be introduced allow the economic performance of proposed projects to be tested against investment criteria and also to be compared with alternative investments.

14.2. CONSTRUCTING A PROJECT CASHFLOW

The construction of a project cashflow requires information from a number of different sources. The principal inputs are typically shown in Table 14.1.

The data gathering process can be lengthy, and each input will carry with it a range of uncertainty. For example, early in the appraisal stage of the field life the range of uncertainty in the reserves and production forecast from the field may be $\pm 50\%$. As further appraisal data is gathered this range will reduce, but at the decision point for proceeding with a project, uncertainties of $\pm 25\%$ are common.

The uncertainty may be addressed by constructing a *base case* which represents the most probable outcome, and then performing sensitivities around this case to determine which of the inputs the project is most vulnerable to. The most influential parameters may then be studied more carefully. Typical sensitivities are considered in Section 14.6.

Table 14.1 Elements of a project cashflow

Sources	Information
Petroleum engineering	Reserves
	Production forecasts – oil, sales gas
Drilling engineering	Drilling and completion costs
Facilities engineering	Capital costs
	Platform structures
	Transportation (e.g. pipelines)
	Production facilities (e.g. separators,
	compressors, pumps)
Operations and maintenance	Operating costs
engineering	Maintenance
	Workover
	Manpower requirement
Human resources	Manpower costs
	Operators
	Technical staff
	Support staff
	Overheads
Host government	Fiscal system
	Tax and royalty rate
	Royalty payment method (in cash or in kind)
	Production sharing agreement
	Company status (e.g. newcomer)
	Project status (e.g. ring fenced)
	Licence terms (duration, final liabilities)
	Decommissioning requirements
Corporate planning	Forecast oil and gas prices
	Discount rates, hurdle rates
	Exchange rates
	Inflation forecast
	Market factors
	Political risk, social obligations

It is therefore important when collecting the data from the various sources that the current range of uncertainty is also requested. In particular, when estimating operating costs it is desirable for the operations and maintenance engineers to estimate the cost of these activities based on the particular facilities and equipment types being proposed in the engineering design. For example, the cost of operating and maintaining an unmanned remote controlled platform will be significantly different to a conventional manned facility.

For any one case, say the base case, the project cashflow is constructed by calculating, on an annual basis, the *revenue items* (the payments received by the project) and then subtracting the *expenditure items* (the payments made by the project: CAPEX,

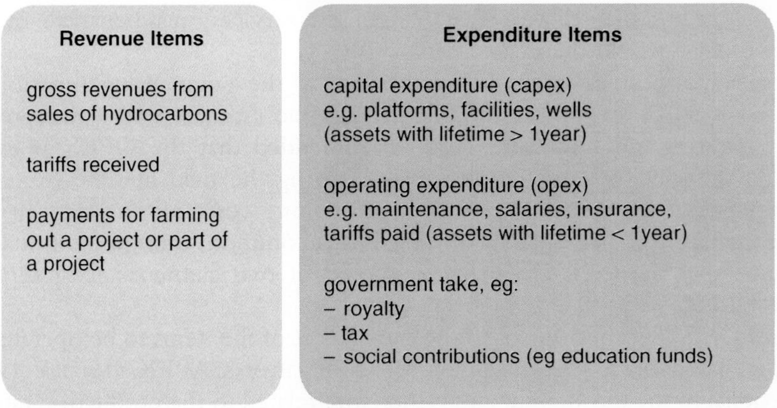

Revenue Items	Expenditure Items
gross revenues from sales of hydrocarbons	capital expenditure (capex) e.g. platforms, facilities, wells (assets with lifetime > 1year)
tariffs received	operating expenditure (opex) e.g. maintenance, salaries, insurance, tariffs paid (assets with lifetime < 1year)
payments for farming out a project or part of a project	government take, eg: − royalty − tax − social contributions (eg education funds)

Figure 14.2 Typical revenue and expenditure items.

OPEX and host government take). For each year the balance is the project net cashflow (or just project cashflow). Hence, on an annual basis

Project net cashflow = Revenue−expenditure

Typical revenues and expenditure items are summarised in Figure 14.2.

14.2.1. Revenue items

In most cases the revenues will be due to the sale of hydrocarbons. In determining these gross revenues, oil and/or gas prices must be assumed. The *oil price forecast* is often based on a flat real terms (RT) price (i.e. increasing in price at the forecast rate of inflation) or flat money of the day (MOD) price (i.e. price stays the same and is thus declining in RT). Both the level and method of price forecast are a matter of taste, and the industry analysts have in the past been notoriously poor at predicting oil price. Oil price is often linked to a regional marker crude such as Brent crude in the North Sea; the specific crude price is adjusted for specific conditions such as crude quality and geographic location. A *gas price forecast* may be indexed to the crude market price or be taken as the result of a negotiated price with an identified customer. A peculiarity of some gas contracts is that a fixed gas price is agreed for a very long period of time, possibly the lifetime of the field, which may result in disparities if the oil price and prevailing gas price change dramatically. Such contracts will often partially index gas price to the market price of the crude, and to other energy forms such as electricity prices.

14.2.2. Expenditure items

14.2.2.1. OPEX and CAPEX

The treatment of expenditures will be specified by the fiscal system set by the host government. A typical case would be to define expenditure on items whose useful life exceeds 1 year as CAPEX, such as costs of platforms, pipelines, wells. Items whose useful life is less than 1 year (e.g. chemicals, services, maintenance, overheads, insurance costs) would then be classed as OPEX.

The capital cost estimates are generated by the engineering function, often expressed as 50/50 (or p50) estimates, meaning an estimate with equal probability of cost overrun and underrun. It is recommended that the OPEX is estimated based on the specific activities anticipated during the field lifetime (e.g. number of workovers, number of replacement items, cost of forecast manpower requirements). In the absence of this detail it is common, though often inaccurate, to assume that the OPEX will be composed of two elements: *fixed OPEX* and *variable OPEX*.

Fixed OPEX is proportional to the capital cost of the items to be operated and is therefore based on a percentage of the cumulative CAPEX. *Variable OPEX* is proportional to the throughput and is therefore related to the production rate (oil or gross liquids). Hence

$$\text{Annual OPEX} = [A(\%) \times \text{cumulative CAPEX(\$)}] + \left[B\left(\frac{\$}{\text{bbl}}\right) \times \text{production}\left(\frac{\text{bbl}}{\text{year}}\right) \right]$$

Any OPEX estimate should not ignore the cost of overheads which the project attracts, especially for example, the cost of support staff and office rental which can form a significant fraction of the total OPEX, and does not necessarily reduce as production declines.

The sum of OPEX and CAPEX is sometimes termed the *technical cost* or total cost. OPEX may be referred to as a *lifting cost*, while CAPEX can be referred to as a *development cost*.

14.2.2.2. Host government take

'Fisc' is an Old English word for taxman. A *fiscal system* refers to the manner in which the host government claims an entitlement to income from the production and sale of hydrocarbons on behalf of the host nation. The simplest and more traditional fiscal system is the *tax and royalty scheme*, such as that applied to income from hydrocarbon production in the United Kingdom.

Royalty is normally charged as a percentage of the gross revenues from the sale of hydrocarbons, and may be paid in cash or in kind (e.g. oil). The prevailing oil price is used.

$$\text{Royalty} = \text{Royalty rate (\%)} \times \text{production (bbl)} \times \text{oil price}\left(\frac{\$}{\text{bbl}}\right)$$

In addition to royalty, one or more profit taxes may be levied (such as a special petroleum tax, plus the usual corporation tax on company profits).

Prior to the calculation of tax, certain allowances may be made against the gross revenue before applying the tax rate. These are called *fiscal allowances* and commonly include the royalty, OPEX and capital allowances (which are explained later in this section). Fiscal allowances may also be referred to as *deductibles*.

$$\text{Fiscal allowances} = \text{Royalty} + \text{OPEX} + \text{capital allowances (\$)}$$

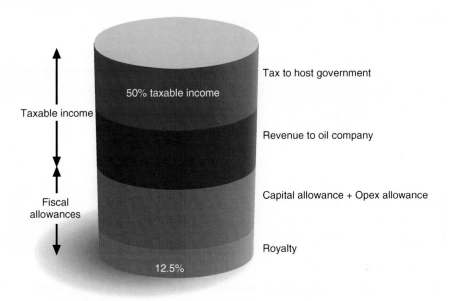

Figure 14.3   Split of the barrel under a typical tax and royalty system.

These are deducted from the gross revenues prior to applying the tax rate.

$$\text{Taxable income} = \text{Revenues} - \text{fiscal allowances (\$)}$$

$$\text{Tax payable} = \text{Taxable income (\$)} \times \text{tax rate (\%)}$$

Royalty is charged from the start of production, but tax is only payable once there is a positive taxable income. At the beginning of a new project the fiscal allowances may exceed the revenues, giving rise to a negative taxable income. Whether the project can take advantage of this depends on the fiscal status of the company and the project. A '*ring-fenced*' *project* would not be able to claim a negative taxable income as a rebate, whereas a '*non-ring fenced project*' may be able to claim a rebate for its negative taxable income by offsetting it against taxable income from another project. It is normally the host government which decrees the fiscal status of the project (Figure 14.3).

14.2.2.3. Capital allowances

Fiscal allowances for investment in capital items (i.e. CAPEX) are made through *capital allowances*. The method of calculating the capital allowance is set by the fiscal legislation of the host government, but three common methods are discussed below.

It should be noted that a capital allowance is not a cashflow item, but is only calculated to enable the taxable income to be determined. The treatment of capital allowance for this purpose is a petroleum economics approach, used to calculate the tax payable. Capital allowance may differ from depreciation, which is a calculation made by the accountant when calculating net book values and annual profit.

Table 14.2   Straight line capital allowance

Year	CAPEX	Capital Allowance			
		1st Year	2nd Year	3rd Year	Total
1	100	20			20
2	400	20	80		100
3	200	20	80	40	140
4		20	80	40	140
5		20	80	40	140
6			80	40	120
7				40	40
8					
	700	100	400	200	700

*14.2.2.3.1. **Straight line capital allowance method.*** This is the simplest of the methods, in which an allowance for the capital asset is claimed over a number of years in equal amounts per year, for example 20% of the initial CAPEX per year for 5 years.

Capital allowances may be accepted as soon as the capital is spent or may have to wait until the asset is actually brought into use. In the case of the newcomer company or the ring-fenced project the allowance may only be applied once there is revenue from the project.

A newcomer company is a company performing its first project in the country, and therefore has no revenues against which to offset capital allowances.

A project is ring-fenced if, for fiscal purposes, its fiscal allowances can only be offset against revenues earned within that ring fence (Table 14.2).

*14.2.2.3.2. **The declining balance method.*** Each year the capital allowance is a fixed percentage of the unrecovered value of the asset at the end of the previous year. The same comments about when the allowance can start apply (Table 14.3).

At the end of the project life a residual unrecovered asset value will remain. This is usually accepted in full as a capital allowance in the final year of the project. Hence the total asset value is fully recovered over the life of the field, but at a slower rate than in the straight line method.

*14.2.2.3.3. **The depletion method or unit of production method.*** This method attempts to relate the capital allowance to the total life of the assets (i.e. the field's economic lifetime) by linking the annual capital allowance to the fraction of the remaining reserves produced during the year. The capital allowance is calculated from the unrecovered assets at the end of the previous year, times the ratio of the current year's production to the reserves at the beginning of the year. As long as the ultimate recovery of the field remains the same, the capital allowance per barrel of

Table 14.3   Declining balance capital allowance

Year	CAPEX	Declining Balance Capital Allowance @ 20% p.a.	
		Unrecovered Assets at Year End	Capital Allowance
1	100	100	20
2	400	480	96
3	200	484	117
4		467	93
5		374	75
6		299	60
7		239	48
8		191	38
9		153	31
10		122	122
	700		700

Table 14.4   Depletion capital allowance

Year	CAPEX ($million)	Annual Production (MMb)	Reserves (MMb)	Depletion Factor	Unrecovered Assets ($million)	Capital Allowance ($million)
1	100	0	250		100	
2	400	0	250		500	
3	200	25	250	0.10	700	70
4		40	225	0.18	630	112
5		50	185	0.27	518	140
6		50	135	0.37	378	140
7		40	85	0.47	238	112
8		25	45	0.56	126	70
9		15	20	0.75	56	42
10		5	5	1.00	14	14
	700	250				700

production is constant. However, this is rarely the case, making this method more complex in practice.

In Table 14.4, where the ultimate recovery remains unchanged throughout the field life, the capital allowance rate remains a constant factor of $700/250 = \$2.8/\text{bbl}$.

Figure 14.4 shows the relative timings of the capital allowance in which the straight line method gives rise to the fastest capital allowance, followed by the declining balance method, and finally the depletion method. From the investor's point of view, earlier capital allowance is preferable since this gives rise to more fiscal

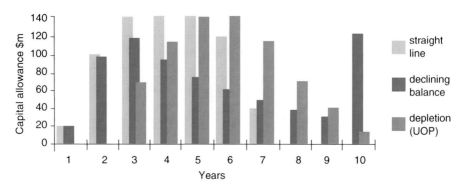

Figure 14.4   Comparison of capital allowance methods.

allowances and thus less tax payable in the early years of the project. The scheme for claiming capital allowance is however set by the host government.

14.2.2.4. Project net cashflow

Having discussed the elements of the cashflow calculation, remember that in any one year this can be calculated from gross revenues and expenditure as follows (for a tax and royalty fiscal system).

$$\text{Project net cashflow} = \text{Gross revenue} - \text{expenditure}$$
$$= \text{Gross revenue} - \text{CAPEX} - \text{OPEX} - \text{royalty} - \text{tax}$$

Project net cashflow may also be referred to as project cash surplus/deficit.

Note that capital allowances do not appear in the expression since they are not items of cashflow. Capital allowances are calculated in order to determine the fiscal allowances and thus the amount of tax payable.

Below is an example of the calculation of the net cashflow for just 1 year of the project.

Suppose in any particular year

Production = 12 MMbbl	CAPEX = $80 million
Oil price = $50/bbl	OPEX = $15 million
Royalty rate = 10%	
Tax rate = 50%	

Assume that the only previous CAPEX had been $120 million, spent in the previous year, with 25% straight line capital allowance, thus capital allowance in this year = $0.25 \times \$120$ million $+ 0.25 \times \$80$ million $= \$50$ million.

Revenue	= Production × oil price
	= 12 MMbbl × $50/bbl = $600 million
CAPEX	= $80 million

OPEX	= $15 million
Technical cost	= $95 million
Royalty	= Revenues × royalty rate = $600 million × 0.10 = $60 million
Fiscal allowances	= Royalty + OPEX + capital allowance
	= $60 million + $15 million + $50 million
	= $125 million
Taxable income	= Revenue−fiscal allowances
	= $600 million−$125 million = $475 million
Tax	= Tax rate × taxable income
	= 0.50 × $475 million
	= $237.5 million
Project net cashflow	= Revenues−CAPEX−OPEX−royalty−tax
	= $600−80−15−60−237.5 million
	= $207.5 million
Host government take	= Tax+royalty
	= $207.5 + 60 million
	= $267.5 million

The above calculation has been demonstrated for just 1 year of the project. In practice, the *project net cashflow* is constructed by performing the calculation for every year of the project life. A typical project cashflow is shown in Figure 14.5, along with a *cumulative net cashflow* showing how cumulative revenue is typically split between the CAPEX, OPEX, the host government (through tax and royalty) and the investor (say the oil company). The cumulative amount of money accruing to the company at the end of the project is the *cumulative net cashflow* (Figure 14.6).

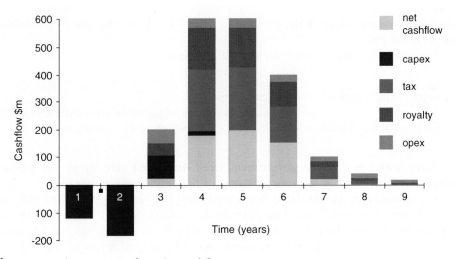

Figure 14.5 Components of a project cashflow.

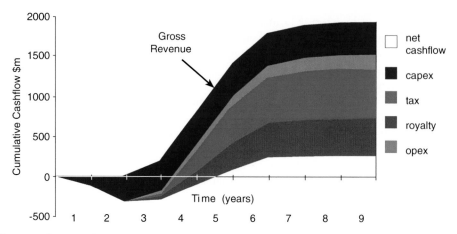

Figure 14.6 Cumulative cashflow.

14.2.2.5. Production sharing contracts

While tax and royalty fiscal systems are common, another prevalent form of fiscal system is the PSC, also referred to in some regions as a production sharing agreement (PSA), as introduced in Chapter 2. In these arrangements, the investor (e.g. oil company) enters into an agreement with the host government to explore and potentially appraise and develop an area. The investor acts as a contractor to the host government, who retains the title of any produced hydrocarbons.

Typically, the contractor carries the cost of exploration, appraisal and development, later claiming these costs from a tranche of the produced oil or gas (*cost oil*), should the venture be fortunate enough to result in a field development. If the cost oil allowance is insufficient to cover the annual costs (CAPEX and OPEX), excess costs are usually deferred to the following year. After the deduction of royalty (if applicable) the remaining volume of production (called *profit oil*) is then split between the contractor and the host government. The contractor will usually pay tax on the contractor's share of the profit oil. Figure 14.7 shows the split of production for a typical PSC.

In terms of cashflow items, for the oil company

Contractor net cashflow = Revenues−expenditures
                         = Cost oil recovery+net of tax profit oil−CAPEX−OPEX
Government net cashflow = Royalty + tax + government share of profit oil

This illustrates that the government need not invest directly in the project, which is not necessarily the case for a tax and royalty system either.

Many variations on the above theme exist, and the percentages applied will vary from country to country and from contract to contract. Some PSC systems do not contain royalty, but adjust the percentage split of profit oil and the amount of cost oil available instead to achieve their requirements. The general terms of the PSC are usually outlined by the government at the bidding stage, but then refined through

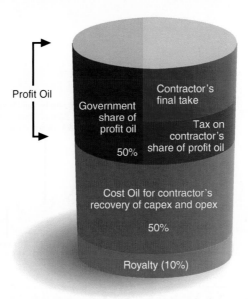

Figure 14.7   Split of production for a typical PSC.

negotiation between the government and the oil company. This can result in a protracted process and require significant effort and patience on behalf of both parties. The final agreement will, by definition, be acceptable to both parties; the government securing a commitment to exploration, appraisal and potential development with a reasonable government take of the revenues, and the oil company gaining a contract with a potential return which meets internal company investment criteria.

In general, the oil company return from a PSC is lower than that of a tax and royalty system. This is because the governments of many of the recently developed basins have elected to set up PSCs (e.g. Algeria, Angola, Kazakhstan). In less mature basins, with larger prospective undiscovered oil and gas targets, the government is able to secure better terms than for a mature area, where traditionally tax and royalty systems were established.

An advantage of PSCs is that the agreement includes a time schedule and fiscal terms for exploration, appraisal and development, and production periods (see Section 2.3, Chapter 2). As terms of the PSC are fixed, this reduces some of the uncertainties associated with tax and royalty systems where the level of royalty and tax rates may vary over the field lifetime. Figure 14.8 gives an indication of how frequently the tax and royalty terms have changed in the UK sector of the North Sea – such risks need to be considered when investing in what may be a politically stable country, but has proved to be fiscally unpredictable.

14.2.2.6. Economic indicators from the cashflow
From the net cashflow and cumulative net cashflow some basic *economic indicators* can be determined. The net cashflow determines the *economic lifetime* of the field. When

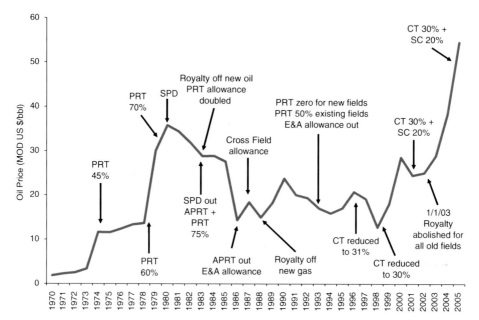

Figure 14.8   UKCS fiscal system changes.

the net cashflow turns permanently negative due to decreasing revenues (e.g. revenues are less than royalty plus OPEX in a tax/royalty system) then the project should be halted, and decommissioning planned. The *first oil* date is important because it indicates the point at which gross revenues commence. For most projects this is the point at which a positive annual net cashflow starts.

The most negative point on the cumulative net cashflow indicates the *maximum cash exposure* of the project. If the project were to be abandoned at this point, this is the greatest amount of money the investor stands to lose, before taking account of specific contractual circumstances (such as penalties from customers, partner claims, contractors' claims). It also represents the funds which are required to finance the project – if the maximum exposure is greater than the company's capacity to raise capital then the investor may consider farming out a portion of the project to a joint investor.

The point at which the cumulative net cashflow turns positive indicates the *payback time* (or payout time). This is the length of time required to receive accumulated net revenues equal to the investment. Payback time is primarily an indicator of risk – the longer the payback the more risky the project, but it says nothing about the net cashflow after the payback time and does not consider the total profitability of the investment opportunity. Some smaller industries use payback as a primary criterion for investment – it is not a bad measure as one can argue that one of the first rules of investment is to get the investment funds back – payback time indicates how long this will take.

The *cumulative net cashflow* accrues to the investor at the end of the economic lifetime of the project. Of the indicators mentioned so far, this is probably the most important as it measures the final prize.

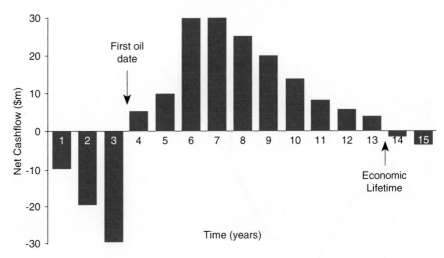

Figure 14.9 Indicators from the annual net cashflow.

The *profit-to-investment ratio* (PIR) may be defined in many ways, and is most meaningful when the net cashflow has been discounted (more of this shortly). On an undiscounted basis, the PIR may be defined as the ratio of the cumulative net cashflow to the capital investment. This is the first of the indicators of project efficiency, indicating the return on capital investment of the project. If capital is constrained, as it often is, then the investor should aim to maximise the return per unit of capital investment, which PIR measures very well. It is simple to calculate, but does not reflect the timing of the investment or the income stream.

Caution is required in the use of the simple cashflow indicators, since they fail to take account of changing general price levels or the cost of capital (discussed in Section 14.4). It is always recommended that the definition of the indicators is quoted for clarity of understanding.

Figures 14.9 and 14.10 show the indicators on the net cashflow and cumulative net cashflow diagrams, while Table 14.7 compares what these indicators tell the investor about the nature of the project economics.

14.3. Calculating a Discounted Cashflow

The project net cashflow discussed so far follows a pattern typical of E&P projects; a number of years of expenditure (giving rise to cash deficits) at the beginning of the project, followed by a series of cash surpluses. The annual net cashflows now need to incorporate the *timing* of the cashflows, to account for the effect of the *time value of money*. The technique which allows the values of sums of money spent at different times to be consistently compared is called *discounting*. This is particularly necessary for typical E&P projects because they are spread over many years. For very short projects, discounting would be less relevant.

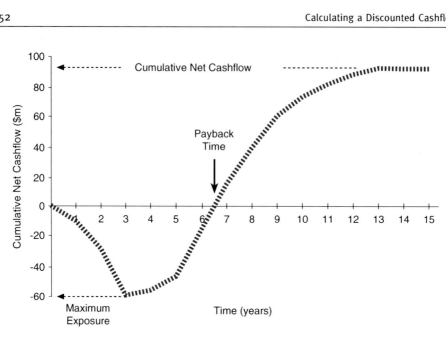

Figure 14.10   Indicators from the cumulative net cashflow.

14.3.1. Discounting: the concept

Suppose you have to meet an obligation to pay a bill of £10,000 in 5 years time. If you could be guaranteed a compound interest rate in your bank of 7% per annum (after tax) over each of the next 5 years, then the sum which you would have to invest today to be able to meet the obligation in 5 years time would be

$$\text{Investment today} \times (1.07)^5 = £10,000$$

$$\text{i.e. Investment today} = \frac{£10,000}{(1.07)^5} = £7130$$

The PV today of the sum of £10,000 in 5 years time is £7130, assuming a *discount rate* of 7% per annum.

What we have calculated is the PV, at a particular reference date, of a future sum of money, using a specified discount rate. In any discounting calculation, it is important to quote the *reference date* and the discount rate.

If you were offered £7130 today, or £10,000 in exactly 5 years time, you should be indifferent to the options, unless you could find an alternative investment opportunity which yielded a guaranteed interest rate better than the bank, in which case you should accept the money today and take the alternative investment opportunity.

In other words, given the constraints on investment opportunities, the £7130 today and the £10,000 in 5 years time are equivalent. The only difference between them is the timing and the opportunity to invest at 7% interest rate.

14.3.2. Setting the discount rate

In the above example, the discount rate used was the annual compound interest rate offered by the bank. In business investment opportunities the appropriate discount rate is the *cost of capital* to the company. This may be calculated in different ways, but should always reflect how much it costs the oil company to borrow the money which it uses to invest in its projects. This may be a weighted average of the cost of the share capital and loan capital of a company.

If the company is fully self-financing for its new ventures, then the appropriate discount rate would be the rate of return of the alternative investment opportunities (e.g. other projects) since this opportunity is foregone by undertaking the proposed project. This represents the *opportunity cost of the capital*. It is assumed that the return from the alternative projects is at least equal to the cost of capital to the company, otherwise the alternative projects should not be undertaken.

14.3.3. Discounting: the procedure

Once the concept of discounting is accepted, the procedure becomes mechanical. The general formula for discounting a flow of money *occurring in t years time* (c_t) to its PV (c_o) assuming a *discount rate* (r) is

$$c_o = \frac{c_t}{(1+r)^t} \quad \text{or} \quad c_o = c_t(1+r)^{-t}$$

the factor $1/(1+r)^t$ is called the *discount factor*.

Since this is a purely mechanical operation it can be performed using the above equation, or by looking up the appropriate discount factor in discount tables, or by setting the problem up on a spreadsheet. Two types of discount factors are presented for *full year* and *half year* discounting.

If the reference date is set at the beginning of the year (e.g. 1.1.2008) then full year discount factors imply that t is a whole number and that cashflows occur in lump sums at the end of each year. If the cashflow occurs uniformly throughout the year and the reference date is the beginning of the year then mid-year discount factors are more appropriate, in which case the discounting equation would be

$$c_o = \frac{c_t}{(1+r)^{t-0.5}}$$

$1/(1+r)^{t-0.5}$ is the mid-year discount factor.

Discounting can also be performed, of course, using a programmable calculator or a spreadsheet such as Lotus 1-2-3 or Microsoft Excel. The 'NPV' function is worth noting. In Excel, for example, a string of numbers (net cashflows in our case) can be discounted at a chosen discount rate (r) by using the function

$= \text{NPV}(r\%, \text{ cell references for net cashflows}) \text{ e.g.} = \text{NPV}(7\%, \text{ B2}:\text{B20})$

By default, Excel will assume the net cashflows occur at the end of each year. If mid-year discounting is required, this can be adapted as follows

$= \text{NPV}(7\%, \text{ B2}:\text{B20}) \times (1+r)^{0.5}$

14.3.4. Discounting a net cashflow

The net cashflow discussed in Section 14.2 did not take account of the time value of money, and was therefore an undiscounted net cashflow. The discounting technique discussed can now be applied to this net cashflow to determine the PV of each annual net cashflow at a specified reference date.

The following example generates the discounted cashflow (DCF) of a project using 20% mid-year discount factors (Table 14.5).

The total undiscounted net cashflow is $190 million. The *total discounted net cashflow* ($24.8 million) is called the NPV of the project. Since in this example the discount rate applied is 20%, this figure would be the 20% NPV also annotated NPV(20) or NPV_{20}. This is the PV at the beginning of Year 1 of the total project, assuming a 20% discount rate.

The example just shown assumed one discount rate and one oil price. Since the oil price is notoriously unpredictable, and the discount rate is subjective, it is useful to calculate the NPV at a range of oil prices and discount rates. One presentation of this data would be in the form of a matrix. The appropriate discount rates would be 0% (undiscounted), say 10% (the cost of capital), and say 20% (the cost of capital plus an allowance for risk). The range of oil prices is again a subjective judgement (Table 14.6).

Table 14.5 Calculating a discounted cashflow

Year	Net Cashflow ($million)	Discount Factor (Mid-Year)	Discounted Cashflow ($million)
1	−100	0.913	−91.3
2	−120	0.761	−91.3
3	+60	0.634	+38.0
4	+200	0.528	+105.6
5	+120	0.440	+52.8
6	+30	0.367	+11.0
Total	+190		+24.8

Table 14.6 NPV under various scenarios

Oil Price ($/b)	Discount Rate (%)		
	0	10	20
	NPV ($million)		
50	200	100	50
40	120	50	−5
20	70	10	−20

The benefit of this presentation is that it gives an indication of how sensitive the NPV is to oil price and discount rate. For example if the 20%, $40/bbl NPV is positive, but the 20%, $20/bbl NPV is negative, it indicates that at an oil price somewhere between $20 and $40/bbl the project breaks even (i.e. has an NPV of $0 at 20% discount rate).

If 10% is the cost of capital to the company, then the NPV(10) represents the real measure of the project value. That is, whatever positive NPV is achieved after discounting at the cost of capital, is the net value generated by the project. The 20% discount rate sensitivity is applied to include the risks inherent in the business, and would be a typical discount rate used for screening projects. Screening is discussed in more detail in Section 14.5.

14.3.5. The impact of discount rate on NPV

As the discount rate increases then the NPV is reduced. The following diagram shows the cashflow from the previous example (assuming an oil price of $20/bbl and ignoring the effect of inflation) at four different mid-year discount rates (10, 20, 25, 30%).

At a specific discount rate the NPV is reduced to zero. This discount rate is called the *internal rate of return* (IRR).

The higher the IRR, the more robust the project is, that is the more risk it can withstand before the IRR is eroded down to the level of the cost of capital. If the project IRR does not meet the cost of capital, then the project is unable to repay the cost of financing (assuming it is funded at the normal cost of capital to the company). IRR is therefore often used as a screening criterion, discussed in Section 14.5.

One way of calculating the IRR is to plot the NPV against discount rate, and to extrapolate/interpolate to estimate the discount rate at which the NPV becomes zero, as in the present value profile in Figure 14.11. The alternative method of calculating IRR is by using spreadsheet software (@ IRR function on Lotus 1-2-3, = IRR function on Microsoft Excel). This function uses an iteration

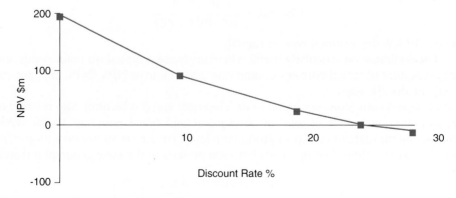

Figure 14.11   The present value profile.

technique to calculate what discount rate reduces the NPV of the net cashflow to zero. Care should be taken when the net cashflow has more than one change of sign (such as a phased project with a delayed significant investment), as multiple solutions for IRR will exist.

So, with a final economic indicator added, it is worth summarising the measures of project attractiveness in the following section.

14.4. ECONOMIC INDICATORS

In Section 14.2, a number of economic indicators were derived from the annual net cashflow; the most useful being the *economic life of the project*, determined when the annual net cashflow becomes permanently negative.

The cumulative net cashflow was used to derive *ultimate cash surplus* – the final value of the cumulative net cashflow; *maximum exposure* – the maximum value of the cash deficit; *payback time* – the time until cumulative net cashflow becomes positive.

The shortcoming of the maximum exposure and payout time is that they say nothing about what happens after the net cashflow becomes positive (i.e. the investment is recouped). Neither do they give information about the return on the investment in terms of a ratio, which is useful in comparing projects.

We have discussed the derivation and importance of NPV, often considered as the most important indicator in the upstream business. It is appealing in its simplicity but is one-dimensional in that it does not test efficiency of investing a constrained amount of capital.

A common ratio which indicates the efficiency with which the project creates profit is the

$$PIR = \frac{Cumulative\ net\ cashflow}{Total\ capital\ expenditure}$$

This may be more useful if the net cashflow items are discounted, for example

$$10\%\ PIR = \frac{10\%\ NPV}{10\%\ PV\ CAPEX}$$

where 10% is the assumed cost of capital.

This indicator is particularly useful where investment capital is a main constraint. It is a measure of capital efficiency, sometimes referred to as NPV/NPC (net present cost), or the *PV ratio*.

Per barrel costs (costs per barrel of development and production), also referred to as *unit costs*, *unit technical costs* or *development and lifting costs*, are useful when production throughput or export production levels are the constraint on a project, or when making technical comparisons between projects in the same geographical area.

$$Per\ barrel\ cost = \frac{CAPEX + OPEX}{Production} \quad \left(\frac{\$}{bbl}\right)$$

Table 14.7 Profitability indicators

Indicator	Unit	Value	Efficiency	Risk	Timing
Economic life	year	×	×	✔	✔
Maximum exposure	$	×	×	✔	×
Payback time	year	×	✔	✔	✔
Cumulative net cashflow	$	✔	×	×	×
PIR, (NPV/NPC)	%	×	✔	✔	×
Unit technical cost (UTC)	$/bbl	×	✔	✔	×
NPV	$	✔	×	×	×
IRR	%	×	✔	✔	×

It is often more useful to use the discounted values, to allow for the time effect of money, hence

$$\text{Perbarrel PV cost} = \frac{\text{PV CAPEX} + \text{PV OPEX}}{\text{PV production}} \quad \left(\frac{\$}{\text{bbl}}\right)$$

Within the same geographical area (e.g. water depth, weather conditions, distance to shore, reservoir setting) this is a useful tool for comparing projects to check that the appropriate development concept is being applied. If the indicators vary significantly then the reasons should be sought.

Unit costs vary dramatically by region – in the order of $10–20/bbl in the North Sea, deep water GoM, Russia, North Slope of Alaska, but just $2–5/bbl in the Middle East. This is a reflection of the location, climate and reservoir productivity.

In the case of a country whose output is constrained, perhaps by a pipeline capacity but more commonly by an OPEC production quota, it makes sense to minimise per barrel PV cost to produce the quota level as cheaply as possible. While this can lead to some inefficiency in development planning, the initial attractiveness of this simple approach is appealing.

In conclusion, Table 14.7 compares the aspects of the project highlighted by the economic indicators discussed so far. It demonstrates that no single indicator can paint a complete picture of the attractiveness of the project, and therefore a combination of these indicators is normally used to make an investment decision. Which indicator is of prime importance depends on the situation of the investor. With no limitations, NPV would probably be the primary indicator. In a capital constrained environment, the PIR would be very important, and if cashflow was a critical issue then payback or IRR would be looked at keenly.

14.5. Project Screening and Ranking

Project screening involves checking that the predicted economic performance of a project passes a prescribed threshold, or 'hurdle'. It is used to quickly sift out

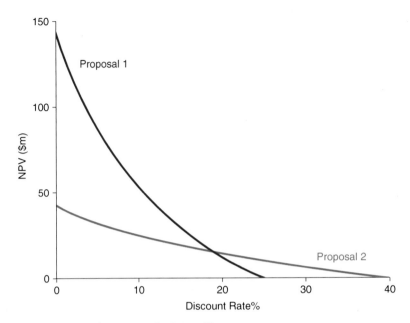

Figure 14.12   Project ranking using the PV profile.

interesting projects from non-starters. Investors commonly use IRR as a screening criterion by testing the project IRR against a minimum hurdle rate, for example, 20% IRR at $30/bbl. Providing that the project IRR exceeds the hurdle rate, then the project is considered further, otherwise it is rejected in current form.

With unlimited resources, the investor would take on all projects which meet the screening criteria. *Project ranking* is necessary to optimise the business when the investor's resources are limited and there are two or more projects (which both pass the screening criterion) to choose between.

The PV Profile can be used to select the more attractive proposal at the appropriate discount rate if the primary indicator is NPV. Figure 14.12 illustrates that the outcome of the decision may change as the discount rate changes.

At discount rates less than 18%, Proposal 1 is more favourable in terms of NPV, whereas at discount rates above 18%, Proposal 2 is more attractive. NPV is being used here as a ranking tool for the projects. At a typical cost of capital of, say, 10%, Proposal 1 generates the higher NPV, despite having the lower IRR.

The typical procedure would be to screen the projects on offer against the hurdle IRR, say 20%. In the above example, both projects pass the test. The next step is to rank the projects on the basis of NPV at the cost of capital. This would then rank Project 1 higher.

Choosing between projects on the basis of IRR alone risks rejecting higher value projects with a more modest, yet still acceptable rate of return.

Again, the comparison of project indicators (Table 14.7) must be kept in mind – one needs to be aware of what criteria are important to the company at the time.

14.6. SENSITIVITY ANALYSIS

As discussed in Section 14.2, the technical, fiscal and economic data gathered to construct a project cashflow carry uncertainty. An economic base case is constructed using, for example, the most likely values of production profile and the 50/50 cost estimates, along with the 'best estimate' of future oil prices and the anticipated production agreement and fiscal system.

In order to test the economic performance of the project to variations in the base case input data, sensitivity analysis is performed. This indicates how robust the project is to variations in one or more parameters, and also highlights which of the inputs the project economics is more sensitive to. These inputs can then be addressed more specifically. For example, if the project economics is highly sensitive to a delay in first production, then the scheduling should be more critically reviewed.

Changing just one of the individual input parameters at a time gives a clearer indication of the impact of each parameter on NPV (the typical indicators under investigation), although in practice there will probably be a combination of changes. The combined effect of varying individual parameters is usually closely estimated by adding the individual effects on project NPV.

Typical parameters which may be varied in the sensitivity analysis are

Technical parameters

- CAPEX
- OPEX (fixed and/or variable)
- reserves and production forecast
- delay in first production.

Economic parameters

- timing of fiscal allowances (e.g. ring-fencing)
- discount rate
- oil price
- inflation (general and specific items).

If the fiscal system is negotiable, then sensitivities of the project to these inputs would be appropriate in preparation for discussions with the host government.

When the sensitivities are performed the economic indicator which is commonly presented is NPV at the discount rate which represents the cost of capital, say 10%, this being considered as the true value of the project.

The results of the sensitivity analysis may be represented in tabular form, but a useful graphical representation is a plot of the change in NPV(10) against the percentage change in the parameter being varied, as shown in Figure 14.13. This is sometimes called a *spider diagram*.

The plot immediately shows which of the parameters the 10% NPV is most sensitive to – the one with the steepest slope. Consequently the variables can be ranked in order of their relative impact.

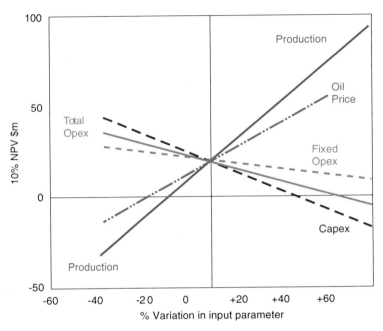

Figure 14.13   Sensitivity diagram for 10% NPV.

It is useful to truncate the lines at the extreme values which are considered likely to occur, for example oil price may be considered to vary between −40% and +20% of the base case consumption. This presentation adds further value to the plot.

14.7. INCORPORATING INFLATION

Inflation is a factor which is usually taken into account in economic evaluations, since it has become a fact of life in most economies in recent decades. Inflation increases the price of goods and erodes the purchasing power of money over time. When making estimates of the future amount of money required to pay for materials and services for our project development, the best we can usually do to is to establish the cost at the reference date, which we call the *base year cost* (BYC), and then predict how this will escalate in the future due to inflation and other specific market factors. This allows us to estimate the *money of the day* (MOD) *cashflow* – the amount of cash which will change hands on the specific date in the future. This escalation may be applied to costs and also to prices of the product if we believe that oil or gas prices will increase in the future.

The economic calculations of capital allowance, royalties and taxation are performed in MOD, yielding a net cashflow in MOD. In order to estimate what this is worth at the reference date, this net cashflow is then deflated back to real terms (RT). Usually, the deflation and the discounting discussed in Section 14.3 are combined into one step. It is important to realise however that the deflation brings

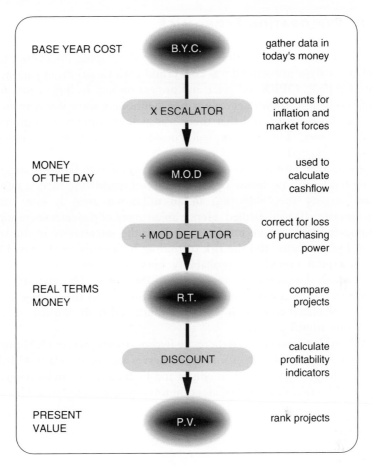

Figure 14.14   Handling inflation – types of money.

the MOD back to RT, while the discounting is done to reflect the cost of capital of the project, correcting future net cashflows to the PV.

The process can be captured in the above diagram (Figure 14.14).

RT refers to the purchasing power of money. Suppose $100 today would buy 100 loaves of bread at $1/loaf. If inflation increased the cost of a loaf to $1.10 in a year's time, then the same $100 in a year's time would only be able to buy 100/1.1 (i.e. 91) loaves of bread. The RT value of the MOD $100 in 1 year's time is only $91. Anyone whose income rises slower than inflation will be aware of the loss of the RT value of their income.

For completeness, when we calculate the project IRR, this is usually performed by working out what discount rate makes the MOD net cashflow equal to zero, as discussed in Section 14.3. The real rate of return (RRoR), as used by some commercial analysts, is the discount rate which sets the RT net cashflow equal to zero. At low inflation rates, the RRoR is lower than the IRR by approximately the rate of inflation.

14.8. EXPLORATION ECONOMICS

So far, we have discussed the economics of developing discovered fields, and the sensitivity analysis introduced was concerned with variations in parameters such as reserves, CAPEX, OPEX, oil price and project timing. In these cases, the risk of there being no hydrocarbon reserves was not mentioned, since it was assumed that a discovery had been made, and that there was at least some minimum amount of recoverable reserves. This section will consider how exploration prospects are economically evaluated.

In exploration economics, we must consider exploration failure – the possibility of spending funds with no future returns. A typical worldwide success rate for rank exploration activity (i.e. exploring in an unknown area) is one commercial discovery for every 10 wells drilled. Hence an estimate of the reserves resulting from exploration activity must take into account both the uncertainty in the volume of recoverable hydrocarbons and the risk of finding hydrocarbons (Figure 14.15).

Recall a typical cumulative probability curve of reserves for an exploration prospect in which the probability of success (POS) is 30%. The 'success' part of the probability axis can be divided into three equal bands, and the average reserves for each band is calculated to provide a low, medium and high estimate of reserves, *if* there are hydrocarbons present.

From this expectation curve, *if* there are hydrocarbons present (30% probability), then the low medium and high estimates of reserves are 20, 48 and 100 MMstb. The NPV for the prospect for the low, medium and high reserves can be determined by estimating engineering costs and production forecasts for three cases. This should be

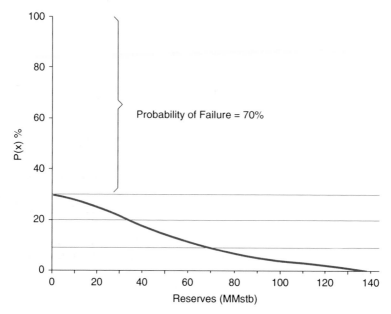

Figure 14.15   Cumulative probability curve for an exploration prospect.

performed not simply by scaling, but by tailoring an engineering solution to each case assuming that we would know the size of reserves before developing the field. For example, the low case reserves may be developed as a satellite development tied into existing facilities, whereas the high case reserves might be more economic to develop using a dedicated drilling and production facility.

We define the EMV of the exploration prospect as

$$\text{EMV} = \text{Unrisked NPV} \times \text{POS} - \text{PV exploration costs}$$

where POS is the probability of success of an economic development; unrisked NPV is the mean of the H, M, L NPVs (without any consideration of exploration and appraisal costs); PV exploration costs are the discounted cost of the exploration activity.

The POS is estimated using the techniques discussed in Chapter 3.

An alternative way of considering EMV is by presenting the outcomes on a decision tree. Figure 14.16 is an example of a decision tree which uses the above values.

It is assumed that the cost of the exploration activity is $10 million. The NPV of developing the high, medium and low case reserves are assumed to be $200, $80 and $5 million respectively, so the low case actually makes a loss when taking into account the exploration costs. With an equal probability of low, medium and high cases occurring, and assuming that the low case would be developed to make the small gain, the EMV of the prospect is $85 million, again assuming an exploration cost of $10 million. The above problem has used software from Palisade, called 'Precision Tree' (Figure 14.16).

In very simple terms, evaluating an exploration opportunity means weighing up the potential prize (multiplied by the probability of winning it) against the certain loss of the exploration cost. Figure 14.17 is a representation of this risk-reward calculation.

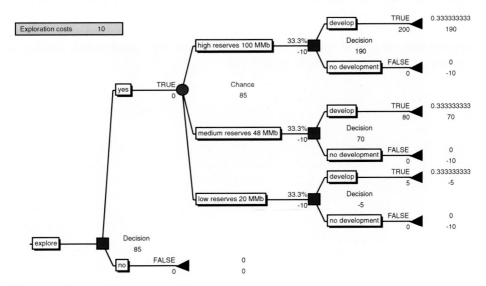

Figure 14.16   Decision tree example.

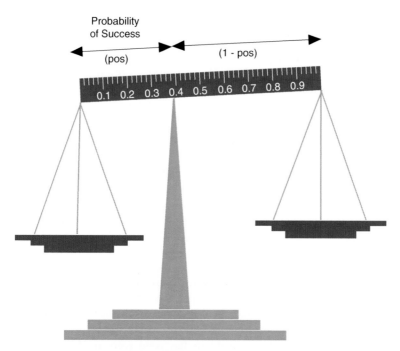

Figure 14.17   Weighing up the risks of exploration.

Even if the EMV of an undrilled prospect (after deducting exploration costs) is positive, the investor still needs to determine whether the prospect is significant. For example, would a prospect with an EMV of $50 million be attractive if the exploration cost is $25 million. Such an opportunity would have a 'risk cover' of 2. In other words, one would spend a guaranteed $25 million to win an expected net prize of $50 million. This may not be attractive to investors who have other, better opportunities to pursue. In this case a farm-out may be considered to involve an investor with a different attitude to such risk.

CHAPTER 15

RISK ANALYSIS

Introduction and Commercial Application: We have established that the oil and gas business involves major investments in all stages of the field life cycle. During the gaining access, exploration and appraisal stages, the expenditure does not guarantee a return, and at the development stage major investments are made in the anticipation of returns over a long period of time. Payback periods are typically long, and the project is subject to large fluctuations in key variables such as oil and gas price and cost of services during the producing life of the asset. For these reasons, it is important that careful technical and commercial risk analysis is performed when making decisions on investment in the industry. In addition, the nature of oil and gas production is intrinsically hazardous, and the potential impact on the people involved is carefully analysed and reduced to acceptable levels by design of plant and operating procedures.

This section will summarise the risk analysis techniques already introduced in the areas of exploration and appraisal and HSE, and will cover some methods used in managing risk in development projects.

15.1. RISK DEFINITION AND UNIT OF MEASURE

Whereas project uncertainties refer to the range of possible values for project variables, project risk can conveniently be defined as the impact of the outcomes on the stakeholders. For example, oil price is an input in project economic analysis, and the risk of price variation will be measured in terms of project NPV. Of course the actual outcome may be better or worse than the base case, so risk can have a positive side as well as a negative side. We usually think of risk in terms of a negative impact, but we should remember that there is often a positive element of risk (we may call it opportunity) and this upside may be worth pursuing.

The dimensions of risk in the above example are clearly in monetary terms, say dollars. However it is not just the investment which is at risk in a project – the company also places its reputation, its people and the environment at risk. Although a somewhat contentious statement, the risks in these other aspects can also be measured in terms of dollars; actuaries provide estimates of the value of injury to people, environmental damage often incurs fines and remediation costs and loss of good reputation can limit the company's ability to gain access to new regions. Of course, positive performance in safety and environment can enhance reputation, reduce costs and add value. In general, monetary terms is a convenient measure of risk, allowing comparisons to be made, and justifying expenditure to reduce risks with a potential negative impact.

In some cases it is worth defining risk as the product of impact and probability, still measured in dollars. An event with high impact but low probability, such as a major plant upset or disaster, will therefore be considered in terms of the product of the two, which may be considerable, and worth reducing through design effort and expenditure. This is a technique used in quantitative risk assessment (QRA).

15.2. Summary of Risk Analysis Techniques in Exploration and Appraisal

In the exploration phase, the key uncertainties are the presence of a petroleum system through which hydrocarbons could be accumulated in a reservoir, and the volume of those hydrocarbons, if present. These two uncertainties are combined into a risked volume by multiplying together the POS and the volumetric range, often represented by an expectation curve, as presented in Chapters 7 and 14.

In summary the risked volume of hydrocarbon reserves can be calculated as follows

Probability of Success (POS)	Volumes of Recoverable Hydrocarbons
POS =	Reserves =
p(source)	gross rock volume (GRV)
×	×
p(migration)	net:gross ratio
×	×
p(sealed trap)	porosity
×	×
p(reservoir)	hydrocarbon saturation
×	×
p(timing)	shrinkage
	×
	recovery factor
Risked reserves = POS × reserves	

As described in Chapter 7, it is common to generate a distribution curve for the range of uncertainty in volume of hydrocarbon reserves, often using Monte Carlo simulation techniques to combine the uncertainty in each input parameter. This distribution curve is multiplied by the POS to yield a range of *risked reserves*, as shown in Figure 15.1.

In this example, the p90, p50, p10 reserves are approximately 10, 50 and 100 MMstb, and the POS is 30%, so the p90, p50, p10 risked reserves are approximately 3, 15 and 30 MMstb. The diagram is useful in indicating that the probability of exceeding any level of reserves, for example the probability of exceeding 100 MMstb

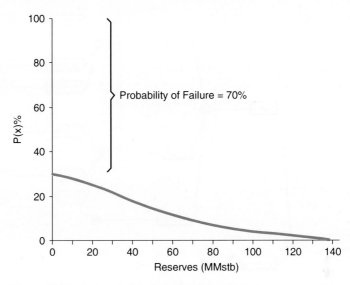

Figure 15.1 Range of risked reserves for an exploration prospect.

is approximately 3% and the probability of exceeding 50 MMstb is approximately 15%. If a commercial threshold for development can be provided by the economists, say 50 MMstb, then the *probability of commercial success* can be quoted, in this case 15% – this is sometimes referred to by the abbreviation POSc.

The simplest way in which to represent the risked reserves range is to multiply the p90, p50, p10 reserves volume by the POS to quote the estimated p90, p50, p10 risked reserves.

The EMV of the project is most elegantly calculated by estimating the NPV of the project for a discrete number of reserves volumes, typically the p90, p50 and p10 cases. These are then combined with the cost of exploration (say E) to calculate the EMV, using the decision tree analysis approach introduced in Chapter 14.

The example below shows a decision tree for an exploration prospect, drawn up in PrecisionTree® software, which is an add-in to Microsoft Excel. The cost of exploration is $5 million, and the NPVs of a p90, p50, p10 oil and gas discovery are $−20, 100, 200 and −20, 60, 140 million respectively. The value of finding water is of course zero. Each leaf on the end of the branches which follow exploration includes the cost of exploration, and the p90, p50, p10 values have been weighted at 30, 40, 30% respectively. The POS for oil is estimated at 16% and the POS for gas 10%, so the POS of finding neither is 74%. The EMV of the prospect is $16.04 million, assuming the exploration cost of $5 million.

The blue numbers on the end of the leaves are calculated net payoffs and the cumulative probability of occurrence (as determined by the software package). The risk weighted outcomes include the assumed cost of exploration.

If exploration is successful, the next phase of the field life cycle would involve considering appraisal of the discovery.

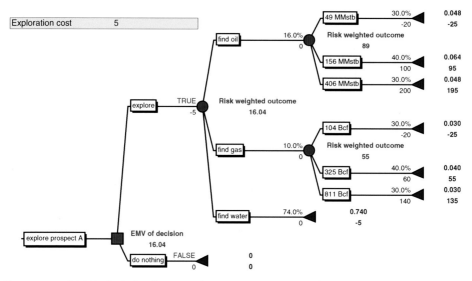

Figure 15.2   EMV calculation for an exploration prospect using decision tree analysis.

During the appraisal phase the key driver is to efficiently reduce uncertainty by data gathering in order to size the development facilities appropriately. Data gathering typically involves shooting seismic, drilling wells and performing production tests. The value of information (VoI) represents the maximum value of the appraisal data and is equal to the value of the project with the information less the value of the project without the information. Again, a decision tree is a convenient way to estimate the VoI. With this in mind an appraisal programme can be designed to reduce uncertainty in the key parameters.

In the above example, the exploration prospect introduced in Figure 15.2 has in fact found oil, and the question now is whether to immediately develop the discovery, or to first appraise. Three discrete reserves values have been taken from the reserves distribution, being the p90, p50 and p10. These have been weighted 30, 40, 30% respectively. The values of the corresponding projects are enhanced after appraisal by right-sizing the facilities ($60, 100 and 200 million if right-sized, as opposed to $40, 100 and 140 million if developed using a facility designed only for the p50 volume). The cost of appraisal has been set at $10 million. With this appraisal cost it is favourable to appraise and then tailor the facility size rather than developing with a single facility size and only after production finding out the true value of reserves. A sensitivity analysis can be performed to determine the maximum cost of appraisal before it would be better to just commit to development without knowing the true reserves size. This cost would be the maximum VoI, and spending more than this on the appraisal information cannot be justified. In this example the maximum VoI is $24 million (Figures 15.3 and 15.4).

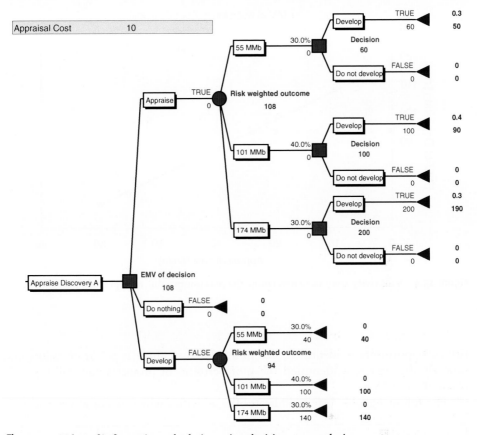

Fig. 15.3 Value of information calculation using decision tree analysis.

15.3. RISK ANALYSIS FOR MAJOR CAPITAL INVESTMENTS IN PROJECTS

Risks can represent potential negative impact or upside opportunity, and are identified in order to plan mitigation against those with a potential negative impact and to take advantage of upsides identified.

For major capital projects, it is common to perform risk analysis at several stages of the development planning, so that risk items can be identified early and actions planned accordingly. A general criticism of the industry is that companies progress too far with planning of projects which eventually prove to be unfeasible – early identification of this is clearly more efficient. To assist this, *a stage-gate process* is used by many companies.

The stage-gate process breaks the project into phases, with an approval required at each stage before progressing through this 'gate' to the next stage. The earlier a

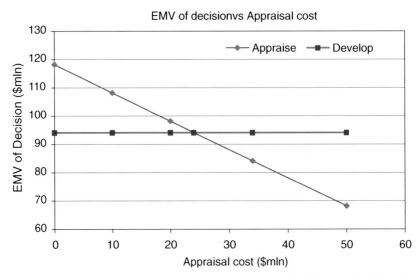

Figure 15.4 Sensitivity analysis to determine the maximum value of appraisal information.

risk or opportunity is identified, the more potential there is to create value, and similarly the later it is identified the more erosion of project value will typically occur (Figure 15.5).

The following table briefly indicates the activities which occur at each stage.

Stage	Activity
Identify	Identify the opportunity for investment.
Assess	Consider options for development and demonstrate that at least one is economically feasible. Determine whether further appraisal activity is required to improve the project value.
Select	Choose between the options generated, based on economic, environmental and safety criteria.
Define	Perform detailed design of the chosen concept for development.
Execute	Procure suppliers and agree the contracts for constructing the facilities, processing equipment and drilling. Fabricate, install and commission the plant.
Operate	Operate the facility.

The *Assess stage* can also be called the Appraise stage in some companies, and the stage-gate process clearly has its roots in the traditional planning and control approach discussed in Section 13.2, Chapter 13.

There are useful risk analysis tools which can be applied at the above stages, and these will now be introduced.

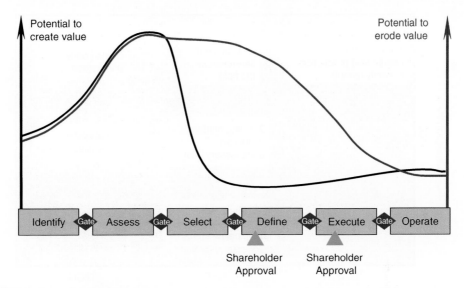

Figure 15.5   A typical stage-gate process.

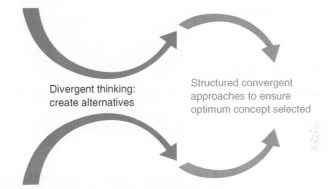

Figure 15.6   Schematic view of the brainstorming technique.

15.3.1. Brainstorming

The objective of the *Assess Stage* is to ensure that all reasonable alternatives have been considered and screened in a consistent fashion before deciding upon a 'solution' too quickly. It is human nature to resort to the known and to exclude the unfamiliar, and this can lead to lost opportunity. A useful technique to use at this stage is that of *brainstorming*. No limits are placed on the team as this is done, and a multidisciplinary approach works best. Some of the most original ideas come from non-discipline team members who are unhindered in challenging what the discipline may consider as paradigms. This stage corresponds to the divergent thinking shown below. Nothing is impossible at this stage! The unfeasible options will be screened out later (Figure 15.6).

Of course VoI techniques are applicable at this stage in determining if further information gathering is justifiable.

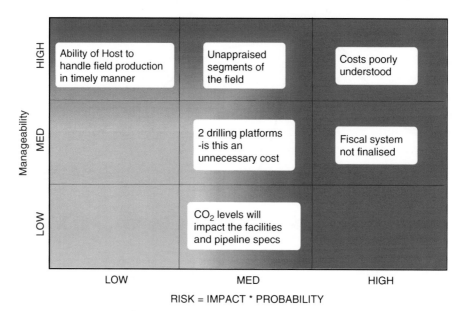

Figure 15.7 Project risk matrix.

15.3.2. Project risk matrices and risk registers

The period of divergent thinking in Figure 15.6 is followed by a convergent process to prune down the feasible options to take forward to the Select Stage, in which multiple options will need to be narrowed down to the favoured approach.

At the Select Stage, it is useful to define the risks which each option could incur. This can be recorded on a Risk Matrix which places the issues on a scale of risk (= impact × probability) versus manageability, that is the ability to influence or control (Figure 15.7).

The scale for risk is usually calibrated by the scale of the project. As discussed in the introduction to this section, the unit of risk is monetary terms, being the result of direct spending, project delay, reputation, environment or potential harm to personnel (Figure 15.8).

Once the items have been evaluated and posted on a project risk matrix, the main items are then recorded on a project *Risk Register* which details the item by discipline, and includes an action for risk mitigation and a responsible party who will follow-up on each item.

A risk register may be drawn up for each project option being investigated during the Select Stage, allowing a comparison of the risks and opportunities associated with each option. This will help to select between which option moves forward to the Define Stage. During the more detailed design of the project, the risk matrix and risk register will be used again, but then focused on one particular design concept (Figure 15.9).

Suggested Definitions

	Occurence probability	Impact time schedule	Impact on Budget $ million
high	> 75%	> 3 months	> 30
medium	30 - 75%	1 - 3 months	10 - 30
low	10 - 30%	< 1month	1 - 10
negligible	< 10%	< 1week	< 1

Points		Impact			
		negligible	low	medium	high
Occurence probability	high	4	8	12	16
	medium	3	6	9	12
	low	2	4	6	8
	negligible	1	2	3	4

Ability to influence (or Manageability)

Ability to influence	high	
	medium	
	low	

Figure 15.8   Outline of scale and scoring for a project risk matrix.

15.3.3. Sensitivity analysis

During the Define Stage, project optimization will be carried out, and sensitivity analysis is a tool which helps to home in on what parameters can most significantly influence project value. This technique and the spider diagram were introduced in Chapter 14, but an alternative format is the tornado diagram. These presentations are summarized in Figures 15.10 and 15.11.

The spider diagram shows the impact of the anticipated range of each parameter on the project indicator, usually the NPV of the project, shown on the y-axis. It is the result of a relatively crude sensitivity analysis in which each input variable is varied independently. The scale of the x-axis is the percentage variation of each input variable away from the base case (often the p50) assumption (Figure 15.10).

The left hand *tornado plot* (Figure 15.11) is usually generated as a result of Monte Carlo simulation, introduced in Section 7.2.4, Chapter 7. This shows the covariance of the output value (the numbers on the y-axis) and each individual input, ranked in order. The benefit of this approach is that it can accommodate any shape of distribution for the input parameter, rather than single changes in the input, and through the Monte Carlo model can also handle dependencies between input variables. It is useful in sorting those parameters of considerable influence from those whose effect on the project value can be ignored. The right hand tornado plot is more deterministic and shows the absolute impact on NPV of

I.D.	Category	Description	Probability	Impact	Points	Ability to Influence	Potential Mitigations	Responsible
C1	Commercial	Fiscal system not finalised	H	M	12	M	Discuss PSC terms with Oil Ministry	Comm. Mgr
C2	Commercial	Lack of local skilled workforce	H	M	12	H	Local recruitment and training	HR
C2	HSE	Environment policies still immature - flaring may not be an option	M	M	9	M	Discuss with govt as soon as possible - opportunity to influence legislation	Env. Mgr
F1	Facilities	Costs poorly understood	M	H	12	H	Consider refining the cost estimates	Cost engr.
F2	Facilities	CO_2 noted in some production tests	M	M	9	H	Sample fluids in next test and consider sweet service metallurgy	Prod Chem.
F3	Facilities	Current capex high - no alternative development concepts yet considered	H	H	16	H	Consider alternative concepts for development - FPSO, subsea tie-backs	Proj. Eng.
F4	Facilities	Ability of Host to handle field production in timely manner	L	L	4	H	Plan timing of tie-back to Host	Comm. Mgr
F5	Process	Potential waxing in tie-back pipeline	M	H	12	H	Investigate use of wax inhibition	Process Eng.
W2	Wells	2 drilling platforms - is this an unnecessary cost	M	M	9	H	Investigate extended reach drilling from one drill centre	Drlg. Mgr
W5	Wells / Fac	CO_2 levels will impact the facilities and pipeline specs	M	M	9	L	The tubulars (and liners) should be designed (and costed) with the CO_2 levels possible. Facilities can be specified once CO_2 levels are confirmed.	Process Eng.
G1	G&G	Fault seal uncertainty affecting compartmentalisation	M	H	12	M	Shoot 3D seismic. Look at fault sealing issues and sealing trends (e.g. faults of a certain orientation) in offset and/or analogue fields. Assess in-situ stress parameters and influence on faults. Mitigate by drilling more wells.	Geologist
G2	G&G	Unappraised segments of the field	H	L	8	H	Consider further well appraisal	Geologist
G3	G&G	P90 estimates currently uneconomic, may affect partner buy in.	H	H	16	M	Review current P90 parameters and assess they're not overly pessimistic. Change where justifiable.	Res. Eng.
RE1	RE	Water injection down dip may not possible	M	H	12	H	Consider further well appraisal down dip - investigate diagenesis potential	Geologist
RE2	RE	Fault seal along spine of field - if not sealing can reduce well count	M	H	12	H	Consider an interference test in appraisal wells or early production wells	Geol/Res Eng
RE3	RE	Recovery from water flood with poorly connected reservoirs may be low	M	H	12	M	Consider dedicated producers and injectors - may require more wells than planned	Res. Eng.

Figure 15.9   Example of a Risk Register.

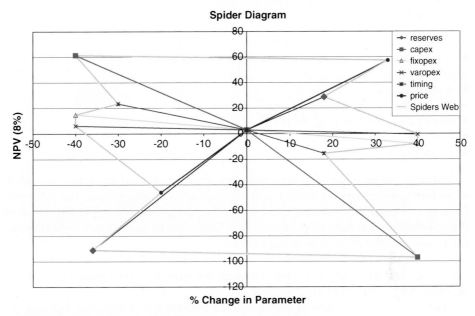

Figure 15.10 Sensitivity analysis shown on a spider diagram.

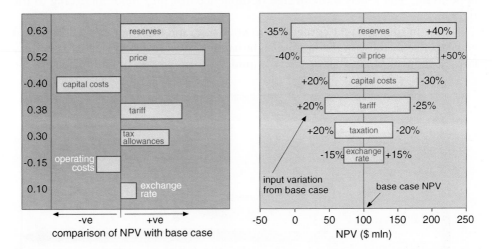

Figure 15.11 Forms of tornado plot.

specific changes in input parameter – usually determined by changing one input parameter at a time.

15.3.4. Stakeholder analysis

Once the development team has selected the preferred development plan, a common stumbling block is inadequate alignment with other stakeholders, both within and outside the company. Stakeholders may include

- company management, employees and shareholders
- government and regulatory bodies (national and local)
- partners
- neighbours
- NGOs, for example environmental, human rights
- financial institutions (e.g. World Bank)
- suppliers and contractors
- competitors.

Primary stakeholders are those who are directly affected by a project, while secondary stakeholders are those not directly affected by the project, but who may have an influence, interest or expertise to offer. Identification and classification of the stakeholders can help to prioritise their involvement and identify opportunities for potential collaboration or partnership.

It is worthwhile considering these stakeholders and plotting them on the following chart in order to develop a strategy for managing the opportunity or risk that they might represent (Figure 15.12).

The y-axis is a measure of the degree of interest the stakeholder has and the support (above the centre line for positive support, and below the centre line for opposition). The x-axis is the degree of influence that they hold, increasing to the right.

Those in the top right corner are on your side, and their support is assumed to be guaranteed. Those in the bottom right hand corner are deadly opposed and may not be influenced. It is those in the middle areas whose support needs to be

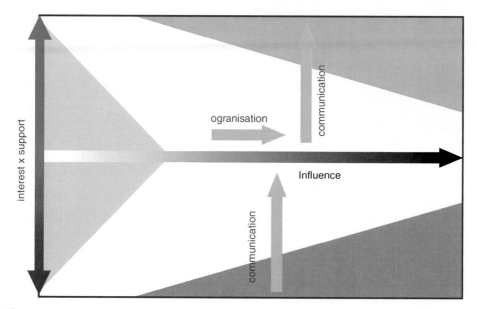

Figure 15.12   Stakeholder analysis plot.

strengthened through an effective communication plan. Those with little influence may become valuable if they are brought together through organisation.

The tools that exist to win over stakeholders include face-to-face meetings, workshops, visits, presentations, briefing papers and written information, or a mixture of these, and the medium needs to be tailored to the target audience. The position of stakeholders on the plot may change over time, and so keeping this plot updated for movements and any new entrants is important.

15.3.5. More complex problems: specific subsurface considerations

The spider diagrams and tornado plots are useful in displaying the impact of input assumptions when the problems are relatively simple and there is a linear relationship between inputs and outputs. In more complex cases (e.g. dependency of reserves on permeability distribution) there are several ways of analysing the risk on the subsurface side. Below we suggest three alternative approaches to *subsurface uncertainty handling*; the most appropriate one will depend on the amount of data and the complexity of the problem. In each case, we are trying to capture the range of uncertainty in the subsurface descriptions, which will involve all subsurface disciplines. These can be termed multiple realisations – in other words possible alternatives (Figure 15.13).

The three approaches suggested to creating the multiple realisations are

- Build a limited number of cases, anchored around a preferred 'Best Guess' model
- Use statistical methods to build a 'Multiple Stochastic' model, constrained by imposed boundary conditions
- Build 'Multiple-Deterministic' models which are designed manually, based on discrete alternative concepts, and can be assigned a relative probability.

In each case, we are trying to build a range of possible reservoir descriptions, all of which honour the limited amount of raw data available (Figure 15.14). A useful initial step is to take a team-based approach to identify the key inputs which affect

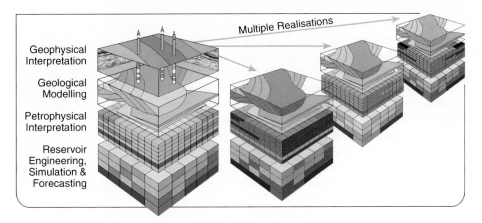

Figure 15.13   Multiple realisations of the subsurface description.

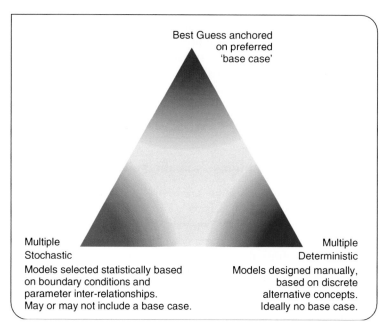

Figure 15.14   Alternative approaches to subsurface uncertainty handling.

the result by

- *Listing the perceived uncertainties in the field.* This is done for both the static and dynamic realms.
- *Rank the uncertainties.* This can be done by assigning a low, medium or high impact label to each of the uncertainties after a round table discussion involving the entire subsurface team, or by building some very simple models to determine the relative impact of each input listed.

If suitable simple models can be built for this initial investigation, the spider diagram is a useful way of representing the impact of the ranges of uncertainty of the input parameters. The ones which need to be captured in the risk analysis are those which move the result furthest from the base case (Figure 15.15).

Alternatively the result of the ranking exercise could be as simple a table listing the input parameters and flagging them as having high, medium or low impact.

In the base case-dominated approach (Figure 15.16), the models are 'anchored' around a favoured interpretation of the raw data, and variations on this model are created by adjusting the values of those input variables which rank highest in the exercise above. This method will typically yield a low, base and high case model. Its advantage is that it is relatively quick to generate, is broader than a single case and is easily understood. The disadvantage is that it may not be broad enough to capture the true range of uncertainty, and may yield a false sense of accuracy in the models derived.

A multiple stochastic approach (Figure 15.17) uses statistical techniques to infill data between fixed points, or boundary conditions, such as well depths, top reservoir

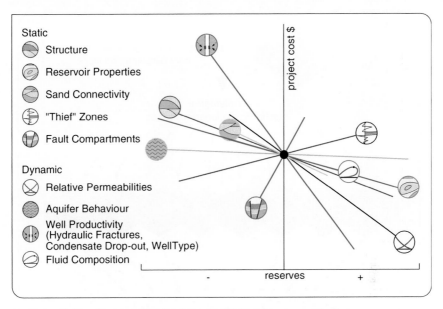

Figure 15.15   Spider diagram showing impact of subsurface uncertainties.

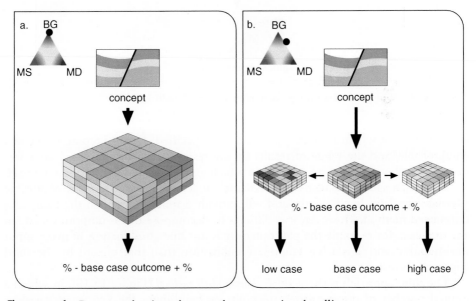

Figure 15.16   Base case–dominated approach to uncertainty handling.

surfaces, or petrophysical values such as net:gross ratios or porosity values measured at the wells. The technique can accommodate inter-dependencies between parameters such as permeability dependency on porosity. By varying the interpolation methods, boundary conditions and the dependencies, many realisations can be

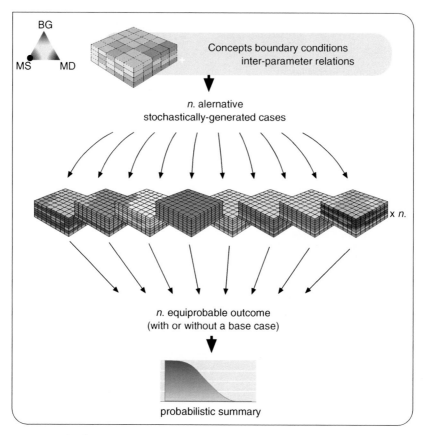

Figure 15.17 Multiple stochastic approach to uncertainty handling.

created, each of which honours the fixed data specified. Each outcome is deemed equiprobable, and the range of results (in say reserves) can be presented on a frequency distribution, shown below, from which a p90, p50 and p10 case can be read.

The advantage of this technique is that it can generate a large number of possible alternatives relatively efficiently, and will provide a fuller range than the base case-dominated method. However, it is difficult to decompose the assumptions made in any one case, for example the p90 outcome is just one combination of many input uncertainties, and it is a less tangible combination than that offered by the third approach, below.

The third approach is the *multiple-deterministic method* (Figure 15.18), in which a limited number of discrete cases are built, by combining the key input variables identified, in specific combinations. In the example below, the key variables are the structural faulting and the reservoir properties. These are differentiated as discrete concepts and then combined together in a limited number of combinations. Alternatives can be narrowed down by dynamic modeling techniques such as material balance, or well deliverability models which are 'history matched' against observations made in the field. In the picture below the check is made against the

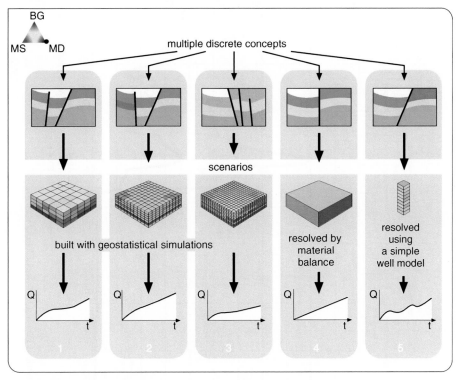

Figure 15.18   Multiple deterministic, or scenario-based approach.

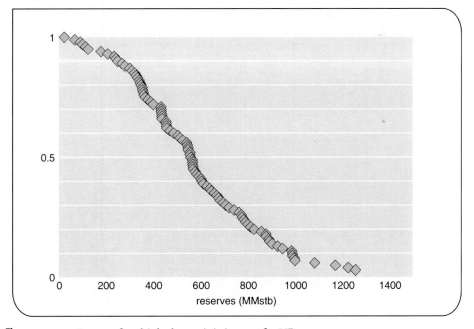

Figure 15.19   Range of multiple deterministic cases for UR.

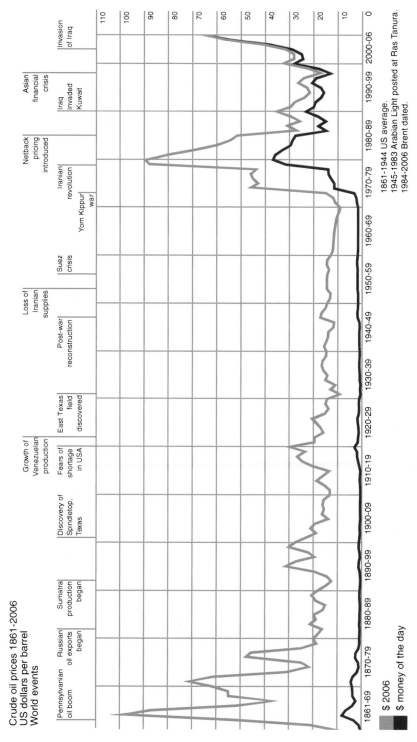

Figure 15.20 History of oil prices over a century period (*source:* BP Statistical Review of World Energy 2006).

field production profile. If no such field data are available, then the scenarios would be considered as viable alternative forecasts.

The advantages of this approach is that any one case can be fully described in terms of the specific assumptions, making it more tangible than the stochastic approach. It is also strongly driven by a limited number of realistic scenarios. The individual cases can be weighted if required and displayed on a cumulative probability curve (Figure 15.19), showing the range of uncertainty, and allowing, say, a p90 case to be selected, if required.

15.4. MANAGING COMMERCIAL RISK

This subject is of course very broad, and only a few key issues will be addressed in this section. One of the main commercial risks faced in a project investment is that of the price of the product. The following diagram shows a history of oil prices in real terms and money of the day for the past century (Figure 15.20).

These large fluctuations in price are a result of both the perception of the market makers and varying supply and demand, which is influenced by factors such as rapid industrialization, war, natural disasters and other perceived threats.

It is basic practice is to test the project economics using a low case price forecast to check that the project is still economic at the company's forecast of the lowest long-term price – clearly a rather individual choice given the fluctuations shown above. Hedging mechanisms exist for taking the impact of price variation out of the project risks. In a hedging arrangement the seller agrees with a buyer to sell oil in the future at a price agreed at the date of striking the hedge agreement. This deal can be brokered, with the broker taking a commission on the deal. Of course in the future the company may gain or lose compared with the prevailing price at the date of sale, but at least the price uncertainty has been eliminated.

In gas projects, it is common to sell the gas forward at an agreed price, and this is often necessary to reduce exposure to price when investing in a large capital project with a relatively small margin, such as an LNG plant, as discussed in Section 9.3, Chapter 9.

A natural risk reduction can be achieved by a company diversifying its investments into a portfolio of projects, rather than placing all of its faith in one project. Large International Oil Companies (IOCs) and NOCs do this as a matter of course.

CHAPTER 16

MANAGING THE PRODUCING FIELD

Introduction and Commercial Application: During the production phase of the field life, the operator will apply field management techniques aimed at maximising the profitability of the project and realising the economical recovery of the hydrocarbons, while meeting all contractual obligations and working within certain constraints. Physical constraints include the reservoir performance, the well performance and the capacity and operability of the surface facilities. The company will have to manage internal factors such as manpower, cashflow and the structure of the organisation. In addition, the external factors such as agreements with contractors and the NOC or government, environmental legislation and market forces must be managed throughout the production lifetime (Figure 16.1).

Some of the approaches and techniques for measuring performance and managing the constraints of the subsurface and surface facilities, and the internal and external factors will be discussed in this section.

First we will look at the constraints in the above groupings, but they are most effectively managed in an integrated approach, since they all act simultaneously on the profitability of the producing field. This requires careful planning and control by a centralised, integrated team, which will also be discussed.

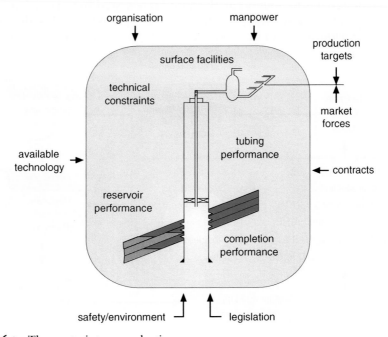

Figure 16.1   The constraints on production.

16.1. MANAGING THE SUBSURFACE

16.1.1. The reservoir performance

At the development planning stage, a *reservoir model* will have been constructed and used to determine the optimum method of recovering the hydrocarbons from the reservoir. The criteria for the optimum solution will most likely have been based on profitability and safety. The model is initially based on a limited data set, perhaps a seismic survey, five exploration and appraisal wells, and will therefore be an approximation of the true description of the field. As development drilling and production commence, further data is collected and used to update both the geological model which comprises the description of the structure, environment of deposition, diagenesis and fluid distribution and the description of the reservoir under dynamic conditions or the reservoir model.

A programme of *monitoring* the reservoir is carried out, in which measurements are made and data are gathered. Figure 16.2 indicates some of the tools used to gather data, the information which they yield and the way in which the information is fed back to update the models and then used to refine the ongoing reservoir development strategy.

The reservoir model will usually be a computer-based simulation model, such as the 3D model described in Chapter 9. As production continues, the monitoring programme generates a database containing information on the performance of the field. The reservoir model is used to check whether the initial assumptions and description of the reservoir were correct. Where inconsistencies between the predicted and observed behaviour occur, the model is reviewed and adjusted until a new match or so-called *history match*, is achieved. The updated model is then used to predict future performance of the field, and as such is a very useful tool for generating

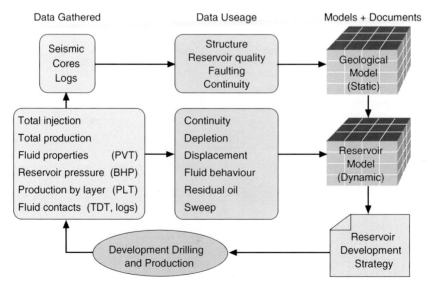

Figure 16.2   Updating the reservoir development strategy.

production forecasts. In addition, the model is used to predict the outcome of alternative future development plans. The criterion used for selection is typically profitability, but the operating company may state other specific objectives.

Some specific examples of the use of data gathered while monitoring the reservoir will now be discussed.

If the original FDP was not based on a *3D seismic survey*, which is a commonly used tool for new fields, then it would now be normal practice to shoot a 3D survey for development purposes. The survey would help to provide definition of the reservoir structure and continuity of faulting and extent of reservoir sands, which is used to better locate the development wells. In some cases time lapse 3D seismic, '4D', surveys carried out a number years apart (see Chapter 3), are used to track the displacement of fluids in the reservoir.

The data gathered from the *logs and cores* of the development wells are used to refine the correlation, and better understand areal and vertical changes in the reservoir quality. Core material may also be used to support log data in determining the residual hydrocarbon saturation or the residual oil saturation to water flooding left behind in a swept zone.

Production and injection rates of the fluids will be monitored on a daily basis. For example, in an oil field we need to assess not only the oil production from the field (which represents the gross revenue of the field), but also the GOR and water cut. In the case of a water injection scheme, a well producing at high water cut would be considered for a reduction in its production rate or a change of perforation interval (see well performance below) to minimise the production of water, which not only causes more pressure depletion of the reservoir but also gives rise to water disposal costs. The total production and injection volumes are important to the reservoir engineer to determine whether the *depletion policy* is being carried out to plan. Combined with the pressure data gathered, this information is used in *material balance* calculations to determine the contribution of the various drive mechanisms such as oil expansion, gas expansion and aquifer influx.

Fluid samples will be taken in selected development wells using downhole sample bombs or the MDT tool to confirm the PVT properties assumed in the development plan, and to check for areal and vertical variations in the reservoir. In long hydrocarbon columns, of about 1000 ft, it is common to observe vertical variation of fluid properties due to *gravity segregation*.

Reservoir pressure is measured in selected wells using either permanent or non-permanent bottom hole pressure gauges or wireline tools in new wells (RFT, MDT, see Section 6.3.6, Chapter 6) to determine the profile of the pressure depletion in the reservoir. The pressures indicate the continuity of the reservoir and the connectivity of sand layers. They are used in material balance calculations and in the reservoir simulation model to confirm the volume of the fluids in the reservoir and the natural influx of water from the aquifer. The following example shows an RFT pressure plot from a development well in a field which has been producing for some time (Figure 16.3).

Comparing the RFT pressures to the original pressure regime in the reservoir yields information on both the reservoir continuity and the depletion. The disconti-nuities in pressure indicate that there is a shale or a fault between sands A and B which

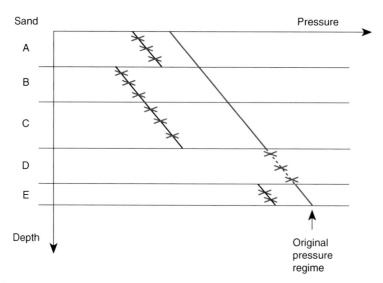

Figure 16.3   RFT pressure plot in a development well.

is at least partly sealing. The shale layers (or faults) between sands C and D and between D and E must be fully sealing, since sand D is still at the original pressure. The vertical pressure communication of the reservoir is therefore limited by these features. Assuming that the reservoir in this example is being produced by natural depletion, then it can be seen that production from layers B and C (which are in vertical pressure communication) is faster than from the other sands, indicating either that the sands have better permeability, or are limited in extent. Meanwhile, no production is occurring from the D sand in this area, since the pressure remains undepleted. The RFT data can therefore be used to derive more than simply a pressure. The modern equivalent of the RFT is the MDT which is a Schlumberger tool.

Monitoring the *reservoir pressure* will also indicate whether the desired reservoir depletion policy is being achieved. For example, if the development plan was intended to maintain reservoir pressure at a chosen level by water injection, measurements of the pressure in key wells would show whether all areas are receiving the required pressure support, and may lead to the redistribution of water injection or highlight the need for additional water injectors. If the chosen reservoir drive mechanism was depletion drive, then reservoir pressure in key wells will indicate if the depletion is evenly distributed around the field. A relatively undepleted pressure would indicate that the area around that well is not in pressure communication with the rest of the field, and may lead to the conclusion that more wells are required to drain this area to the same degree as the rest of the field. The presence of an active *natural aquifer* can also be detected by measuring the reservoir pressure and the produced volumes; the contribution of the aquifer support for the reservoir pressure would be calculated by the reservoir engineer using the technique of material balance (Section 9.1, Chapter 9).

In a reservoir consisting of layers of sands, the sweep of the reservoir may be estimated by measuring the *production rate of each layer* using the PLT. This is a tool

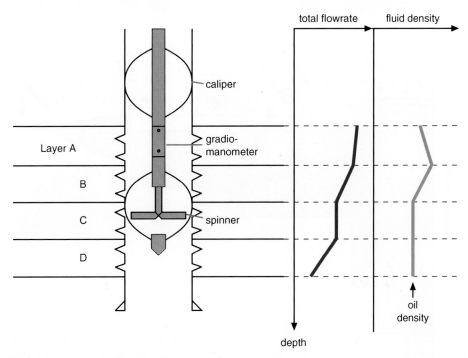

Figure 16.4 The production logging tool (PLT).

run on electrical wireline, and contains a spinner and gradiomanometer which can determine the production rate flowing past the tool as well as the density of that fluid. By passing the tool across a series of flowing layers, the flowrate and fluid type of each producing layer can be determined. This is useful in confirming how much of the total flowrate measured at surface is contributed by each layer, as well as indicating in which layers gas or water breakthrough has occurred (Figure 16.4).

The above example reveals that layer C is not contributing to flow as demonstrated by the zero increase in total production as the tool passes this layer, and that a denser fluid, such as water, is being produced from layer B, which is also a major contributor to the total flowrate in the well. These results would be interpreted as showing that water breakthrough has occurred earlier in layer B than in the other layers, which may give reason to shut-off this layer, as discussed below. The lack of production from layer C may indicate ineffective perforation, in which case the interval may be re-perforated. The lack of production may be because layer C has a very low permeability, in which case little recovery would be expected from this layer.

Hydrocarbon-water contact (HCWC) movement in the reservoir may be determined from the openhole logs of new wells drilled after the beginning of production, or from a *thermal decay time (TDT)* log run in an existing cased production well. The TDT is able to differentiate between hydrocarbons and saline water by measuring the TDT of neutrons pulsed into the formation from a source in the tool. By running the TDT tool in the same well at intervals of say 1 or 2 years (*time lapse TDTs*), the

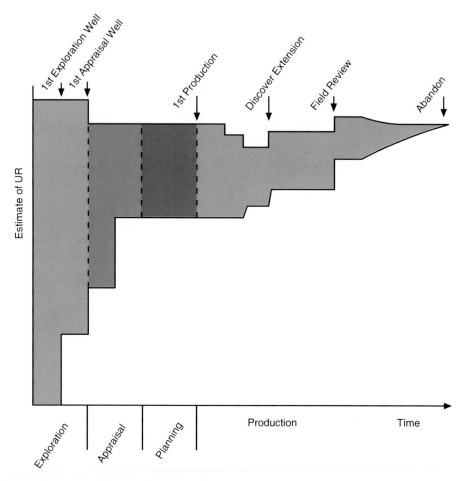

Figure 16.5   Typical change of estimate of UR during the field life.

rate of movement of the HCWC can be tracked. This is useful in determining the displacement in the reservoir, as well as the encroachment of an aquifer.

During the producing life of the field, data is continuously gathered and used to update the reservoir model, and reduce the uncertainties in the estimate of STOIIP and UR. The following diagram indicates how the range of uncertainty in the estimate of UR may change over the field life cycle. An improved understanding of the reservoir helps in selecting better plans for further development, and may lead to increases in the estimate of UR. This is not always the case; the realisation of a more complex reservoir than previously described, or parts of the anticipated reservoir eroded for example, would reduce the estimate of UR (Figure 16.5).

16.1.2. The well performance

The objective of managing the well performance in the scheme shown in Figure 16.1 is to reduce the constraints which the well might impose on the production of the hydrocarbons from the reservoir. The well constraints which may limit the reservoir potential may be split into two categories; the *completion interval* and the *production tubing*.

The following table indicates some of the constraints

Completion Interval Constraints	Production Tubing Constraints
• Damage skin	• Tubing string design
• Geometric skin	– size
• Sand production	– restrictions to flow
• Scale formation	• Artificial lift optimisation
• Emulsion formation	• Sand production
• Asphaltene drop-out	• Scale formation
• Producing unwanted fluids	• Choke size

To achieve the potential of the reservoir, these well constraints should be reduced where economically justified. For example, *damage skin* may be reduced by acidising, while *geometric skin* is reduced by adding more perforations, as described in Section 10.2, Chapter 10. *Scale formation* may occur when injection water and formation water mix together, and can be precipitated in the reservoir as well as on the inside of the production tubing; this could be removed from the reservoir and tubing chemically or mechanically scraped off the tubing.

Unwanted fluids are those fluids with no commercial value, such as water, and non-commercial amounts of gas in an oil field development. In layered reservoirs with contrasting permeabilities in the layers, the unwanted fluids are often produced firstly from the most permeable layers, in which the displacement is fastest. This reduces the actual oil production, and depletes the reservoir pressure. Layers which are shown by the PLT or TDT tools to be producing unwanted fluids may be '*shut-off*' by *recompleting* the wells. The following diagrams show how layers which start to produce unwanted fluids may be shut-off. An underlying water zone may be isolated by setting a *bridge plug* above the water bearing zone; this may be done without removing the tubing by running an inflatable *through-tubing bridge plug*. An overlying gas producing layer may be shut-off by squeezing cement across the perforations or by isolating the layer with a casing patch called a scab liner, an operation in which the tubing would firstly have to be removed. This would be termed a *workover* of the well and would require a rig or at least a hoist, for shallower wells with simple completions (Figures 16.6 and 16.7).

Workovers may be performed to repair downhole equipment or surface valves and flowlines, and involve shutting in production from the well, and possibly retrieving and re-running the tubing. Since this is always undesirable from a production point of view, workovers are usually scheduled to perform a series of tasks simultaneously, for example renewing the tubing at the same time as changing the producing interval.

Tubing corrosion due to H_2S (sour corrosion) or CO_2 (sweet corrosion) may become so severe that the tubing leaks. This would certainly require a workover.

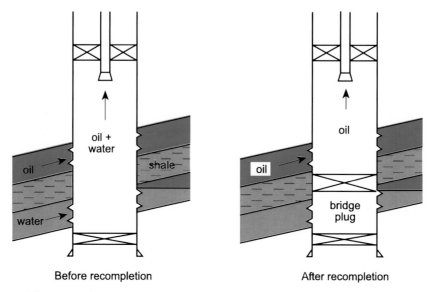

Before recompletion   After recompletion

Figure 16.6   Upwards recompletion of a well.

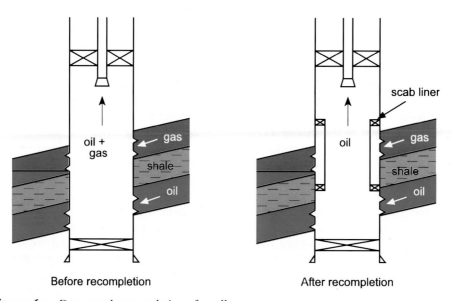

Before recompletion   After recompletion

Figure 16.7   Downwards recompletion of a well.

Monitoring of the tubing condition to track the rate of corrosion may be performed to anticipate tubing failure and allow a tubing replacement prior to a leak occurring.

The *tubing string design* should minimise the restrictions to flow. *Monobore completions* aim at using one single conduit size from the reservoir to the tubing head

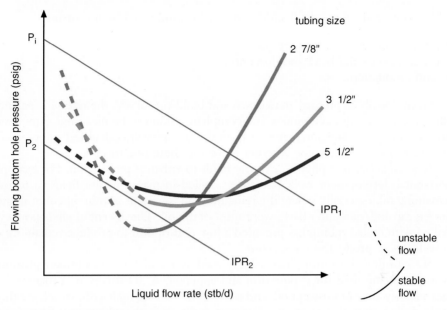

Figure 16.8 Tubing size selection.

to achieve this. The tubing size should maximise the potential of the reservoir. The example shown in Figure 16.8 shows that at the beginning of the field life, when the reservoir pressure is P_i, the optimum tubing size is $5\frac{1}{2}$ in. However, as the reservoir pressure declines, the initial tubing is no longer able to produce to surface, and a smaller tubing ($2\frac{7}{8}$ in.) is required. Changing the tubing size would require a workover. Whether it would be better to install the smaller tubing from the beginning (initially choking the flowrate but not requiring the later workover) is an economic decision.

The relationship between the tubing performance and reservoir performance is more fully explained in Section 10.5, Chapter 10.

Artificial lift techniques are discussed in Section 10.8, Chapter 10. During production, the operating conditions of any artificial lift technique will be optimised with the objective of maximising production. For example, the optimum gas–liquid ratio will be applied for gas lifting, possibly using computer assisted operations (CAO) as discussed in Section 12.2, Chapter 12. Artificial lift may not be installed from the beginning of a development, but at the point where the natural drive energy of the reservoir has reduced. The implementation of artificial lift will be justified, like any other incremental project, on the basis of a positive NPV (see Section 14.4, Chapter 14).

Sand production from loosely consolidated formations may lead to erosion of tubulars and valves and sand-fill in both the sump of the well and surface separators. In addition, sand may bridge off in the tubing, severely restricting flow. The presence of sand production may be monitored by in-line detectors. If the quantities of sand produced become unacceptable then downhole sand exclusion should be considered (Section 10.7, Chapter 10).

During production, *the 'health' of the well* is monitored by measuring

- production rates – oil, water, gas
- pressures – tubing head and downhole
- sand production.

From downhole pressure drawdown and build-up surveys the reservoir permeability, the well productivity index and completion skin can be measured. Any deviation from previous measurements or from the theoretically calculated values should be investigated to determine whether the cause should be treated.

New technology is applied to existing fields to enhance production. For example, horizontal development wells have been drilled in many mature fields to recover remaining oil, especially where the remaining oil is present in thin oil columns after the gas cap and/or aquifer have swept most of the oil. The advent of *multilateral wells* drilled with coiled tubing has provided a low cost option to produce remaining oil as well as low productivity reservoirs.

3D seismic is becoming increasingly used as a tool for development planning, as well as being used for exploration and appraisal. A 3D survey in a mature field may identify areas of unswept oil, and is useful in locating *infill wells*, which are those wells drilled after the main development wells with the objective of producing remaining oil.

16.2. MANAGING THE SURFACE FACILITIES

The purpose of the surface facilities is to deliver saleable hydrocarbons from the wellhead to the customer, on time, to specification, in a safe and environmentally acceptable manner. The main functions of the surface facilities are

- gathering, for example manifolding together producing wells
- separation, for example gas from liquid, water from oil, sand from liquid
- transport, for example from platform to terminal in a pipeline
- storage, for example oil tanks to supply production to a tanker.

The surface facilities used to perform these functions are discussed in Section 11.1, Chapter 11, and are installed as a sequence or train of vessels, valves, pipes, tanks etc. This section will concentrate on the optimisation of the production system designed and installed in the development phase. The system needs to be managed during the production period to maximise the system's *capacity* or possible throughput and *availability* or the fraction of time for which the system is available.

16.2.1. Capacity constraints

During the design phase, the hardware items of equipment or facilities are designed for operating conditions which are anticipated based on the information gathered during field appraisal, and on the outcome of studies such as the reservoir simulation.

The design parameters will typically be based on assessments of

- fluid flowrates (oil, water, gas) and their variations with time
- fluid pressures and temperatures and their variations with time
- fluid properties (density, viscosity)
- the required product quality.

During the production period of the field, managing the surface facilities involves optimising the performance of existing production systems. The operating range of any one item of equipment will depend on the item type, for example liquid–gas separator, and its selection at the design stage, but there will be maximum and minimum operating conditions, such as throughput. The minimum throughput may be described by the *turndown ratio*

$$\text{Turndown ratio} = \frac{\text{Minimum throughput}}{\text{Design throughput}} \times 100\%$$

Below the minimum throughput an equipment item such as a gas compressor will not function. The process must therefore be managed in a way which keeps production above that of the minimum throughput.

Often a more common concern is the *maximum* capacity of the item of equipment, since optimising performance usually means maximising possible production. For an individual equipment item such as a separator, increases in the maximum capacity may be achieved by *monitoring* the operating conditions, such as temperature, pressure, weir height, and *fine-tuning* these conditions to optimise the throughput. This fine-tuning of specific items of equipment is ongoing, since the properties of the feed change over time, and is performed by the process engineer and the operator. *Records* of the operating conditions of the equipment items are kept to help to determine optimum conditions, and to indicate when the equipment is performing abnormally.

The surface production system consists of a series of equipment items, such as that illustrated below, which shows the maximum oil handling capacity of the items. The maximum capacity of the system is determined by the component of the system with the smallest throughput capacity.

This very simplified example indicates that the export pump is limiting the system throughput to 45 Mb/d, although the production potential of the wells is 50 Mb/d. If the pump was upgraded or a duplicate pump was installed in parallel to a new capacity of, say 80 Mb/d, then the system capacity would become limited by the separator. Identifying and then uprating the item which is limiting the capacity is called *de-bottlenecking*. It is common to find that solving one restriction in the capacity leads on to the identification of the next restriction, as in the above example. Whether or not de-bottlenecking is economically worthwhile can be determined by treating it as an *incremental project* and calculating its NPV. The operators and engineers should constantly try to identify opportunities to de-bottleneck the production system. A de-bottlenecking activity may be as simple as changing a valve size, or adjusting the weir height in a separator.

The above example is a simple one, and it can be seen that the individual items form part of the chain in the production system, in which the items are dependent

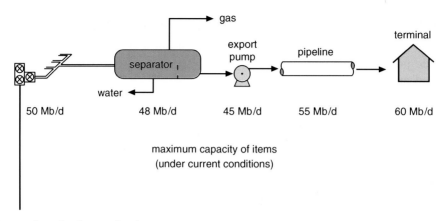

Figure 16.9   Surface production systems.

on each other. For example, the operating pressure and temperature of the separators will determine the inlet conditions for the export pump. *System modelling* may be performed to determine the impact of a change of conditions in one part of the process to the overall system performance. This involves linking together the mathematical simulation of the components, for example the reservoir simulation, tubing performance, process simulation and pipeline behaviour programmes. In this way the dependencies can be modelled, and sensitivities can be performed as calculations prior to implementation.

De-bottlenecking is particularly important when the producing field is on plateau production, because it provides a means of earlier recovery or acceleration of hydrocarbons, which improves the project cashflow and NPV.

Figure 16.9 may be characterised by an alternative diagram, called a *choke model* in which equipment items are represented as chokes in the system. Again a system model can be built around this to identify the current constraints and hence opportunities for increasing throughput or availability.

16.2.2. Availability constraints

Availability refers to fraction of time which the facilities are able to produce at full capacity. Figure 16.10 shows the main sources of *non-availability* of an equipment item.

An equipment item is designed to certain operating standards and conditions, beyond which it should not be operated. To ensure that the equipment is capable of performing safely at the design limit conditions, it must be periodically *inspected* and/or *tested*. For example, a water deluge system for fire-fighting would be periodically tested to ensure that it starts when given the appropriate signal, and delivers water at the designed rate. If equipment items have to be shutdown to test or inspect them, for example inspecting for corrosion on the inside of a pressure vessel, this will make the equipment temporarily unavailable. If the equipment item is a main process system item, such as one of those shown in Figure 16.9, then the complete production train would be shutdown. This would also be the case in testing

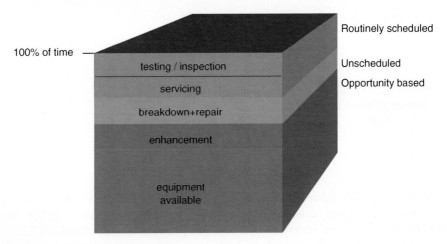

Figure 16.10 Availability of equipment.

a system that was designed to shutdown the process in the case of an emergency. This causes a loss of production. Where possible, *inspection* and *testing* is designed to be performed on-line to avoid interrupting production, but otherwise such inspections are scheduled to coincide. The periods between full function testing of process equipment is sometimes set by legislation.

Servicing of items is a routinely scheduled activity which is managed in the same way as inspection, and the periods between services will depend on the design of the equipment. The periods may be set on a calendar basis, that is every 24 months, or on a service hours basis such as every 10,000 operating hours.

Breakdown and subsequent repair is clearly non-scheduled, but gives rise to non-availability of the item. Some non-critical items may actually be maintained on a breakdown basis, as discussed in Section 12.3, Chapter 12. However, an item which is critical to keeping the production system operating will be designed and maintained to make the probability of breakdown very small, or may be backed up by a stand-by unit.

Enhancements to the process may be required due to sub-optimal initial design of the equipment, or to implement new technology or because an idea for improving the production system has emerged. De-bottlenecking would be an example of an enhancement, and while making the changes for the enhancement, the system becomes temporarily unavailable.

All of the above activities reduce the total availability of items, and possibly the availability of the production system. Managing the availability of the system hinges upon *planning and scheduling* activities such as inspection, servicing, enhancements and workovers, to minimise the interruption to producing time. During a *planned shutdown*, which may be for 1 or 2 months every 2 or 3 years, as much of this type of work as possible is completed. Reducing the non-availability due to breakdown is managed through the initial design, maintenance and back-up of the equipment. If the planned shutdowns are excluded, then a typical up-time, the time which the system is available, should be around 95%.

16.2.3. Managing operating expenditure

During the producing life, most of the money spent on the field will be on OPEX. This includes costs such as

- maintenance of equipment offshore and onshore
- transport of products and people
- salaries of *all* staff in the company, housing, schooling
- rentals of offices and services
- payment of contractors
- training.

In Section 14.2, Chapter 14, it was suggested that OPEX is estimated at the development planning stage based on a percentage of cumulative CAPEX (fixed OPEX) plus a cost per barrel of hydrocarbon production (variable OPEX). This method has been widely applied, with the percentages and cost per barrel values based on previous experience in the area. One obvious flaw in this method is that as oil production declines, so does the estimate of OPEX, which is not the common experience; as equipment ages it requires more maintenance and breaks down more frequently.

Figure 16.11 demonstrates that despite the anticipation of an incremental project, for example gas compression during the decline period, the *actual OPEX* diverges significantly from the estimate during the decline period. Underestimates of 50–100% have been common. This difference does not dramatically affect the NPV of the project economics when discounting back to a reference date at the development planning stage, because the later expenditure is heavily discounted. However, for a company managing the project during the decline period, the difference is very real; the company is faced with actual increases in the expected OPEX of up to 100%.

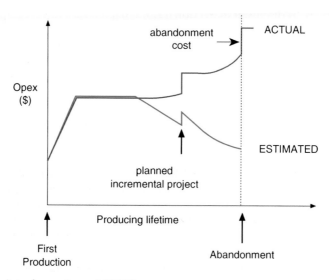

Figure 16.11   Actual vs. estimated OPEX.

Such increases in planned expenditure may threaten the profitability of a project in its decline period; the OPEX may exceed the cost oil allowance under a PSC.

A more sophisticated method of estimating OPEX is to *base the calculation on actual activities expected* during the lifetime of the field. This requires estimates of the cost of operating the field based on planning what will actually be happening to the facilities, and the manpower forecasts throughout the lifetime of the field. This means involving petroleum engineering, drilling, engineering, maintenance, operations and human resources departments in making the activity estimates, and basing the costs on historical data. This *activity-based costing* technique is much more onerous than the simple economic approach (see Section 14.2, Chapter 14), but does allow a more accurate and auditable assessment of the true OPEX of the development.

Often the divergence in costs shown in Figure 16.11 does occur, and must be managed. The objective is to maintain production in a safe and environmentally responsible manner, while trying to *contain or reduce costs*. The approach to managing this problem is through reviewing

- use of new technology
- effective use of manpower and support services − automation, organisational setup, supervision
- sharing of facilities between fields and companies, for example pipelines, support vessels, terminal
- improved logistics − supplying materials, transport
- reduction of down-time of the production system
- improving cost control techniques − measurement, specifications, quality control.

It is worth noting that typically personnel and logistics represent 30–50% of operating costs while maintenance costs represent 20–40% of operating costs. These are particular areas in which *cost control* and reduction should be focused. This may mean reviewing the operations and maintenance philosophies discussed in Chapter 12, to check whether they are being applied, and whether they need to be updated.

16.3. Managing the External Factors

Production levels will be influenced by external factors such as agreed production targets, market demand, the level of market demand for a particular product, agreements with contractors and legislation. These factors are managed by planning of production rates and management of the production operation.

For example, a *production target* may be agreed between the oil company and the government. An average production rate for the calendar year will be agreed, at say 30,000 stb/d, and the actual production rates will be reviewed by the government every 3 months. To determine the maximum realisable production level for the forthcoming year, the oil company must look at the *reservoir potential*, and then all of the constraints discussed so far, before approaching the government with a proposed production target. After technical discussions between the oil company and the government, an agreed production target is set. Penalties may be incurred if the target is not met within a tolerance level of typically 5%.

The oil company will also be required to periodically submit *reports* to the NOC or government, and to partners in the venture. These typically include

- well proposals
- FDPs
- annual review of remaining reserves per field
- six-monthly summary of production and development for each field
- plans for major incremental projects, for example implementing gas lift.

Market forces determine the demand for a product, and the demand will be used to forecast the sales of hydrocarbons. This will be one of the factors considered by some governments when setting the production targets for the oil company. For example, much of the gas produced in the South China Sea is liquefied and exported by tanker to Japan for industrial and domestic use; the contract agreed with the Japanese purchaser will drive the production levels set by the NOC.

The demand for domestic gas changes seasonally in temperate climates, and production levels reflect this change. For example, a sudden cold day in Northern Europe causes a sharply increased requirement for gas, and gas sales contracts in this region will allow the purchaser to demand an instant increase (up to a certain maximum) from the supplier. To safeguard for seasonal swings, imported gas is frequently stored in underground reservoirs during summer months in salt caverns or depleted gas fields and then withdrawn at times of peak demand.

Contracts made between the oil company and supply or service companies are a factor which affects the cost and efficiency of development and production. This is the reason why oil companies focus on the types of contract which they agree. Types of contract commonly used in the oil industry are summarised in Section 13.5, Chapter 13.

Legislation in the host country will dictate work practices and environmental performance of the oil company, and is one of the constraints which must be managed. This may range from legislation on the allowable concentration of oil in disposal water, to the maximum working hours per week by an employee, to the provision of sickness benefits for employees and their families. The oil company must set up an internal organisation which passes on the current and new legislation to the relevant parts of the company, for example to the design engineers, operators, human resources departments. The technology and practices of the company must at least meet legislative requirements, and often the company will try to anticipate future legislation when formulating its development plan.

One particular common piece of legislation worth noting is the requirement for an *EIA* to be performed prior to any appraisal or development activity. An EIA is used to determine what impact an activity would have on the natural environment including flora, fauna, local population, and will be used to modify the activity plan until no negative impact is foreseen. More details of the EIA are given in Section 5.3.1, Chapter 5.

16.4. Managing the Internal Factors

During production, the oil company will need to structure its operation to manage a number of internal factors, such as

- organisational structure and manpower
- planning and scheduling
- reporting requirements
- reviews and audits
- funding of projects.

In order to function effectively, the *organisational structure* should make the required flow of information for field development and management as easy as possible. For example, in trying to co-ordinate daily operations, information is required on

- external constraints on production-target rates
- planned production shutdowns
- budget availability
- delivery schedules to the customer
- injection requirements
- workover and maintenance operations
- routine inspection schedules
- delivery times for equipment and supplies
- manpower schedules and transport arrangements.

There is no single solution to the organisational structure required to achieve this objective, and companies periodically change their organisation to try to improve efficiency. The above list shows information required for daily operations, and a quite different list would be drawn up for development planning. Often the tasks required for production and development are split up, and this is reflected in the organisation. The following structure is one example of just part of a company's organisation (Figure 16.12).

This structure is organised by function – members of a technical function are grouped together. An alternative to the function-*based organisation* is an *asset-based organisation*, in which a multidisciplinary team is grouped together within an asset. The asset may be a producing field, a group of fields or an area of exploration interest.

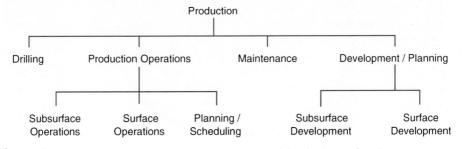

Figure 16.12   Organisational structure for operations and development planning.

Planning is carried out to steer the company's business and operations, and sets out what activities the company wants to perform. Typically there will be a 5-year *business plan* setting out the long-term objectives, a 1 year *operations plan* for operations activities and a 3 month *operations schedule* setting out the timing of the work. From the 3-month plan, a 30-day *schedule* of *when* the activities will be performed is made firm, running into detail such as the production expected from each well, and any wireline operations and maintenance work and the co-ordination of surface and subsurface operations. Even within this 30-day schedule, there will be some flexibility, but the first week of the 30-day period will be programmed by the *production programmers* in detail, determining, for example bean size for wells or production target per well. Each of these plans will involve a budget which describes the proposed expenditure.

In addition to the external reporting requirements mentioned in Section 16.3, there will be *internal reports* generated to distribute information within the organisation. These will include

- monthly reports of producing fields – production, injection, workover, development drilling
- management briefs on field progress
- safety performance statistics
- monthly budget summaries.

One of the important reasons for internal reporting is to provide a *database* of the activities which can be analysed to determine whether improvements can be made. Although the process of reviewing progress and implementing improvements should be ongoing, there will be periodic *audits* of particular areas of the company's business. Audits are often targeted at areas of concern and provide the mechanism for a critical review of the process used to perform business. This is simply part of the cycle of learning, which is one of the basic principles of management.

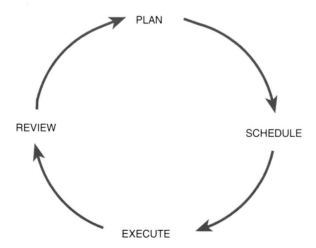

Figure 16.13   One of the basic principles of management.

The *audit team* may be formed on an ad-hoc basis, pulling in the individuals with the relevant experience, or could be a full-time team dedicated to the task, roving from project to project. A popular approach to this form of *quality assurance* is termed a peer review. A peer review team is typically a group of professionals working on a similar asset who are taken into the peer review for a short, dedicated period to apply their knowledge and experience to test the assumptions made by the asset team being audited (Figure 16.13).

The *funding* of the activities of the company is managed by the finance department, but the spending of the funds is managed by the technical managers. The budget reports are the mechanism by which the manager keeps track of how the actual revenue and expenditure is performing against the plan as laid out in the budget. The budget will be planned on an annual basis, split into 3-month periods and will be updated each quarter.

CHAPTER 17

MANAGING DECLINE

Introduction and Commercial Application: The *production decline* period for a field is usually defined as starting once the field production rate falls from its plateau rate. Individual well rates may, however, drop long before field output falls. This section introduces some of the options that may be available, initially to arrest production decline, and subsequently to manage decline in the most cost-effective manner.

The field may enter into an *economic decline* when either income is falling (production decline) or costs are rising, and in many cases both are happening. Whilst there may be scope for further investment in a field in economic decline, it should not tie up funds that can be used more effectively in new projects. A mature development must continue to generate a positive net cashflow and compete with other projects for funds. The options that are discussed in this chapter give some idea of the alternatives that may be available to manage the inevitable process of economic decline, and to extend reservoir and facility life.

17.1. INFILL DRILLING

Oil and gas reservoirs are rarely as simple as early maps and sections imply. Although this is often recognised, development proceeds with the limited data coverage available. As more wells are drilled and production information is generated, early geological models become more detailed and the reservoir becomes better understood. It may become possible to identify reserves which are not being drained effectively and which are therefore potential candidates for *infill drilling*. Infill drilling means drilling additional wells, often between the original development wells. Their objective is to produce yet unrecovered oil (Figure 17.1).

Hydrocarbons can remain undrained for a number of reasons:

- attic/cellar oil may be left behind above (or below) production wells
- oil or gas may be trapped in isolated fault blocks or layers
- oil may be bypassed by water or gas flood
- wells may be too far apart to access all reserves.

In the case of *attic/cellar oil* and *isolated fault blocks or layers*, it is clear that hydrocarbon reserves will not be recovered unless accessed by a well. The economics of the incremental infill well may be very straightforward; a simple comparison of well costs (including maintenance) against income from the incremental reserves. Reserves which have been bypassed by a flood front are more difficult to recover. Water will take the easiest route it can find through a reservoir. In an inhomogeneous sand, injected water or gas may reach producing wells via high-permeability layers without sweeping poorer sections. In time, a proportion of the oil in the bypassed sections

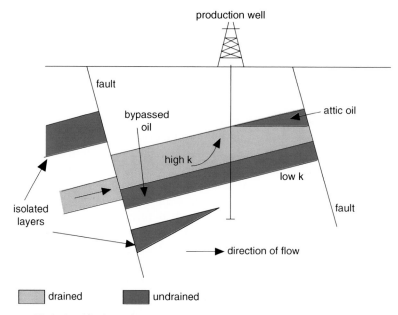

Figure 17.1   Undrained hydrocarbons.

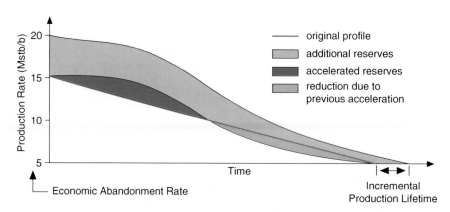

Figure 17.2   Additional and accelerated reserves.

may be recovered, though inefficiently in terms of barrels produced per barrel injected. Drilling an infill well to recover bypassed oil will usually generate extra reserves as well as some accelerated production (of reserves that would eventually have been recovered anyway). To decide whether to drill additional wells it is necessary to estimate both the extra reserves recovered, as well as the value of accelerating existing reserves (Figure 17.2).

In a completely homogenous unfaulted reservoir, a single well might, in theory, drain all the reserves, though over a very long period of time. FDPs address the compromise between well numbers, production profiles, equipment life and the

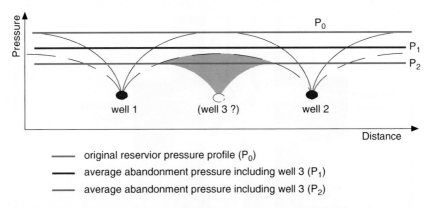

—— original reservior pressure profile (P_0)
—— average abandonment pressure including well 3 (P_1)
—— average abandonment pressure including well 3 (P_2)

Figure 17.3   Influence of an infill drainage point.

time value of money. Compared to the base case development plan, additional wells may access reserves which would not necessarily be produced within the field economic lifetime, simply because the original wells were too far apart. This is illustrated in Figure 17.3 by considering the pressure distribution in a reservoir under depletion drive. A third well in this situation could recover additional reserves before the wells reach their abandonment pressure. The additional well would have to be justified economically; the incremental recovery alone does not imply that the third well is attractive.

17.2. WORKOVER ACTIVITY

Wells are 'worked over' to increase production, reduce operating cost or reinstate their technical integrity. In terms of economics alone (neglecting safety aspects), a workover can be justified if the NPV of the workover activity is positive (and assuming no other constraints exist). The appropriate discount rate is the company's cost of capital (Figure 17.4).

Well *production potential* is the rate at which a well can produce with no external constraints and no well damage restricting flow. Actual well production may fall below the well potential for a number of reasons, which include

- mechanical damage such as corroded tubing or stuck equipment
- formation productivity impairment around the wellbore
- flow restriction due to sand production or wax and scale deposition
- water or gas breakthrough in high-permeability layers
- cross flow in the well or behind casing.

If *mechanical damage* is severe enough to warrant a workover, the production tubing will normally have to be removed, either to replace the damaged section or gain access for a casing repair. Such an operation will require a rig or workover hoist, and on an offshore platform may involve closing in neighbouring wells for

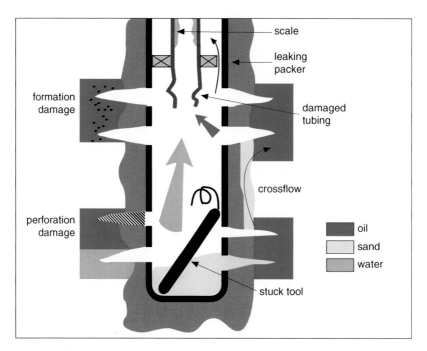

Figure 17.4   A workover candidate.

safety reasons. Where damage is not so severe, it may be possible to use 'through tubing' techniques to install a tubing patch or plug, on *wireline* or *coiled tubing* – both cheaper options.

Formation damage is usually caused by pore throat plugging. It may be a result of fine particles such as mud solids, cement particles or corrosion products invading the formation. It can also be caused by emulsion blocking or chemical precipitation. Impairment can sometimes be bypassed by deep perforating or fracturing through the damaged layer, or removed by treatment with acids. *Acid treatment* can be performed directly through production tubing or by using coiled tubing to place the acid more carefully (Figure 17.5).

Normally acid would be allowed to soak for some time and then back-produced if possible along with the impairing products. One of the advantages of using coiled tubing is that it can be inserted against wellhead pressure so the well does not have to be killed, a potentially damaging activity.

Coiled tubing can also be used to remove sand bridges and scale. Sometimes simple jetting and washing will suffice, and in more difficult cases an acid soak may be required. For very consolidated sand and massive scale deposits, a small fluid-driven drilling sub can be attached to the coiled tubing. In extreme cases, the production tubing has to be removed and the casing drilled out. Coiled tubing drilling (CTD) is explained in Section 4.5, Chapter 4.

When only small amounts of sand, wax or scale are experienced, the situation can often be contained using wireline bailers and scrapers, run as part of a well maintenance programme.

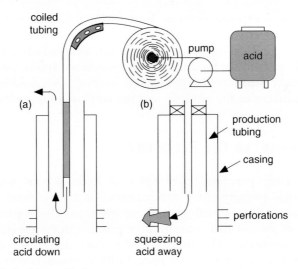

Figure 17.5 Coiled tubing acid placement.

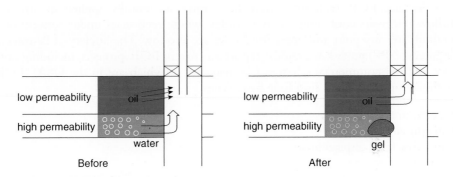

Figure 17.6 Water shut-off with chemicals.

If *water* or *gas breakthrough* occurs (in an oil well) from a high-permeability layer, it can dominate production from other intervals. Problems such as this can sometimes be prevented by initially installing a selective completion string, but in single-string completions on multiple layers, some form of *zonal isolation* can be considered. *Mechanical* options include plugs and casing patches (or 'scab' liners), which can be installed on wireline or pipe, although production tubing has to be pulled unless a well has a monobore completion. These options were illustrated in Section 16.1, Chapter 16. *Chemical* options, which are becoming much more common, work by injecting a chemical, for instance a polymer gel (Figure 17.6) which fills pore spaces and destroys permeability in the more permeable layers. These chemicals can be placed using coil tubing. *Squeezing off* water or gas producing zones using cement is a cheap but often unsatisfactory option.

Cross flow inside the casing can also be prevented by isolating one zone. However, this may still result in reduced production. Installing a selective completion can solve the problem but is an expensive option. To repair cross flow behind casing normally requires a full workover with a rig. Cement has to be either squeezed or circulated behind the production casing and allowed to set, after which cement inside the casing is drilled out, and the producing zones perforated and recompleted.

In very difficult situations the production interval is plugged back, a *side-track* well is drilled adjacent to the old hole and the section completed as a new well.

17.3. ENHANCED OIL RECOVERY

A considerable percentage (40–85%) of hydrocarbons are typically not recovered through primary drive mechanisms, or by common supplementary recovery methods such as waterflood and gas injection. This is particularly true of oil fields. Part of the oil that remains after primary development is recoverable through EOR methods and can potentially slow down the decline period. Unfortunately, the cost per barrel of most EOR methods is considerably higher than the cost of conventional recovery techniques, so the application of EOR is generally much more sensitive to oil price.

Generally, EOR techniques have been most successfully applied in onshore, shallow reservoirs containing viscous crudes, where recoveries under conventional methods are very poor and operational costs are also low. The Society of Petroleum Engineers (SPE) publishes a regular report on current EOR projects, including both pilot and full commercial schemes (the majority of which are in the USA). EOR methods can be divided into four basic types:

- steam injection
- in situ combustion
- miscible fluid displacement
- polymer flooding.

In the North Sea, which is more representative of large, offshore, capital-intensive projects developing lighter hydrocarbon reservoirs, it has been estimated that around 4 billion barrels are theoretically recoverable using known EOR techniques, which is equivalent to 15% of the estimated recoverable oil from existing North Sea fields. This represents a considerable target. Therefore, EOR research also continues into methods more suited to this type of environment, such as waterflooding with viscosified injection water (polymer-augmented waterflood).

The physical reasons for the benefits of EOR on recovery are discussed in Section 9.8, Chapter 9, and the following gives a qualitative description of how the techniques may be applied to manage the production decline period of a field.

17.3.1. Steam injection

Steam is injected into a reservoir to reduce oil viscosity and make it flow more easily. This technique is used in reservoirs containing high-viscosity crudes, where conventional methods only yield very low recoveries. Steam can be injected in a cyclic process in which the same well is used for injection and production, and the steam is allowed to *soak* prior to back-production (sometimes known as 'Huff and Puff'). Alternatively, steam is injected to create a *steam flood*, sweeping oil from injectors to producers much as in a conventional waterflood. In such cases, it is still found beneficial to increase the residence (or relaxation) time of the steam to heat treat a greater volume of reservoir.

Steam injection is run on a commercial basis in a number of countries (such as the USA, Germany, Indonesia and Venezuela), though typically on land, in shallow reservoirs where well density is high (well spacings in the order of 100–500 ft). There is usually a trade-off between permeability and oil viscosity, that is higher permeability reservoirs allow higher viscosity oils to be considered. Special considerations associated with the process include the insulation of tubing to prevent heat loss during injection, and high production temperatures if steam residence times are too low. Safety precautions are also required to operate the equipment for generating and injecting high-temperature steam.

17.3.2. In situ combustion

Like steam injection, in situ combustion is a thermal process designed to reduce oil viscosity and hence improve flow performance. Combustion of the lighter fractions of the oil in the reservoir is sustained by continuous air injection. Though there have been some economic successes claimed using this method, it has not been widely employed. Under the right conditions, combustion can be initiated spontaneously by injecting air into an oil reservoir. However, a number of projects have also experienced explosions in surface compressors and injection wells.

17.3.3. Miscible fluid displacement

Miscible fluid displacement is a process in which a fluid, which is miscible with oil at reservoir temperature and pressure conditions, is injected into a reservoir to displace oil. The miscible fluid (an oil-soluble gas or liquid) allows trapped oil to dissolve in it, and the oil is therefore mobilised.

The most common solvent employed is carbon dioxide gas, which can be injected between water spacers, a process known as 'water alternating gas' (WAG). In most commercial schemes, the gas is recovered and re-injected, sometimes with produced reservoir gas, after heavy hydrocarbons have been removed. Other solvents include nitrogen and methane (Figure 17.7).

17.3.4. Polymer-augmented waterflood

The three previous methods tend to yield better economics when applied in reservoirs containing heavy and viscous crudes, and are often applied either after or in conjunction with secondary recovery techniques. However, polymer-augmented waterflood is best considered at the beginning of a development project and is not restricted to viscous crudes. In this process, polymers are used to thicken the injected water to improve areal and vertical sweep efficiency by reducing the

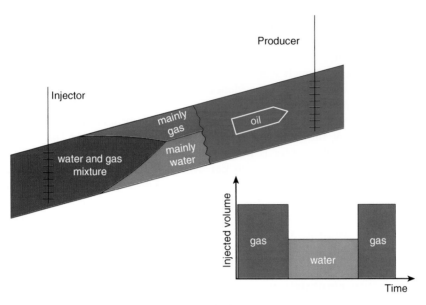

Figure 17.7   Water alternating gas injection (WAG).

tendency for oil to be bypassed. As with conventional flooding, once oil has been bypassed it is difficult to recover efficiently by further flooding.

One problem facing engineers in this situation, where the process is applied from waterflood initiation, is how to quantify the incremental recovery resulting from the polymer additive.

17.4. Production De-Bottlenecking

As introduced in Section 16.2, Chapter 16, bottlenecks in the process facilities can occur at many stages in a producing field life cycle. A process facility bottleneck is caused when any piece of equipment becomes overloaded and restricts throughput. In the early years of a development, production will often be restricted by the capacity of the processing facility to treat hydrocarbons. If the reservoir is performing better than expected it may pay to increase plant capacity. If, however, it is just a temporary production peak such a modification may not be worthwhile (Figure 17.8).

As a field matures, bottlenecks may appear in other areas, such as water treatment or gas compression processes, and become factors limiting oil or gas production. These issues can often be addressed both by surface and subsurface options, though the underlying justification remains the same – the NPV of a de-bottlenecking exercise (net cost of action vs. the increase in net revenue) must be positive.

This seems obvious, but it is not always easy to predict how a change in one part of a processing chain will affect the process as a whole (there will always be a bottleneck somewhere in the system). In addition, it may be difficult to estimate the cost in terms of extra manpower and maintenance overheads, where an increase in capacity demands additional equipment. To be able to make a decision, it is

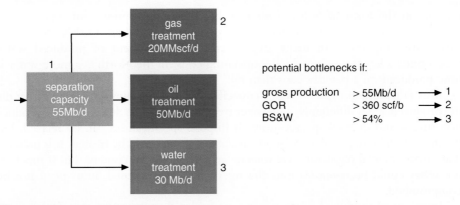

Figure 17.8 Potential facility bottlenecks.

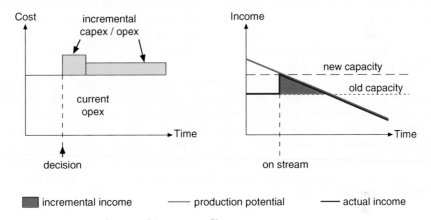

Figure 17.9 Incremental cost and income profiles.

important to have realistic incremental cost and revenue profiles, to judge the consequence of either action or no action (Figure 17.9).

The types of facilities bottleneck which appear late in field life depend upon the reservoir, development scheme and facilities in place. Two of the most common capacity constraints affecting production include

- produced water treatment
- gas handling.

17.4.1. Produced water treatment

Both the issues above are more difficult to manage offshore than on land, where space and load-bearing capacity are less likely to represent restrictions. Produced water treatment is a typical case, as extra tankage or other low maintenance options are usually too heavy or take up too much room on an offshore platform. Additional capacity in the form of *hydrocyclones* may be a technical option, but will increase existing operating and maintenance costs, at a time when OPEX control is particularly important. In many mature areas, the treatment of produced water is becoming a key factor in reducing operating costs. In the North Sea more water is now produced on a per day basis than oil!

If extra treatment capacity is not cost-effective, another option may be to handle the produced water differently. The water treatment process is defined by the production stream and disposal specifications. If disposal specifications can be relaxed, less treatment will be required, or a larger capacity of water could be treated. It is unlikely that environmental regulators will tolerate an increase in oil content, but if much of the water could be *re-injected* into the reservoir, environmental limits need not be compromised.

Injection of produced water is not a new idea, but the technique initially met resistance due to concerns about reservoir impairment (solids or oil in the water may block the reservoir pores and reducing permeability). However, as a field produces at increasingly high water cuts, the potential savings through reduced treatment costs compared with the consequences of impairment become more attractive. Local legislation has become the catalyst for produced water re-injection (PWRI) in some areas.

Rather than attempting to treat increasing amounts of water, it is possible in some situations to reduce water production by *well intervention* methods. If there are several wells draining the same reservoir layer, water cut layers in the 'wettest' wells can sometimes be isolated with bridge plugs or 'scab' liners. Unless a well is producing nothing but water, high water cut wells will also reduce oil production which may not be made up elsewhere. Similar operations can be considered in water injectors to shut-off high-permeability zones if water is being distributed inefficiently (Figure 17.10).

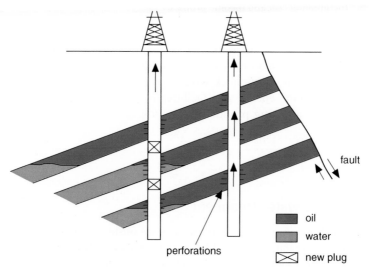

Figure 17.10   Well intervention to reduce water cut.

A promising technique currently under development is *downhole separation* whereby a device similar to a hydrocyclone separates oil and water in the wellbore. The water is subsequently pumped into a zone beneath the producing interval and only the oil is produced to surface.

In stacked reservoirs, such as those found in deltaic series, it is common to find that some zones are not drained effectively. *Through-casing logs* such as thermal neutron and GR spectroscopy devices can be run to investigate whether any layers with original oil saturations remain. Such zones can be perforated to increase oil production at the expense of wetter wells.

In high-permeability reservoirs, wells may produce dry oil for a limited time following a shut-in period, during which gravity forces have segregated oil and water near the wellbore. In fields with more production potential than production capacity, wells can be alternately produced and shut-in (*intermittent production* or *cycling*) to reduce the field water cut. This may still be an attractive option at reduced rates very late in field life, if redundant facilities can be decommissioned to reduce operating costs.

17.4.2. Gas handling

As solution gas drive reservoirs lose pressure, produced GORs increase and larger volumes of gas require processing. Oil production can become constrained by gas handling capacity, for example by the limited compression facilities. It may be possible to install additional equipment, but the added operating cost towards the end of field life is often unattractive, and may ultimately contribute to increased abandonment costs.

If gas export or disposal is a problem, *gas re-injection* into the reservoir may be an alternative, although this implies additional compression facilities. Gas production may be reduced using well intervention methods similar to those described for reducing water cut, though in this case up-dip wells would be isolated to cut back gas influx. Many of the options discussed under 'water treatment' for multilayered reservoirs apply equally well to the gas case.

In some undersaturated reservoirs with non-commercial quantities of gas but too much to flare, gas has be used to *fuel* gas turbines and generate electricity for local use.

17.5. Incremental Development

Most oil and gas provinces are developed by exploiting the largest fields first, since these are typically the easiest to discover. Development of the area often involves installing a considerable infrastructure of production facilities, export systems and processing plant. As the larger fields decline, there may be considerable working life left in the infrastructure which can be exploited to develop smaller fields that would be uneconomical on a stand-alone basis. If a *satellite development* utilises a proportion of the existing process facilities (and carries the associated operating costs), it may allow the abandonment rate of the mature field to be lowered and extend its economic life.

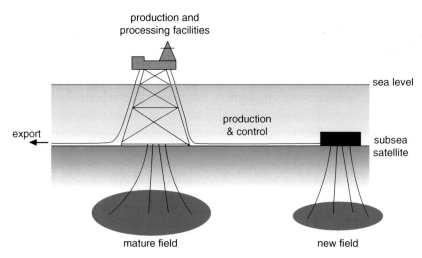

Figure 17.11   Satellite development.

Whether on land or offshore, the principle of satellite development is the same. A new field is accessed with wells, and an export link is installed to the existing (host) facility. Development is not always easier on land, as environmental restrictions mean that some onshore fields have to be developed using directional drilling techniques (originally associated with offshore developments). A vertical well can be drilled offshore away from the host facility, and the well completed using a *subsea wellhead* (Figure 17.11).

The role that a *host facility* plays in an incremental development project can vary tremendously. At one extreme, all production and processing support may be provided by the host (such as gas lift and water treatment). On the other, the host may just become a means of accessing an export pipeline (if a production and processing facility is installed on the new field).

17.5.1. Extended reach development

One form of incremental development is *ERD*, to access either remote reserves within an existing field or reserves in an adjacent accumulation. Provided the new hydrocarbons are similar to those of the declining field, production can be processed using existing facilities without significant upgrading. If no spare drilling slots are available, old wells may have to be plugged and abandoned to provide slots for new extended reach wells (Figure 17.12).

In such cases, the development scheme for the original reserves may have to be modified to make processing capacity available for the new hydrocarbons. The economics of such a scheme can be affected negatively if substantial engineering modifications have to made to meet new safety legislation. For more background to ERD refer to Section 4.5, Chapter 4.

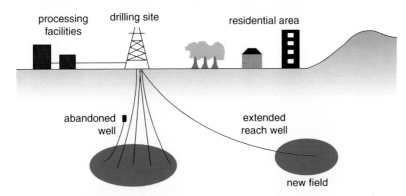

Figure 17.12 Extended reach development.

17.5.2. Satellite development

Handling production from, and providing support to, a satellite field from an older facility is at first glance an attractive alternative to a separate new development. However, whilst savings may be made in capital investment, the operating cost of large processing facilities may be too much to be carried by production from a smaller field.

Initially, if operating costs can be divided based on production throughput, the satellite development project may look attractive. However, the unit costs of the declining host field will eventually exceed income and the satellite development may not be able to support the cost of maintaining the old facilities. If the old facilities can be *partly decommissioned*, and provision made for part of the abandonment cost, then the satellite development may still look attractive. The satellite development option should always be compared to options for independent development.

In an offshore environment, development via a subsea satellite well can be considered in much the same way as a wellhead on land, although well maintenance activity will be more expensive. However, if a simple self-contained processing platform is installed over a new field and the host platform is required only for 'peak shaving' or for export, a number of other development options may become available. The host platform may actually cease production altogether and develop a new role as a pumping station and accommodation centre, charging a tariff for such services. There may be significant construction savings gained for the new platform if it can be built to be operated unmanned. The old reservoir may even in some cases be converted into a water disposal centre or gas storage facility.

Whatever form of incremental development is considered, the benefits to the host facility should not be gained at the expense of reduced returns for the new project. Incremental and satellite projects can in many situations help to extend the production life of an old field or facilities, but care must be taken to ensure that the economics are transparent.

CHAPTER 18

DECOMMISSIONING

Introduction and Commercial Application: Eventually every field development will reach the end of its economic lifetime. If options for extending the field life have been exhausted, then decommissioning will be necessary. Decommissioning is the process by which the operator of an oil or natural gas installations will plan, gain approval and implement the removal, disposal or re-use of an installation when it is no longer needed for its current purpose.

The cost of decommissioning may be considerable, and comes of course at the point when the project is no longer generating funds. Some source of funding will therefore be required, and this may be available from the profit of other projects, from a decommissioning fund set up during the field life or through tax relief rolled back over the late field production period.

Decommissioning is often a complex and risky operation. The five key considerations are the potential impact on the environment, potential impact on human health and safety, technical feasibility, costs of the plan and public acceptability.

Decommissioning may be achieved in different ways, depending on the facilities type and the location. This section will also briefly look at the ways in which decommissioning can be deferred by extending the field life, and then at the main methods of well abandonment and facilities decommissioning.

18.1. LEGISLATION

National governments play an extensive role in assessing and licensing decommissioning options. Most countries which have offshore oil and natural gas installations have laws governing decommissioning.

The prime global authority is the International Maritime Organisation (IMO). The IMO sets the standards and guidelines for the removal of offshore installations. The guidelines specify that installations in less than 75 m of water with substructures weighing less than 4000 tons be completely removed from the site. Those in deeper water must be removed to a depth of 55 m below the surface so that there is no hazard to navigation. In some countries the depth to which structures have to be removed has already been extended to 100 m.

The planning of decommissioning activities involves extensive periods of consultations with the relevant authorities and interested parties, such as fishing and environmental groups.

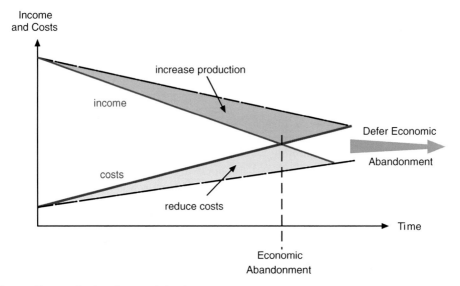

Figure 18.1 Deferring decommissioning.

 ## 18.2. ECONOMIC LIFETIME

The economic lifetime was introduced in Section 14.3, Chapter 14, and was defined as the point at which the annual net cashflow turned permanently negative. This is the time at which income from production no longer exceeds the costs of production, and marks the point when decommissioning should occur, since it does not make economic sense to continue to run a loss-making venture. Technically, the production of hydrocarbons could continue beyond this point but only by accepting financial losses. There are two ways to defer decommissioning (Figure 18.1)

- reduce the operating costs, or
- increase hydrocarbon throughput.

Of course the operator will strive to use both of these means of deferring abandonment.

In some cases, where production is subject to high taxation, tax concessions may be negotiated, but generally host governments will expect all other means to have been investigated first.

18.2.1. Reducing operating costs

Operating costs represent the major expenditure late in field life. These costs will be closely related to the number of staff required to run a facility and the amount of hardware they operate to keep production going. The specifications for product quality and plant up-time can also have a significant impact on running costs.

Operating strategies and *product quality* should be carefully reassessed to determine whether less treatment and more down-time can be accommodated and what cost saving this could make. Many facilities are constructed with high levels of built-in redundancy to minimise production deferment early in the project life. Living with periodic shutdowns may prove to be more cost effective in decline. Intermittent production may also reduce treatment costs by using gravity segregation in the reservoir to reduce water cuts or gas influx, as mentioned in Section 17.4, Chapter 17.

18.2.2. Increasing hydrocarbon production

As decommissioning approaches and all well intervention opportunities to increase productivity have been exploited, *enhanced recovery processes* may be considered as a means of recovering a proportion of the remaining hydrocarbons. However, such techniques are generally very sensitive to the oil price, and whilst some are common in onshore developments they can rarely be justified offshore.

When production from the reservoir can no longer sustain running costs but the operating life of the facility has not expired, opportunities may be available to develop *nearby reserves* through the existing infrastructure. This is becoming increasingly common as a method of developing much smaller fields than would otherwise be possible.

Companies which own process facilities and evacuation routes, but no longer have the hydrocarbons to fill them, can continue to operate them profitably by *renting the extra capacity* or by charging *tariffs* for the use of export routes.

18.3. Decommissioning Funding

Management of the cost of decommissioning is an issue that most companies have to face at some time. The cost can be very significant, typically 10% of the cumulative CAPEX for the field. On land sites, wells can often be plugged and processing facilities dismantled on a *phased* basis, thus avoiding high spending levels just as hydrocarbons run out. Offshore decommissioning costs can be very significant and less easily spread as platforms cannot be removed in a piecemeal fashion. How provision is made for such costs depends partly on the size of the company involved and the prevailing tax rules.

If a company has a number of projects at various stages of development, it has the option to pay for decommissioning with cash generated from younger fields. A company with only one project will not have this option and may choose to build up a *decommissioning fund* which is invested in the market until required. In both cases cash has to be made available. However, in the first situation the company is likely to prefer to use cash generated from the early projects to finance new ventures, on the assumption that investing in projects generates a better return than the market. A combination of using cash generated from profitable projects and money drawn from a decommissioning fund is then likely.

The fiscal treatment of decommissioning costs is a very live issue in many mature areas where the decommissioning of the early developments has already begun. Energy or industry departments within governments, in their role as custodian of the national (hydrocarbon) asset, have a responsibility to ensure that the recovery of oil and gas is maximised. Operating companies have a responsibility to their shareholders to generate a competitive return on investment. The preferred point of decommissioning will therefore be viewed differently by oil company and host government.

In some areas it is obligatory for the oil company to contribute to a decommissioning fund throughout the producing life of the field. The cost of decommissioning is usually considered as an operating cost, for which a fiscal allowance is made. This is typically claimed in the final year of the field life. Complex arrangements exist for dealing with decommissioning costs which exceed the gross revenue in the final year of the field life. For instance costs may be expensed and carried back for a number of years against either revenue, taxation or royalties paid.

18.4. Decommissioning Methods

The basic aim of a decommissioning programme is to render all wells permanently safe and remove most, if not all, surface (or seabed) signs of production activity. How completely a site should be returned to its 'green field' state, is a subject for discussion between government, operator and the public.

18.4.1. Well abandonment

Whether offshore or on land an effective well abandonment programme should address the following concerns

- isolation of all hydrocarbon bearing intervals
- containment of all overpressured zones
- protection of overlying aquifers
- removal of wellhead equipment.

A traditional abandonment process begins with a *well killing* operation in which produced fluids are circulated out of the well, or pushed (bull headed) into the formation, and replaced by drilling fluids heavy enough to contain any open formation pressures. Once the well has been killed the Christmas tree is removed and replaced by a blowout preventer, through which the production tubing can be removed.

Cement is then placed across the open perforations and partially squeezed into the formation to seal off all production zones. Depending on the well configuration it is normal to set a series of cement and wireline plugs in both the liner and production casing (see Figure 18.2), to a depth level with the top of cement behind the production casing.

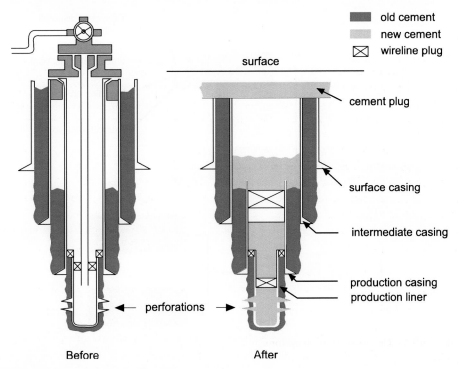

Figure 18.2 A well before and after abandonment.

The production casing is cut and removed above the top of cement, and a *cement plug* positioned over the casing stub to isolate the annulus and any formation which may still be open below the intermediate casing shoe. If the intermediate casing has not been cemented to surface then the operation can be repeated on this string. Alternatively the remaining casing strings will be cut and removed close to surface and a cement plug set across the casing stubs. On land the wellsite may be covered over and returned to its original condition.

Traditional well abandonment techniques are being reviewed in many areas. In some cases wells are being abandoned without rig support by perforating and squeezing without cutting and pulling tubing and casing.

18.4.2. Pipelines

All pipelines will be circulated clean and those that are buried, or on the seabed, left filled with water or cement. Surface piping will normally be cut up and removed. Flexible subsea pipelines may be 'reeled-in' onto a lay barge and disposed of onshore.

18.4.3. Offshore facilities

There are currently more than 6500 oil and gas installations located on the continental shelves of some 53 countries. About 4000 of these are in the US GoM, 950 in Asia, 700 in the Middle East and 400 in Europe.

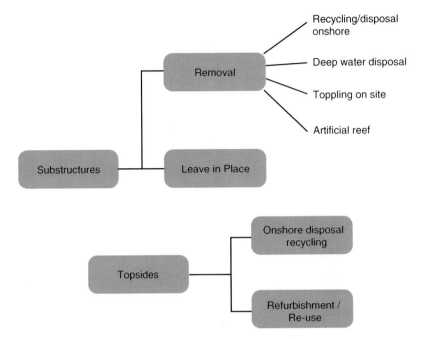

Figure 18.3 Decommissioning options.

Each of the main facility types, for example steel jacket, gravity structure, tension leg and floating platform, have different options for decommissioning. The main factors which need to be considered and which will impact on costs are type of construction, size, distance from shore, weather conditions and the complexity of the removal, including all safety aspects. Figure 18.3 shows the options available.

Tension leg and floating platforms can easily be released and towed away for service elsewhere, which is cheap and attractive. In the case of the fixed platforms, the topside modules are removed by lift barge and taken to shore for disposal. Gravity-based structures can in theory be deballasted and floated away to be re-employed or sunk in the deep ocean, and steel jackets cut and removed at an agreed depth below sea level. In some areas jackets are cleaned and placed as artificial reefs on the seabed. The largest 'rigs to reefs' programme involving more than 90 decommissioned installations has been implemented in the GoM.

Subsea facilities are easily decommissioned as they are relatively small and easy to lift. However, subsea manifolds and templates can weigh in excess of 1000 tons and will require heavy lift barges for removal.

18.4.4. Land facilities

Onshore processing facilities, and modules brought onshore, have to be cleaned of all hazardous compounds and scrapped. Cellars of single wells, drilling pads, access roads and buildings will have to be removed. If reservoir compaction affects the surface area above the abandoned field future land use may be prevented, in particular in coastal or low land environments.

The land under the facilities may also have to be reconditioned if pollutants have been allowed to escape during operation. The return of industrial sites to green field conditions has proved very expensive for many companies in the USA, and a number of law suits are currently outstanding, brought by local authorities and environmental groups.

It is no longer acceptable in most countries to treat decommissioning as an issue that can be ignored until the end of a project. Increasingly operators are being required to return industrial sites to their original condition after use. Many operators now perform a *base line survey* before they build on an area so that the impact of operations can be quantified, and in some cases so that they are not held responsible for the pollution of previous site owners.

References and Bibliography

> ## Selected Bibliography for Further Reading

Allen, P. A. and Allen, J. R., 2006. *Basin Analysis: Principles and Applications*. Blackwell Science, 480p.

Amyx, J. W., Bass, D. M. and Whiting, R. L., 1960. *Petroleum Reservoir Engineering: Physical Properties*. McGraw-Hill.

Archer, J. S. and Wall, C. G., 1999. *Petroleum Engineering: Principles and Practice*. Kluwer Academic.

Azar, J. and Samuel, R., 2007. *Drilling Engineering*. Pennwell, 486p.

Beggs, D., 2003. *Production Optimization using Nodal Analysis*. OGCI, 411p.

Bentley, M. R. and Smith, S. A., 2007. Scenario-based reservoir modelling and the need for more determinism. In: *The Future of Geological Modelling in Hydrocarbon Development*. Geological Society of London Special Publication.

Bentley, M. R. and Woodhead, T. J., 1998. Uncertainty Handling Through Scenario-Based Reservoir Modelling. SPE 39717.

Bernstein, P. L., 1998. *Against the Gods: The Remarkable Story of Risk*. Wiley, 400p.

Bradley, H., 1987. *Petroleum Engineering Handbook*. Society of Petroleum Engineers.

Consentino, L., 2001. *Integrated Reservoir Studies*. Edition Technip, 360p.

Dake, L., 1983. *Fundamentals of Reservoir Engineering*. Elsevier Science, 462p.

Dake, L., 2001. *The Practice of Reservoir Engineering*, revised ed. Elsevier Science, 572p.

Dake, L. P., 1978. *Fundamentals of Reservoir Engineering*, Elsevier.

Economides, M., Watters, L. and Dunn-Norman, S., 1997. *Petroleum Well Construction*. Wiley, 622p.

Fraser, K., 1991. *Managing Drilling Operations*. Elsevier Science, 246p.

Gluyas, J. and Swarbrick, R., 2003. *Petroleum Geoscience*. Blackwell Science, 376p.

Hilterman, F., 2001. *Seismic Amplitude Interpretation – Distinguished Instructor Short Course No. 4*. SEG-EAGE.

Jack, I., 1998. *Time-Lapse Seismic in Reservoir Management – Distinguished Instructor Short Course No. 1*. SEG.

Johnston, D., 1994. *International Petroleum Fiscal Systems and Production Sharing Contracts*. Pennwell.

Joshi, S. A., 1991. *Horizontal Well Technology*. Pennwell books, 535p.

Kearey, P., Brooks, M. and Hill, I., 2002. *An Introduction to Geophysical Exploration*, 3rd ed. Blackwell Science, 262p.

Lowell, J. D., 1985. *Structural Styles in Petroleum Exploration*. OGCI Publications, 460p.

McAleese, S., 2000. *Operational Aspects of Oil and Gas Well Testing*. Elsevier Science, 352p.

McCain, W. D., Jr., 1973. *The Properties of Petroleum Fluids*. Pennwell.

Miall, A., 1984. *Principles of Sedimentary Basin Analysis*. Springer Verlag, 468p.

North, F. K., 1985. *Petroleum Geology.* Allen & Unwin, 607p.

Perrin, D., 1999. *Well Completion and Servicing.* Institut Francais du Petrole, 325p.

Reynolds, J., 1997. *An Introduction to Applied and Environmental Geophysics.* Wiley, 796p.

Rider, M., 1996. *The Geological Interpretation of Well Logs.* Whittles Publishing, 280p.

Royal Dutch/Shell Group, 1983. *The Petroleum Handbook.* Elsevier.

Scholle, P. A. and Spearing, D., 1982. *Sandstone Depositional Environments.* AAPG Memoir 31, 410p.

Scholle, P. A., Bebout, D. G. and Moore, C. H., 1983. *Carbonate Depositional Environments.* AAPG Memoir 33, 708p.

Serra, O., 1984. *Fundamentals of Well Log Interpretation Volume 1: The Acquisition of Logging Data. Developments in Petroleum Science 15A.* Elsevier, 440p.

Standing, M. B. and Katz, D. L., 1942. *Density of Natural Gases,* Trans AIME.

The International Offshore Oil and Natural Gas Exploration and Production Industry, 1996. *Decommissioning Offshore Oil and Gas Installations: Finding the Right Balance.*

Tiab, D. and Donaldson, E., 1996. *Petrophysics: Theory and Practice of Measuring Reservoir Rock and Fluid Transport Properties.* Gulf Publishing Company, 706p.

Walker, R. G. and James, N. P., 1992. *Facies Models.* Geological Association of Canada, 409p.

Yergin, D., 1991. *The Prize.* Simon & Schuster, 885p.

TERMS AND ABBREVIATIONS

AHBDF	alonghole depth below derrick floor
AI	acoustic impedance
ALARP	as low as reasonably practicable
API	American Petroleum Institute
AVA	amplitude variation with angle
AVO	amplitude variation with offset
BHA	bottom hole assembly
BOP	blowout preventor
BS&W	base sediment and water
BYC	base year cost
CAO	computer-assisted operations
CAPEX	capital expenditure
CD	chart datum
CFC	chlorofluorocarbon
CMP	common midpoint
CO_2	carbon dioxide
CPA	critical path analysis
CPI	corrugated plate interceptor
CRA	corrosion resistant alloy
CSEM	controlled source electro-magnetic
CT	computerised tomography
CTD	coiled tubing drilling
DC	drill collar
DCD	declaration of commercial discovery
DCQ	daily contract quantity
DHSV	downhole safety value
DP	dynamic positioning
DST	drill stem test
DSV	diver support vessel
EIA	environmental impact assessment
EM	electro-magnetic
EMV	expected monetary value
EOR	enhanced oil recovery
ERD	extended reach drilling
ESDV	emergency shutdown valve
ESP	electrical submersible pumps
FDP	field development plan
FEED	front end engineering design
FEWD	formation evaluation while drilling

FMI	Fullbore Formation Microimager (Schlumberger)
FMT	formation multitester
FPSO	floating production, storage and offloading
F-T	Fischer–Tropsch
FTHP	flowing tubing head pressure
FWKO	free water knockout vessel
FWL	free water level
GDT	gas down to
GIIP	gas initially in place
GLR	gas–liquid ratio
GLV	gas lift value
GOC	gas–oil contact
GOM	Gulf of Mexico
GOR	gas:oil ratio
GR	gamma ray
GRV	gross rock volume
GTL	gas to liquids
HAZOP	hazard and operability studies
HCIIP	hydrocarbons initially in place
HCWC	hydrocarbon–water contact
HPHT	high pressure high temperature
H_2S	hydrogen sulphide
HSE	health safety and environment
HSP	hydraulic submersible pumps
IMO	International Maritime Organisation
IOC	International Oil Company
IPR	inflow performance relationship
IRR	internal rate of return
JT	Joule Thomson
LCM	lost circulation material
LMV	lower master valve
LNG	liquefied natural gas
LPG	liquefied petroleum gas
LTI	lost time incident
LTS	low temperature separation
LWD	logging while drilling
mb/d	thousand barrels per day
MCS	master control station
MD	measured depth
MDT	modular dynamic tester
MOD	money of the day
MODU	mobile offshore drilling unit
MSL	mean sea level
MSV	multipurpose service vessel
MTF	mean time to failure
MWD	measurement while drilling

N/G	net to gross ratio
NGL	natural gas liquid
NGO	non-government organisation
NMO	normal move out
NMR	nuclear magnetic resonance
NOC	National Oil Company
NOS	net oil sand
NPC	net present cost
NPV	net present value
OBC	ocean bottom cables
OBM	oil-based mud
OBP	overburden pressure
OBS	ocean bottom stations
OD	outside diameter
ODT	oil down to
OPEC	The Organization of the Petroleum Exporting Countries
OPEX	operating expenditure
OUT	oil up to
OWC	oil–water contact
PBR	polished bore receptacle
PDC	polycrystalline diamond compact
PDF	probability density function
PDSG	permanent downhole strain gauge
PFS	process flow schemes
PGOC	possible gas–oil contact
PI	productivity index
PIF	productivity improvement factor
PIR	profit-to-investment ratio
PLT	production logging tool
POOH	pulling out of hole
POS	probability of success
ppm	parts per million
PSA	production sharing agreement
PSC	production sharing contract
PSDM	pre-stacked depth migration
PTW	permit to work (systems)
PV	present value
PVT	pressure volume and temperature
PWRI	produced water re-injection
QRA	quantative risk assessment/analysis
RCI	reservoir characterization instrument
RF	recovery factor
RFT	repeat formation tester
RIF	recordable injury frequency
RMS	root mean square
ROP	rate of penetration

RPM	revolutions per minute
RRoR	real rate of return
RT	real terms
SBHP	static bottomhole pressure
SBM	single buoy mooring
SCAL	special core analysis tests
SEC	Securities and Exchange Commission (US)
SIMOPS	simultaneous operations
SiO_2	silicon dioxide (quartz)
SIPROD	simultaneous production and drilling
SMS	safety management systems
SNG	synthetic natural gas
SOBM	synthetic oil-based mud
SP	spontaneous potential
SPE	Society of Petroleum Engineers
SPM	side pocket mandrel
SSD	sliding side door
STOIIP	stock tank oil initially in place
STP	standard conditions of temperature and pressure
SV	swab valve
SWS	sidewall sampling tool
tcf	trillion cubic feet
TD	total depth
TDT	thermal decay time
TEG	tri-ethylene glycol
TLP	tension leg platform
TPR	tubing performance relationship
TVP	true vapour pressure
TVSS	true vertical depth subsea
TWT	two way time
UMV	upper master valve
UR	ultimate recovery
USD	US dollars
UTC	unit technical cost
UV	ultra violet
VLCC	very large crude carrier
VSP	vertical seismic profiling
WAG	water alternating gas
WBM	water-based mud
WEG	wireline entry guide
WI	Wobbe index
WOB	weight on bit
WOC	wait on cement
WV	wing valve

Subject Index